Joe Blaikie
2/21/04

Wine

# Wine

## A Scientific Exploration

Edited by

Merton Sandler
Imperial College Medical School
London, UK

and

Roger Pinder
Organon Inc.
USA

NEW YORK AND LONDON

First published 2003

Simultaneously published in the UK, USA and Canada by
Routledge
29 West 35th Street, New York NY 10001
and
Routledge
11 New Fetter Lane, London EC4P 4EE

*Taylor & Francis is an imprint of the Taylor & Francis Group*

Reprinted 2003



Typeset in 11/12pt Garamond 3 by Graphicraft Limited, Hong Kong
Printed and bound in Great Britain by The Cromwell Press, Trowbridge, Wiltshire



*British Library Cataloguing in Publication Data*
A catalogue record for this book is available from the British Library

*Library of Congress Cataloging in Publication Data*
A catalog record has been requested

ISBN 0-415-24734-9

There are more old drunkards than old doctors.

Benjamin Franklin (1706–1790)

# Contents

# List of figures

# List of tables

# List of contributors

Michael Aviram
The Lipid Research Laboratory
Rambam Medical Center
Haifa 31096
Israel

Martin Bobak
Department of Epidemiology and Public Health
University College London
London
WC1E 6BT
UK

Glen L. Creasy
Centre for Viticulture and Oenology
Lincoln University
Canterbury
New Zealand

Leroy L. Creasy
Department of Horticulture
Cornell University
Ithaca
New York 14853
USA

Bianca Fuhrman
The Lipid Research Laboratory
Rambam Medical Center
Haifa 31096
Israel

George Gale
Department of Philosophy
University of Missouri-Kansas City
Kansas City
MO 64110
USA

Rosemary George
Master of Wine
Independent wine writer
UK

David M. Goldberg
Department of Laboratory Medicine and Pathobiology
University of Toronto
Toronto
Ontario
Canada
M5G 1L5

Ron S. Jackson
CCOVI
Brock University
St Catherines
Ontario
Canada

Arthur L. Klatsky
Kaiser Permanente Medical Care Program
280 West MacArthur Boulevard
Oakland

CA 94611
USA

Michael Marmot
Department of Epidemiology and Public Health
University College London
London
WC1E 6BT
UK

Carole P. Meredith
Department of Viticulture and Enology
University of California
Davis
CA 95616
USA

Renée S. Moore
Department of Pediatrics
West Virginia University School of Medicine
Morgantown
WV 26506
USA

Philip A. Norrie
Hawkesbury Campus
University of Western Sydney
Sydney
Australia

Thomas O. Obisesan
Section of Geriatrics and Gerontology
Department of Medicine
Howard University Hospital and College of Medicine
Washington DC
USA

Roger M. Pinder
Organon Inc.
375 Mount Pleasant Avenue
West Orange
NJ 07052
USA

Jane M. Renfrew (Lady Renfrew of Kairnsthorn)
Lucy Cavendish College
Cambridge
CB3 0BU
UK

Merton Sandler
Institute of Reproductive and Developmental Biology
Hammersmith Campus
Imperial College Medical School
Du Cane Road
London
W12 0NN
UK

George J. Soleas
Quality Assurance
Liquor Control Board of Ontario
Toronto
Ontario
Canada
M5E 1A4

Martin E. Weisse
Department of Pediatrics
West Virginia University School of Medicine
Morgantown
WV 26506
USA

# Preface

As we both wind down into the second half (we hope) of our lives, we take some consolation in wine. We like wine: we enjoy its spectrum of tastes, its buoyant effect on our state of mind, its mythology. Not that we think of ourselves, for a moment, as wine snobs – but we do have a minimal expertise, yes, and some discrimination, which add to our pleasure. Even so, we are well aware that our good vibes are just that and strictly subjective. Above all, we are scientists and our aim at all times is to be as objective as possible.

Wine carries a heavy baggage. In this book, we hope to winnow the facts, as it were, from the chaff and have assembled top talent from a broad range of scientific interests to assist us. Their authoritative views, presented in this volume, ought, we think, to be required reading for medical practitioners from many different disciplines – cardiologists, neurologists, psychiatrists, ophthalmologists, and microbiologists – as well as for botanists, oenologists and viticulturists, historians of science, and simple wine buffs.

Wine has a long history and was already being made in neolithic times. Early winemakers used the wild Eurasian grape variety *Vitis vinifera sylvestris*, and their wine was resinated in the manner of modern Greek retsina. Domestication of the grapevine to *Vitis vinifera vinifera* was fairly rapid, and this variety became the basis for most subsequent and, indeed, current, wine production. Genetic manipulation of the grapevine and medical applications of wine also have long histories, but only in the past two decades has proper scientific exploration been possible with the application of the modern technologies of analytical chemistry, molecular biology and clinical epidemiology. There has been an explosion of interest and research in grapevine chemistry and genetics, as well as in wine's potential health benefits, making it something of a scientific superstar.

Although the beneficial effects of wine were well recognized by the ancients – the writings of Hippocrates, for instance, are peppered with references and sage advice – the defining moment in modern times came with the publication in 1979 by a British team of researchers of a paper entitled 'Factors associated with cardiac mortality in developed countries with particular reference to the consumption of wine' (St Leger, A. S. *et al.* 1979. *Lancet*: 1, 1017–20). This study, which examined the health histories of men and women aged 55 to 64 years from 18 countries, found that regular alcohol consumption, most particularly of wine, lowers the incidence of heart disease by 30–40%. Confirmation was rapid and extensive, and, although some disagree, the present consensus and that derived from meta-analyses is that coronary heart disease (CHD) mortality and/or morbidity is reduced by daily moderate consumption of wine compared with abstention or heavy drinking.

Despite a high dietary intake of saturated fat, the incidence of CHD in France is low. This is the 'French paradox', first noted by the Irish physician Samuel Black in 1819 (Black, S. 1819. *Clinical and Pathological Reports*. Wilkinson, Newry, pp. 1–47), and it galvanized the debate in 1991 when French epidemiologist and nutritionist Serge Renaud appeared on US television and aired his views on wine and heart disease in the programme *60 Minutes*. The debate continues. Wine bottles sold in the USA continue to carry health warnings, and evidence-based medicine does not seem to apply to moderate wine consumption.

The evidence is less comprehensive but still substantial for a protective effect of wine on the incidence of other forms of vascular disease, particularly stroke. In addition, emerging population-based data also point to protective effects against dementia, some types of cancer and macular degeneration. Wine has antimicrobial and antifungal activity and may play a role in the aetiology of migraine. Red wine may even protect against the common cold (Takkouche, B. *et al*. 2002. Intake of wine, beer and spirits and the risk of the common cold. *American Journal of Epidemiology*: **155**, 853–8). Such an apparent panacea must possess multiple actions, and wine probably acts in many cases through its antioxidant ability.

Wine contains polyphenols from both the flavonoid and stilbene families of chemicals, mostly as grape tannins (about 35%) and anthocyanin pigments (about 20%), with more, by and large, in red than white wines. Although much of the research has emanated from Bordeaux, ordinary table wines are as efficacious in antioxidant activity as are those from the grandest châteaux. In fact, Chilean Merlot may have the highest concentrations of polyphenols, while the stilbene most targeted for its possible health benefits, resveratrol, is significantly more common in wines made from Pinot Noir. Polyphenols in wine affect many factors including blood lipids, platelet aggregation and atherogenic processes. For example, their protective effects upon CHD may reflect antiatherosclerotic and antithrombotic actions in addition to having direct effects upon glucose metabolism.

Genetic modification of the grapevine, and of the yeasts which participate in the fermentation process to produce wine, have also been recognized for many centuries. Many of our modern and most highly regarded wine grapes – Cabernet Sauvignon, Chardonnay, Syrah – are the product of accidental crosses between varieties grown close to each other, probably in the same vineyard, a common practice in medieval times. The parents are often obscure and sometimes humble varieties. The application of modern DNA fingerprinting technology to grapevines only began a decade ago in South Australia, but has subsequently blossomed in both likely (Austria, California, France and Italy) and less likely locations (Crete and Croatia). The major Vitis database for grapevine genetic mapping is held at the University of Crete (http://www.biology.uch.gr/gvd/). It has enabled both the parentage and synonymy of grapevine varieties to be established unequivocally, and has permitted the tracing of their geographical origins.

The technology has also opened up the possibility of genetic engineering, enabling the introduction of genes from grapevines or other organisms into existing grape cultivars, a process already taking place in yeasts. So far restricted to methods to reduce disease loss and pesticide usage, the technology might eventually lead to alteration of the wine attributes of grapevine varieties. Whether for good or ill, we do well to remember that much of the wine that we drink and value today is made from varieties that were already genetically modified many centuries ago.

Wine science is on a roll. It has become the stuff of the popular press, magazines, radio and television. Pundits abound. Some of us are lucky enough to be able to combine our vinous and scientific interests. A little of what you fancy seems to be a good guiding principle for the health benefits of wine, taken daily, preferably red and perhaps Merlot or Pinot Noir. We have used the same principle in selecting the topics and authors to be included in this book.

*Merton Sandler and Roger Pinder, Twickenham and New Providence, August 2002*

# 1 Drinking wine

*R. George*

A simple definition of wine is the fermented juice of the grape. Although so-called wine can be made from other fruits, in this context it is the grape alone that counts. There is no other fruit that can provide such a wonderfully diverse and complex drink with such infinite variations of flavour. It is a many faceted drink, enjoyable in its youth yet capable of considerable longevity. It enhances any social occasion, inducing conviviality and conversation; it can turn a humble picnic into a feast, or a simple dinner into a banquet. Its subtleties can be discussed and pondered over at length, at best around a congenial table, rather than in the clinical atmosphere of a tasting room.

## Tasting

Although it is perfectly possible to drink and enjoy wine without any knowledge of the subject, it is certain that a greater appreciation of it comes with a little information. I have lost count of the number of times that someone has said to me: 'I don't know anything about wine, but I do know what I like'. And that is fair enough. Our sense of taste tends to be our least developed sense as we all tend to take it for granted and do not really think about what we are putting in our mouths. Nevertheless, we are all capable of making an olfactory evaluation of wine, as indeed we do about our food, usually subconsciously, at every meal. Scientific research has recently shown that some of us are endowed with a more than average number of tastebuds, while an unlucky few are quite deficient in them; the great majority of us have a decent number, which we can learn put to good use in the exercise of our critical faculties.

Tasting and drinking wine is, above all, fun, with no right or wrong answers as taste is essentially subjective. However, there is a logical way in which to consider an unfamiliar glass of wine. The first thing is the colour. Colour will tell you much. With white wine, the colour spectrum can range from the very palest almost watery hue, through shades of yellow to deep gold, and from there to the brown-amber tones of old sherry. Colour in white wine is an indication of dryness or sweetness, and also of age. A sweet wine, such as a Sauternes, is always likely to be richly golden in colour. A young dry wine tends to be very pale, while an older dry wine develops some colour with age. Red wines, too, enjoy an enormous colour variation, from the vibrant red of a young wine to the orange-brick red of a mature tawny port. Again, age is a factor here, for red wine loses colour with age, fading from a vivid red around the rim of the wine to a brick red-tawny colour over the years, while a deep, intense colour indicates a wine of stature and body.

Smells are wonderfully evocative and an intrinsic part of the enjoyment of a wine, to the extent of being the most important sense with which we appreciate it. The initial sniff when the wine is poured gives an indication of the pleasure to come, and a quick swirl of the glass first will release even more aroma so that the senses are engaged. Experienced wine-drinkers may recognize a grape variety or a style of wine that they enjoy. Some grape varieties are more distinctive than others. Sauvignon from New Zealand is sometimes described as cat's pee on a gooseberry bush, while Sauvignon from Sancerre in the Loire Valley is an altogether more subtle thing, with flinty, mineral characteristics. Smells are also very subjective. A wine-drinker makes associations with different fruits, pencil shavings, vanilla, vegetal aromas and so on. The list can be infinite in the attempt to define a seemingly elusive characteristic that defies words, and one taster will not necessarily agree with another. The bouquet that makes one wine-drinker think of strawberries, may be associated with raspberries by another, and of course both are right. It is this subjectivity that leads to all the far-fetched fancy descriptions and flights of fantasy one hears, in an attempt to pin down something that is essentially fleeting and indefinable.

The same applies to the flavour of wine. Here again our appreciation is purely subjective and hampered by the fact that our tongue registers only four primary tastes: sweetness, saltiness, bitterness and sourness. But what we actually taste is so much more, a combination of all the numerous components in wine, a harmonious whole, enhanced by the presence of alcohol, which provides the body and weight in a wine, and also a hint of sweetness. Without alcohol, wine tastes hollow and unbalanced. And when you have swallowed the wine, consider whether it leaves a long-lasting taste, or is the finish short and abrupt. A good wine will leave a lingering flavour in the mouth that invariably makes you want to take another sip.

## What determines taste?

There is no other drink that can provide such a vivid sense of place, that can speak of its origins in the glass. Therein lies part of the charm and appeal of wine. The wines of the south of France smell of the warmth of Mediterranean sunshine, linked with the scents of the herbs of the *garrigue*, the scrubland that covers the hillsides of the Languedoc behind the towns of Montpellier and Béziers. Contrast that with dry, flinty Chablis, from a more severe climate of northern France, where winters are harsh, and summers uncertain. There is a steely austerity about Chablis that is reminiscent of the chalk hills of the vineyards. And that sense of place explains the inevitable link with the food of a region: no other wine goes better with the salty goat's cheese of Chavignol than flinty Sancerre, for example; and there is nothing better than a glass of Chianti, with its sour cherry fruit and typical backbone of astringency of the Sangiovese grape, to offset the rich flavours of Tuscan cuisine, with its generous quantities of luscious olive oil.

So what makes a wine taste as it does? Essentially there are four factors, of which three are very much related to the place. They are: grape variety; soil; climate; and the human hand of the winemaker. The winemaker is the international factor; the other three are determined by the location of the vineyard.

Although grape varieties have travelled, they will only thrive in certain conditions and some are more pernickety than others. There are several thousand different grape varieties used for winemaking, mostly of the *Vitis vinifera* species, but fewer than 50 are of international significance, and among those only a handful or so actually appear on a wine label or have travelled from their origins in France, Germany, Spain or Italy to become established in the New World vineyards of California, the Antipodes, South America and South Africa. The grape variety, or blend of grape varieties, will determine the basic flavour of a wine; some have more character than others, while Muscat is the one variety that really does taste of grapes. If you drink fresh, chilled Asti Spumante, you immediately think of those luscious Italian Muscat table grapes. Gewürztraminer, Sauvignon and Riesling are other varieties that have an intrinsic flavour of their own that shines through any winemaking process.

The same grape produced in different places will change accordingly in taste. This is where climate comes into play. The weather must not be too hot or cold, nor too dry or wet, so there is a broad band of land with a suitable climate across Europe, roughly between latitudes 50°N and 35°N, between the Rhineland of Germany and the North African coast, and a similar band in the southern hemisphere that includes central Chile and Argentina, the Capelands of South Africa, a small part of Western Australia as well as the south-east corner of that vast continent, and most of New Zealand. There are of course exceptions to every rule, and consequently you can find vines in the most unlikely places. There are only four North American states that do not produce wine – and think how the climate can vary between Texas and Washington State or between Virginia and Ohio. There are vineyards in the foothills of the Peruvian Andes and in the Sahyadri mountains not too far from Bombay; Norway has just one small vineyard and, in sharp climatic contrast, there is a vineyard near Alice Springs in the heart of the Australian outback.

However, vines tend to perform their best on the cooler edge of that climatic band. For the same reasons that English strawberries and apples have more flavour than those grown in Italy, or Tuscan olive oil is finer than

Greek, grapes grown where the climate is less certain and the sunshine less constant, make wines that attain greater depth of flavour and heights of quality. It is as though vines that are assured of endless days of sunshine become bored. They are not stressed in any way, so the wine they produce is consequently bland, insipid and over alcoholic. In contrast, where they are subjected to climatic vagaries, great wines can be made, but not every year. That is why Burgundy and Bordeaux are two of the great red wine regions of the world, with quite different climates, one continental and the other maritime, allowing the production of quite contrasting but superlative grape varieties, Pinot Noir in Burgundy, and Cabernet Sauvignon, along with Merlot and Cabernet Franc in Bordeaux. Again, it is climate that determines New Zealand's growing ability to produce world class Pinot Noir, rivalling and even out-performing that of California and Oregon.

Chardonnay, which is generally deemed to be the most versatile of grape varieties because it has travelled the world from its native Burgundy and is found in a variety of climes and terrains, is a superb example of flavour varying depending on where it has been grown. Contrast the Chablis of northern France, from vines grown in a cool climate of uncertain sunshine hours, with an Australian Chardonnay from warm vineyards, where lack of rain rather than lack of sunshine is the perennial problem. The taste is quite different; one is lean and steely, the other fat and opulent.

The weather also determines the annual variations of the vintage. In regions such as Bordeaux and Burgundy, where there are considerable differences from year to year, this is all-important. Sunshine and rain at the right time are crucial for a great vintage. Spring frosts can affect the size of the crop, as can cool weather or rain at the flowering. Rain at the harvest may swell the grapes, but to the detriment of their quality, diluting their flavour. A fine summer generally means a good vintage, with ripe healthy grapes, but nothing is certain until the last grapes are safely picked. Potentially fine vintages have been ruined by a hailstorm days, even hours, before the harvest. In warmer regions, a drought can also prevent the grapes from ripening properly. And microclimate also comes into play: why is one vineyard better than another, even when they are only separated by a narrow track? Maybe the better vineyard tends to enjoy a little more sunshine, or maybe there is a subtle difference in the soil.

Generally, vines like poor stony soil and do not produce good results from rich fertile land. Many of the world's great vineyards are in places where nothing else but vines will grow. Consider the granite slopes of the Douro valley, or the steep slate hillsides of the Mosel. Compare the production of vines grown on the wild, rugged foothills of the Massif Central in the south of France, with the dull, insipid wine from the coastal plains of the Midi. The soil type affects flavour, and some grape varieties prefer a particular soil. Pinot Noir and limestone is a happy combination in the Côte d'Or; Cabernet Sauvignon and gravel in the Médoc; Syrah and schist in Côte Rôtie; Gamay and granite in Beaujolais.

The French have a wonderful word, *terroir*, which is quite impossible to translate, for it does not simply mean soil, but covers all the other aspects of the vine's environment in the vineyard. It includes altitude (which may temper excessive heat), the degree of the slope (which affects the intensity of sunlight), and the aspect (whether the vineyard is north- or south-facing), or any permutation in between. There are other factors, such as the water table and the prevailing winds, that affect the vine and define the *terroir* and, ultimately, the flavour of the wine.

And finally that brings us to the human hand, to the winemaker who plays a pivotal role in determining the taste of the wine in a glass. As with cooks, the diversity among winemakers is infinite. Take a region like Chablis, where the only grape variety permitted is Chardonnay. The soil is a mixture of clay and limestone and the vineyard area is small enough for there not to be an extreme variation in the climate from one hillside to another. And yet there are as many different nuances of flavour in Chablis as there are producers of Chablis. There may be an underlying similarity in the wines, but no one Chablis will be exactly like another. And the reason is the winemaker. No one winemaker will make his wine exactly like another. They may be talented or careless, experimental or conservative in outlook, favour new oak barrels or stainless steel vats, bottle their wine as early as possible or leave it to age in barrel for months or even years. A good winemaker will be able to redeem an unpromising vintage, while a bad winemaker can spoil perfect grapes. The permutations are endless, and in some regions more variable than others.

The contrast between Europe and the New World is very marked in this context. In the more traditional parts of Europe, the winemaker has to follow what the French *appellation* laws call '*les usages locaux, loyaux et constants*', which basically means that the local viticultural practices and traditions must be firmly adhered to. If you own a château in Bordeaux, you may not plant Pinot Noir, but must remain with Cabernet Sauvignon, Cabernet Franc and Merlot. Traditionally there is a very good reason for this – Pinot Noir does not grow well in the Gironde – but you are not allowed to find this out for yourself. Similarly, if you have an estate in Sancerre, Sauvignon is the only possible choice of white grape variety, even if you would like to satisfy a hunch that Riesling might do rather well there. Local winemaking practices may also feature in the *appellation* regulations. For example, the grapes for Champagne and for Beaujolais must be harvested by hand as the regulations demand the pressing of whole bunches of grapes. But if you have a vineyard in New Zealand or Chile, there is nothing to stop you from planting exactly what you want to wherever you want to. The only constraint might be quarantine laws, which may mean that a particular grape variety is not yet available in your country. While more obscure varieties such as Petit Manseng or Vermentino have yet to find their way to the Antipodes, in recent years New Zealand has seen a sizeable increase in the choice of grape varieties available to growers there, such as Viognier, Roussanne, Petit Verdot or Montepulciano, whereas 20 or even ten years ago the choice was much more limited. Of course, the good wine grower will carefully consider the conditions of the vineyard before deciding what to plant. It would be unrealistic to expect to produce elegant Pinot Noir in some of the hotter vineyards of Australia, or full-bodied Cabernet Sauvignon from the cooler vineyards of New Zealand.

The same grape variety vinified in different ways may give different results, but you will never completely lose the underlying character and taste of the grape, which may then be enhanced or overwhelmed, depending on what the winemaker has done. Sometimes you may be more aware of how the wine was made than of the grape flavour or any regional characteristics. Some grape varieties, such as Chardonnay, are more malleable, responding better to the winemaker's whims than others, such as Gewürztraminer, where the flavour of the grape is the main thing. The use of oak is the most obvious factor determined by the winemaker. An excess of new barrels can completely overwhelm a wine, while their discreet use may just add a certain indefinable extra flavour, in the same way that a touch of garlic may enhance a dish without its presence being obvious to the tastebuds.

There are many other aspects of winemaking that may make an impact on the ultimate flavour that we enjoy in the glass. They are too numerous to discuss in detail here, but include fermentation temperatures, the blending of different grape varieties or vineyard sources, the amount of residual sugar left in the wine, the ageing period before bottling and so on.

## The history of wine

The history of wine and winemaking is as old as civilization itself. Stories abound about how wine was first discovered, and one of the more delightful tells of a mythical Persian king called Jamsheed. At his court, grapes were kept in jars for eating out of season. One jar was discarded because the juice had lost its sweetness and the grapes were deemed to be poisonous. A damsel from the king's hareem was suffering from nervous headaches and tried to take her life with the so-called poison. She fell asleep, to awake later feeling revived and refreshed. She told everyone what she had done and of the miraculous cure, and thereupon 'a quantity of wine was made and Jamsheed and his court drank of the new beverage'. And that is it in a nutshell. Someone, somewhere in Asia Minor, possibly in modern Anatolia or Georgia, put wild grapes in a container, which were pressed by their own weight. The resulting juice began to ferment and a new drink was discovered that was to give untold pleasure to an untold number of people. Of course the liquid would then have gradually turned to vinegar as the effects of oxygen took their toll, coining the phrase that it was Man who made wine, whereas God gave us vinegar.

The great civilizations of Ancient Greece and Rome trace wine back into their pre-history, with similar legends about its discovery. Ancient Egypt has left us wine lists and wall paintings; indeed they even recorded the vintage, vineyard and winemaker on individual jars of wine. The Babylonians instigated laws to regulate the running of a wineshop and wrote vivid descriptions of a magical, jewel-bearing vineyard in the Epic of Gilgamesh.

The first important step was the cultivation of the vine. Archaeologists are able to tell whether grape pips found at the site of ancient settlements are from wild or cultivated grapes. The earliest cultivated grape pips to have been unearthed date back some 7000 years and were found in the Caucasus at the eastern end of the Black Sea in what is now the Anatolian part of Turkey and Georgia, in an area well suited to the cultivation of the grape. The most crucial thing was that the Ancient Greeks and later the Romans made wine an important part of their lives, whereas in other parts of the world where vines grew wild, such as Persia, India and China, wine came and went leaving little trace. However, in the Hellenic world the god Dionysus was an important figure, as was the god Bacchus to the Romans. Dionysus is said to have brought the vine to Greece from Asia Minor, and the Greeks took wine across the Adriatic. In turn, the Romans spread viticulture throughout Europe, taking it to France, England, the area we now call Germany, and the Iberian peninsula. With the spread of the Roman Empire, the cult of Bacchus was important. The Romans defined the best vineyards of southern Italy, which produced wines such as Caecuban, Massic and Falernum, which were rated highly by Pliny the Elder. The great wine port of Ancient Rome was Pompeii, where one wine merchant was rich enough to build both the theatre and the amphitheatre.

Even more significant was the use of wine in the Eucharist as a symbol of Christ's blood and of life itself. There are numerous references to wine throughout the Bible. Noah is attributed with being the first drunkard, and throughout the Old Testament wine features as a symbol of prosperity. Christ's first miracle was turning water into wine at the wedding feast at Cana. The medicinal power of wine was recognized by St Paul, who exhorted Timothy to 'Drink no longer water, but use a little wine for thy stomach's sake and thine often infirmities'. Wine is inextricably linked with Christianity, and also forms a vital part in the ritual of Judaism, whereas, in contrast, it is prohibited in the Muslim religion.

As Christianity spread throughout Western Europe, so did the need for vineyards, and it was this need that ensured that wine production continued throughout the Dark Ages after the fall of the Roman Empire. Wine had become an essential part of the Mediterranean way of life, but north of the Alps, which suffered attacks from hordes of invaders as the Roman empire collapsed, the future of the Roman vineyards might have been less certain had it not been for the Christian Church and its need for wine for the Eucharist. The monastic communities across Europe maintained the vineyards, and as western Europe emerged from the gloom of the Dark Ages it was the monks who did much to improve wine. The mediaeval Cistercians of Burgundy were the first to study the soil of the Côte d'Or, selecting the best sites where the grapes ripened earliest. The monks of the Cistercian abbey of Pontigny played their part in the development of the vineyards of Chablis, while the inmates of another Cistercian monastery, Kloster Eberbach in the Rhineland of Germany, helped develop the vineyards of the Rheingau. They were aware of the benefits of pruning the vines and of selecting vineyard sites where frost was less likely to be a hazard, a factor of immense significance in these northern climes.

The monks needed wine, not only for the Eucharist, but also for their guests, because the monasteries were the three-star hotels of the Middle Ages for wealthy travellers. Wine was also significant as a trading commodity. As such it had been important to the Greeks and Romans, and it regained its importance once again in medieval Europe. The port of Bordeaux flourished as a result of the export of wine to England, Scotland and northern Europe. Ease of transport was crucial, and above all that meant rivers at a time when travel overland was dangerous and roads were often impassable. Wine was transported in barrels, which are heavy and cumbersome, so carriage by water along rivers to the coast was essential. Thus the vineyards around Bordeaux developed, and the Bordelais jealously guarded their privileged position when faced with competition from vineyards further downstream. The wines of the Rhône valley flourished, as did those of the Mosel and the Rhine; Burgundy found its outlet in Paris because that is where its river systems led.

Trade between London and Bordeaux flourished; Edward II of England ordered the equivalent of more than one million bottles to celebrate his wedding to Isabella of France in 1308. Under Elizabeth I almost three centuries later the English were drinking more than 40 million bottles of wine a year, in addition to beer and cider, at a time when the population was just 6.1 million. In effect, wine was a necessity rather than a luxury because water supplies, especially in cities, were often impure, and wine was known to be an antiseptic that would help to alleviate the effects of contaminated water.

It was in the late seventeenth and eighteenth centuries that the connoisseur appeared, along with the realization that wine had an aesthetic appeal, that it was not simply a beverage but also a drink with many life-enhancing qualities. England and France saw the rise of a social group with money and taste who were prepared to spend money on fine wine. In England, men like the first Prime Minister, Robert Walpole, sought out the finest red wines from Bordeaux. Arnaud de Pontiac, president of the *parlement* of Bordeaux around 1660 and owner of Château Haut Brion, is generally accredited with pioneering a quality approach to winemaking, with small crops and careful cellar practices. Other Bordeaux estates, such as Latour, Lafite and Margaux, followed his example. The development of glass bottles allowed wines to be kept, and for sparkling wine, in the form of champagne, to be bottled. The use of cork stoppers was also developing at around the same time that Dom Pérignon was refining the art of blending champagne. He actually tried to eliminate the presence of bubbles in his wine, without much success, but at least the development of stronger glass ensured that the wine could be bottled safely.

Sherry was known of in Elizabethan times, with sherris sack being enjoyed by Shakespeare's Falstaff, while port was turned from a table wine into a fortified wine in the eighteenth century when it was realized that the addition of brandy helped to stabilize it for the long sea journey from Portugal to northern Europe. British merchants with names like Sandeman, Graham and Croft did much to develop port as the drink we know today, and the British also played their part in the development of other fortified wines, such as Marsala and Malaga.

Wine was also affected by politics. The Methuen treaty of 1703 established Portugal as a favoured source of wine, in preference to France, as reflected in a popular jingo of the time, attributed by some to Jonathan Swift:

> Be sometimes to your country true,
> Have once the public good in view;
> Bravely despise champagne at court
> And choose to dine at home with Port.

The future of wine was looking rosy, but then *Phylloxera* was discovered in the south of France in middle of the nineteenth century. This tiny aphid feeds off the roots of vines and ultimately kills them. It was introduced quite by accident from North America when steam ships began to cross the Atlantic quickly enough for the pest to survive on imported plants. From the south of France it spread across Europe, causing considerable devastation in its wake. Many vineyard areas were never replanted, notably the extensive vineyards around Paris, even once the solution of grafting vines on to American rootstock was discovered.

Two other diseases, oidium (or powdery mildew) and downy mildew, also caused considerable damage at that time, but the remedy of sulphur spraying was more easily accomplished. *Phylloxera* remains a constant threat, and there are very few areas in the world – the main two exceptions being Chile and Cyprus – where ungrafted vines survive without there being fear of infection.

The twentieth century saw both lows and highs. Wine consumption was affected by the two world wars, and the economic crisis of the Great Depression in the 1930s also had a dramatic impact. The system of *appellation contrôlée* was developed in France and subsequently imitated in other European countries. It was borne out of an urgent need for some kind of wine regulations in order to protect the consumer from fraud. In 1911 rioters in Champagne protested against wine from outside the region being sold as champagne. Ordinary wines were often passed off as something much finer, and wines were often adulterated. In the south of France it had not been unknown for so-called wine to be made from a mixture of raisins, brandy and hot water. Since 1936 the *appellation* system has gradually been extended to encompass the principal vineyards of France, determining the grape variety or blend of grape varieties, methods of production in vineyard and cellar, yields and so on. The vineyard boundaries are meticulously delimited, as *terroir* is all important here. *Vin délimité de qualité supérieure* forms a lesser category, while *vin de pays* covers a broad range of wines that may not conform to the traditions of the region but nonetheless retain some local identity. In the Midi (south of France), for example, they can include some of the best and some of the worst, but may still be labelled Vin de Pays de l'Hérault, without any quality distinction. The key here, as always, is the producer's name on the label. Below this category is the anonymous *vin de table*, which is of uncertain provenance and is usually sold in bulk without any label by village co-operatives. In Italy they have *Denominazione di Origine Controllata*, in Spain *Denominación de Origen* and in Germany a system of *Qualitätswein*, all based on similar guidelines.

## The New World

Europeans, not unnaturally, took viticulture with them wherever they colonized the New World, for even if they did not come from a wine-producing country, wine was nonetheless still an intrinsic part of their culture. The first to do this were the Spanish missionaries who went initially to Mexico, and then, around 1770, to what is now called California. The Gold Rush of 1849 helped to spread wine-growing throughout the state, and it was realized just how suitable California is for growing grapes. Although Prohibition between 1920 and 1933 wreaked havoc on the industry, it has since flourished in the bountiful climate of the Sunshine State. From Mexico the Spanish missionaries also went south, encouraging viticulture with grape varieties such as Pais and Criolla, which may have originally come from Spain, reaching Peru in the late sixteenth century and then on to Chile and Argentina. In the middle of the nineteenth century, happily before *Phylloxera* had reached Europe, French grape varieties were introduced into Chile, which remains for the moment *Phylloxera*-free, protected by its natural boundaries of desert, mountain, ocean and icepack.

The first European settlers in South Africa planted vines. The Dutch settler and first Governor of the Cape, Jan van Riebeeck, imported grape cuttings and planted a vineyard. His diary for 2 February 1659 records the first harvest of grapes in the southern hemisphere with the words: 'Today praise be to God, wine was pressed for the first time from Cape grapes . . .'. Simon van der Stel, van Riebeeck's successor as governor, was also a wine enthusiast and established vineyards in Constantia with the help of French Huguenot refugees who were skilled in winemaking and viticulture. Today there are numerous French names among the wine estates and winemakers of the Cape, especially in the area of Franschhoek, which translates literally as 'French valley'.

The vine first arrived in Australia from the Cape of Good Hope in 1788 and was planted in Governor Philip's garden in what is now the centre of Sydney. Viticulture gradually spread throughout the south eastern corner of the continent during the nineteenth century, and then to Tasmania and Western Australia, encouraged by the European immigrants who wanted to drink the wines that they had enjoyed at home. Today, Australia is rivalling France as the main source of wine imports into the UK.

A British missionary called Samuel Marsden planted the first vines in New Zealand in 1819. There are no records to show that he actually produced any wine, but the assumption is that he did, if only a tiny amount for the Eucharist. It was James Busby, who is widely regarded as the father of Australian viticulture, who takes the credit for the first recorded New Zealand wine in 1836. Before settling in New Zealand he had toured the vineyards of Europe, selecting cuttings of French and Spanish grape varieties for planting in New South Wales. Many were established in the Sydney Botanical Gardens, some in Busby's own vineyard in the Hunter Valley and others he took with him to the Bay of Islands on New Zealand's North Island. The French explorer d'Urville tasted the wine and recorded that he was given 'a light white wine, very sparkling and delicious to taste, which I enjoyed very much'.

## Choice

The world of wine has grown enormously in the past 20 years, and its international face and accompanying tastes have changed beyond recognition. Twenty years ago France held sway. It produced some of the world's finest wines, and today, without question, still makes wine that winemakers from the New World use as a yardstick and seek to emulate, be it champagne, claret or white Burgundy.

The other main European contributions to the world's wine scene have come from Germany, with elegant Rieslings, from Spain with sherry, and from Portugal with port. Although wines like Chianti and Barolo from Italy, Dão from Portugal and Rioja from Spain were known of 20 years ago, they were not necessarily the wines that they are today. Eastern Europe was dominated by brands like Lutomer Laski Riesling and Bull's Blood, and as for the New World, South Africa produced sherry and Australia port. California and Chile were hard to find on our shelves in 1980, and in New Zealand 1980 marks the first commercial vintage of Montana's Marlborough Sauvignon, the wine that was to convert us to the charms of the Antipodes.

In contrast, look at the choice today. Go into any wine shop in London, which is generally considered to be the wine capital of the world, and you are confronted with a bewildering choice. It could be argued that we are living in a golden age of wine. Even some of the smallest shops provide a showcase of the world's wine production, with wines from every European wine region from England to the toe of Italy, and from the Eastern Europe to the Atlantic seaboard of Portugal, and from all five continents. Asia is represented by Château Musar in the Lebanon, sparkling Omar Khayyám is produced in India, and, incredibly, Thailand can offer two vintages a year, differentiated by the seasons, dry or rainy. The most common French grape varieties have travelled the world, so that you can find Chardonnay in umpteen different countries, from Chile to New Zealand, via over 40 North American states to Austria and Switzerland. Californian Cabernet Sauvignon, Australian Chardonnay or New Zealand Sauvignon are commonplace, but most of the New World countries are also seeking to find the element of originality that is not available in Europe.

Zinfandel is the grape of California. At first it was thought to be unrelated to any European grape variety, but it has been proven to be a relation of the Primitivo of southern Italy. In California it has been decried and taken for granted, but in the right hands it produces wonderfully rich satisfying wines, rather than the anaemic white Zinfandel. And now Primitivo is coming into its own in Puglia.

As for South America, you find Malbeck in Argentina, not to mention Torrontés, a perfumed white grape variety that does not seem to have a European equivalent. Uruguay, with its Basque traditions, is developing Tannat, found more commonly in Madiran, a small *appellation* of the Pyrenees in south west France, and Chile has distinguished Carmenère, one of the original grape varieties of the Gironde, which today is rarely found in Bordeaux.

South Africa has Pinotage, a crossing of Cinsaut and Pinot Noir. Developed there in the 1920s, it was once used for high volume wines, but today it is increasingly vinified as a serious red wine, with satisfying results. Syrah may come from France, but Australia has given it an individuality of its own, with Barossa Valley Shiraz, while Hunter Valley Semillon has a distinctive flavour, quite unlike anything produced in Bordeaux. And then there is the unique and wonderfully luscious Liqueur Muscat from Rutherglen in northeast Victoria. New Zealand has taken Sauvignon and created a brand new flavour quite different from anything we can find in the Loire Valley, while Canada offers the world ice wine from frozen grapes with an annual consistency that is impossible in Germany or Austria.

The New World wine labels are a consumer's delight as they give you the essential pieces of information: the grape variety, and thereby a basic indication of flavour, and the name of the producer, with the indication of quality which that entails. The Old World, in contrast, places much more emphasis on *terroir* and expects you to know that Chablis comes from Chardonnay or Côte Rôtie from Syrah. But the signature of the winemaker, the winery or estate name, is equally important throughout the world.

## Improvements in winemaking

Making wine is a relatively simple process, with the yeast on the grape skins naturally turning the sugar in the ripe grapes into alcohol and carbon dioxide.

Far more complicated is the process of ageing it, allowing it to develop in wooden barrel and glass bottle without detriment from the contact with air. The second half of the twentieth century saw the most enormous developments in winemaking techniques, bringing a vast improvement in taste and flavour. Aspects of vinification that were previously little understood are now mastered and controlled. Malolactic fermentation is a key example of this. When the wine became agitated in the spring as the ambient temperature rose, it was believed to be reacting in sympathy with the sap rising in the vines. Now it is understood that a malolactic fermentation is taking place, with the conversion of malic acid, as in apples, into softer lactic acid, as in milk. At first wine growers thought that they had no control over this phenomenon and that it happened of its own accord. Then they realized that temperature came into play, and that by keeping your cellar warm during the late autumn you could encourage *le malo*, as the French call it, to take place earlier. Nowadays the more interventionist winemaker can even make the fermentation start at will, or indeed prevent it from happening, as is preferable for some white wines.

The importance of temperature represents another fundamental improvement in winemaking practices. Yeast only operates between certain temperatures, about 10°–30°C, up to 35°C at the very most. In the bad old days, if the temperature rose above 30°C, the growers resorted to adding blocks of ice to their vats, or were helpless in the face of the elements. Then various cooling systems were invented, and now most modern wineries can control their fermentation temperatures to the precise degree, with the aid of a computer. It is generally recognized that a fermentation benefits from different temperatures at different stages during its course and that white wines generally prefer a cooler temperature than red wines, making for fresher, fruitier flavours. Temperature is also an important factor for storing wine. One of the perennial problems in the south of France, for instance, is the lack of insulated warehouses or underground cellars as the summer temperatures of the Mediterranean can have a distinctly adverse effect on cartons of wine kept in a warehouse without any airconditioning.

Winemaking equipment has generally become more sophisticated. Presses are no longer the brutal beasts of the past that did nothing more than squash the grapes. The port trade, however, has yet to find an improvement on the human foot as a means of extracting the maximum flavour and extract without imparting bitter tannins from crushed pips and stalks. It is now possible to remove stalks without overly damaging the grapes themselves. Filters are also more sophisticated, and efforts are made to avoid excess pumping of juice and grapes. There is nothing to beat using gravity, with the grapes arriving at a level above the fermentation vats. This is something that the Etruscans, the very first producers of Orvieto, a white wine of central Italy, were aware of in the seventh century BC.

The subtleties of ageing wine in barrel or vat are now better understood. In the days before the development of cement and steel, wood was really the only choice for a storage container. The Romans used amphora, but these were later replaced by sturdier barrels, usually of oak, though chestnut was traditional in central Italy and raule was used in Chile. Tradition also determined barrel sizes. For example, why does the Chablis *feuillette* contain only 132 litres, while the Burgundian *pièce* used in the Côte d'Or is nearly twice the size? No one really quite knows. A port pipe also varies in size, depending on whether it is a Douro pipe or a shipping pipe, and a Bordeaux *barrique*, which is now used throughout the New World, holds 225 litres. In contrast, the south of France and Italy favoured much larger barrels, enormous *foudres* or *botti* of 100 hectolitres or more. These, however, were never used for transporting wine, but for storage, unlike the *barriques* or *feuillettes*.

Today it is the Bordeaux *barrique* that holds sway all over the world as the desirable barrel size for ageing wine. It seems to provide the right ratio of wine to wood surface, allowing for a very gentle, imperceptible element of oxygenation, rather than oxidation, during the ageing process, thereby allowing the flavour of the wine to develop. Not all wines benefit from oak ageing, nor do all wines benefit from new oak barrels, which tend to give quite a strong, almost overpowering flavour to the wine. The success of what the French call *élevage*, a word that means upbringing and could equally be applied to children, depends very much on the quality of wine. One quip has it that there is no such thing as an over-oaked wine and that the problem lies with the response of wines with insufficient body and extract to a heavy-handed use of oak. The obvious contenders for oak ageing are classed growth clarets, *grand cru* Burgundies, the finer wines of the Rhône valley such as Côte Rôtie and Hermitage, and, for white wine, *grand cru* Burgundies, which are usually fermented, as well as aged, in oak, along with the obvious parallels from the New World. However, that list is by no means conclusive and much depends on the winemaker's personal preference and what he or she is trying to achieve with the wine. Nonetheless, the benefit to today's wine-drinker is that the winemaker has a much better understanding of what oak will do to a wine, and so quality is significantly enhanced.

The bottling process is also much more carefully controlled, with meticulous attention paid to hygiene. The joker in the pack, however, remains the cork that is affected by trichloranisole (TCA), which imparts an unpleasantly musty taste to the wine. The cork industry is finally addressing this problem, trying to find a means of detecting cork taint and also of preventing its occurrence. To be fair, TCA does not occur only in cork; there are other sources of the contamination, but cork, as the offending stopper, is the most obvious culprit. There are alternatives to pure cork: plastic corks and composition corks have been developed, but they can bring problems of their own. The Stelvin or screwcap is also favoured by some producers whose wines do not require lengthy ageing in bottle, such the Riesling producers of Australia's Clare Valley and a group of New Zealand Sauvignon producers in Marlborough. But not unnaturally there is consumer resistance. A screw cap may be efficient, but it has 'cheap and nasty' connotations. And who among us can resist that wonderful inimitable sound of a cork being extracted from a bottle? There is no doubt that presentation does count enormously in the enjoyment of wine.

## Presentation

It is true that we can drink wine out of a plastic beaker and essentially it will taste the same as out of a glass, but our tastebuds and impressions are influenced by external circumstances, so in some ways it will not taste the same. It has also been proven that the same wine can taste differently in varying shapes of glass. This has led the Austrian glass producer Georg Riedel to refine his range of wine glasses, devising different glass shapes for different wines with meticulous detail. Apparently a particular shape of glass will result in the liquid hitting your tongue and taste buds in a particular spot, so that you will register one flavour more than another. So if the wine is particularly tannic, such as a rather astringent Sangiovese-based Chianti, the glass shape will propel the liquid away from the part of the mouth that instantly registers tannin to the part of the mouth where the underlying fruit in the wine is more likely to be appreciated.

For most of us, however, this is a fine science; what really matters is a glass that looks attractive on a dinner table. It is perfectly possible to use the same shape of glass for red and white wine, and even for sparkling wine. That does not mean the dumpy Paris goblets favoured by most public houses, but an elegant glass with a bowl that is large enough to contain a reasonable amount of wine without being overly full. A wine glass should never be filled to the brim, but to about two thirds full, so that the glass can be gently shaken to release the bouquet without detriment to a tablecloth. The bowl should be tapered, curving inwards towards the rim, as this also helps to concentrate the bouquet of the wine. The champagne cups supposedly modelled on Marie Antoinette's breasts do a great disservice to champagne as they allow the delicate mousse, on which so much care has been lavished, to escape into the atmosphere. A tall, fluted glass is the ideal here, while smaller glasses are better for dessert and fortified wines. The Jerez region of Spain has a glass of its own, the narrow *copita*, which concentrates the aromas very effectively. Coloured glass, beautiful though it may be, is best avoided as it detracts from the colour of the wine, removing one of the anticipatory pleasures of wine-drinking.

The use of a decanter can also enhance the visual as well as the drinking pleasure of a bottle of wine. There is something enormously appealing about a decanter of red wine glinting, perhaps in candlelight, on the dining table. Apart from the aesthetic appeal, decanting a red wine or port has two purposes. Most good quality red wines that are destined for some bottle-ageing before drinking, such as fine claret, Rhône wines and of course vintage port, will shed some colour pigmentation during the ageing in bottle. Quite simply this is unsightly in the glass and not very appetizing to look at, so the solution is to stand the bottle upright for 24 hours before serving it, allowing the fine particles to settle naturally to the bottom of the bottle. It is then perfectly possible to pour the clear wine into the decanter with the help of a candle or bright light positioned at the neck of the bottle, and leave the deposit in the bottle. But in decanting you are also allowing the wine to breathe. Contact with air, particularly oxygen, brings out the flavours of a wine and in decanting a wine, which may have been in the bottle for as much as 20 years, you are allowing it the opportunity to 'stretch its legs' and to recover its composure before it performs at your dinner table. The catch-22 question, of course, is how long before a meal to decant? It is really only possible to offer nothing better than a guestimate. Suffice it to say, that it is better to err on the side of caution and observe the wine developing during a meal, rather than decant it hours beforehand and find a faded beauty lingering in the glass.

Temperature is another important factor in the enjoyment of wine. The wrong temperature can completely destroy the flavour of a wine, while the right temperature brings out its flavours as a harmonious whole. Perceived wisdom used to say that a red wine should be *chambré*, at room temperature, but that was before the days of central heating. The room temperature of a twenty-first century house is much too warm for the average red wine; opt instead for cool room temperature, but certainly not a cold room. A red wine that is too cold will taste angular and awkward as the cool temperature emphasizes its tannins and acidity while numbing its fruit. On the other hand, a red wine that is too warm will taste flabby and spineless. White wine, and also *rosé*, can be too cold, again numbing the flavour to the extent that the wine tastes of nothing but icy water, but too warm and the wine will be soft and flabby. More full-bodied wines benefit from a slightly higher temperature than light wines. Champagne needs to be colder than, say, white Burgundy, especially as the chill will also control the bubbles and the exit of the cork from the bottle.

## Keeping wines

There are some statistics about the life expectancy of a wine bottle once it has been removed from the supermarket or wine merchant's shelf showing that a very high proportion of wine is drunk within 24 hours, or even minutes, to a week of purchase. And indeed much of the wine on our wine merchant's or supermarket shelves is destined for relatively early drinking. There is usually little virtue, other than curiosity, in keeping Muscadet or Orvieto, Beaujolais or Bardolino, but with more complex wines that have already been given some *élevage* in barrel, or maybe in vat, much is to be gained from extra bottle-ageing. The ageing process allows for the youthful edges of the wine to mellow, for the tannins to soften and for the adolescent wine to become a harmonious adult. The same applies with good white wines, for ageing in bottle allows any overtly oaky flavours from a fermentation in barrel to evaporate and for extra layers of taste to develop. Chablis is a fundamental example of this, with a curious chameleon quality that leads you to think that the wine has been aged in barrel when it has not been near a stave of oak. Young Chablis has a fresh minerally fruit with a firm backbone of acidity; it then goes into a period of adolescent gangly youthfulness, when it can seem plain boring or edgy and unharmonious, but leave it for five years, or longer in the case of *premier cru* and *grand cru* Chablis, and it develops all manner of subtleties that would have been difficult to discern in its youth. What the French call *pierre à fusil*, the mineral gunflint notes, begin to develop and it is these than can lead you to think that the wine has spent months in an oak barrel. Your patience is well rewarded by a startling and immensely pleasurable transformation.

## Labels

The label on the bottle is another aspect of wine-drinking. The label provides the key: it tells us what wine is in the bottle and where it comes from. Various legal conventions dictate what must go on the label, such as country of origin, alcoholic degree, producer, and so on. But the label also does more than give us facts. Deliberately, or sometimes subconsciously, its design conveys an impression. The label is our first point of contact with the bottle. Do you judge a wine by its appearance, as you might do a person you are meeting for the first time? Its appearance on the wine merchant's shelf must say: 'Buy me, drink me! This is what I taste like.' There is no doubt that we are influenced by appearances. The classic elegant design of a claret label leads us to expect a serious wine with stature, whereas a cheerful colourful label automatically implies a simpler, less sophisticated wine. Some producers may attach more importance to the appearance of their bottles, while others merely accept whatever their local printer has to offer them. The ultimate in sophistication is at Château Mouton Rothschild, where a famous contemporary artist is commissioned each year to design the label, so that works by Picasso, Chagall and many others are now collectors' items. Without a doubt, an attractive label on a bottle on the dining table enhances the anticipatory pleasure. And that is really what wine is all about. Pleasure and enjoyment; good company and conviviality.

## Health

The health aspects of wine-drinking are covered at much greater length and in meticulous detail elsewhere in this book. As someone with no medical knowledge whatsoever, I would simply venture to say that wine makes me feel better. There is no greater pleasure at the end of a day, maybe a day that has been beset by stress and mishap, than sharing a bottle of wine around a dinner table with family and friends. A glass of wine helps one to relax and therefore must reduce prescriptions for sleeping pills or antidepressants. I know from personal experience that it helps the digestive juices, stimulating them as an aperitif, while some dishes would be unthinkable without a glass of wine to aid the digestion. I could not envisage a plate of pasta accompanied by water, nor a delicious roast chicken. Indigestion would be sure to follow.

## Wine and culture

Over the centuries wine has become an essential part of the economic life of many towns and villages all over Europe. Towns as diverse as Bordeaux, Oporto, Jerez and Marsala would probably not have flourished without the wine trade to sustain them. More than half a million people are employed in the wine industry in some way or other in France; certainly there are innumerable villages all over France that would have died without viticulture to sustain them, for it is on vines alone that the livelihood of the villagers depends. There are numerous parts of the Mediterranean where little else will grow apart from vines, and maybe olives trees, and where the success of the village wine cooperative is fundamental to the economic and social fabric of the village. Why is Tuchan in the heart of the Corbières hills a bustling thriving village in comparison to its neighbours? The answer is quite simply that it has one of the most dynamic wine cooperatives in the whole of the south of France. There are numerous villages and towns that would be unknown to the outside world but for their fine wine production: what takes people to Chablis, St Jean-de-Minervois or Banyuls, to name just a few, but wine?

Wine has featured extensively in literature. The Greeks and Romans wrote of wine; Homer at the end of the eighth century BC was the earliest Greek author to write about wine, and many others followed his example. The eleventh century Arabian poem *The Rubáiyát of Omar Khayyám* contains many references to wine, including the lines: 'I often wonder what the vintners buy, one half so precious as the goods they sell'. It continues to hold its place in European literature following its translation by Edward FitzGerald in 1859. References to wine in English literature are relatively common from Chaucer onwards, providing an intriguing record of fashion in wine styles and a history of the wines available in the British Isles. In Shakespeare Falstaff drank sherris sack and Sir Toby Belch called for 'a cup of Canary'. Samuel Pepys mentioned wines grown around London and enjoyed Ho Bryan and champagne. Sheridan in *A School for Scandal* wrote of claret that 'women give headaches, this don't'. What we think of as a sturdy red wine, Hermitage, was deemed by the Scottish novelist Tobias Smollett in the eighteenth century as a small wine, and his hero Humphrey Clinker considered that 'life would stink if he did not steep it in Claret'. Samuel Johnson gives an insight into the prodigious quantities of wine consumed in the eighteenth century, though wine was generally much lower in alcohol in those days. Jane Austen included an occasional reference to her characters enjoying wine, but always in genteel quantity. South African Constantia was considered an appropriate restorative for a young lady in *Sense and Sensibility*. Byron recommended 'hock and soda water' as a hangover remedy, and of course the hangover is one of the hazards of an overenthusiastic enjoyment of wine, leading most of us to follow the precepts of moderation.

The authors of the nineteenth century indicate the growing range of wine important in the British Isles. Thomas Love Peacock described wine as both a 'hierarchical and episcopal fluid' and 'the elixir of life'. William Thackeray named Beaune and Chambertin and referred to claret, sherry and Madeira. 'If there is to be Champagne, have no stint of it . . . save on your hocks, sauternes and moselles, which count for nothing' is a quote from *Pendennis*. Robert Louis Stevenson in the *Silverado Squatters* made the first literary reference to the vineyards of California. Charles Dickens mentions punch more than wine.

The twentieth century shows yet more variety and quality. Galsworthy referred to hock in the *Forsyte Saga*; Aldous Huxley observed that 'champagne has the taste of an apple peeled with a steel knife'. The characters of P.G. Wodehouse are enthusiastic imbibers and wine features in the scenario of *Brideshead Revisited*. And finally Ian Fleming gave James Bond rather common tastes in champagne, causing him to mention non-existing vintages of Taittinger or Dom Pérignon. Nonetheless it illustrates the power of pen and the literary inspiration of wine.

And that bring us back to the hazards of enjoying wine. The alcohol content of wine, while part of the enjoyment, inevitably entails an element of intoxication and this has been recognized since the first drinking days. Excessive drinking has always had moral or religious connotations, beginning with the intoxication of Noah after the Flood. Attitudes have varied about what constitutes excess. An early Mesopotamian tale described a man drunk from strong wine thus: 'he forgets his words and his speech becomes confused, his mind wanders and his eyes have a set expression'. The Old Testament contains many warnings against drunkenness alongside the positive benefits of temperate wine-drinking. Ancient Greece also seems ambiguous in its attitudes, with guidelines to curb excess. Plato advised no wine before the age of 18 and moderation until 30. The aim of the all-male drinking party known as the symposium, with poetry, entertainment and debate, was aimed at pleasant intoxication but not the loss of reason. There are stories of legendary Roman drunkenness, and in mediaeval Europe high-living monks earned a reputation for drunkenness, while the discovery of the art of distillation in the twelfth century also increased the scope for drunkenness. By the eighteenth century wine was relatively expensive and those seeking intoxication preferred spirits, as immortalized in Hogarth's cartoon *Gin Lane*, inscribed with the words: 'Drunk for a penny, dead drunk for tuppence'. The Victorian age saw the development of the Temperance movement, while the twentieth century witnessed Prohibition in the USA and the lingering legacy of dry counties. The hangover, the unfortunate and inevitable consequence of an excessive enjoyment of not only wine but any other alcoholic drink, claims many cures, perhaps the most popular of which is the so-called 'hair of the dog'.

So maybe enjoyment entails moderation, and drinking wine with a meal to lessen the impact of alcohol. The flavours of wine are inevitably enhanced by what you eat with them. Much has been written on wine and food combinations, or pairings as they are now called, but really there are no hard and fast rules. It is your own tastebuds that determine whether you are delighted or disgusted by the combinations you choose. Rules are made to be broken, to which end there is no harm in experimenting with flavours and dishes. Perceived wisdom has it that red wine goes with red meat, game and cheese, while white wine goes with fish. However, a red wine with more acidity than tannin will happily accompany fish, while goat's cheese is perfect in combination with white wine. A sweet wine will enhance the salty flavour of a blue cheese, and the classic French combination of Sauternes and foie gras is incomparable. In a restaurant, a wine is often expected to accompany several different dishes, and very frequently it will succeed. Regional combinations also work well: Chianti with spaghetti comes to mind, or Muscadet with shellfish, and in France the local wine with the local cheese. There are some wines that are delicious drunk on their own, unaccompanied, such as champagne, many Rieslings from Germany and some dessert wines that are not sweet enough to accompany a pudding successfully or else too sweet for a simple piece of fruit. And then there are wines that demand food to bring out their true qualities. Italian red wines are above all food wines, especially the better ones, which the Italians call *vini da meditazione*, wines for meditating over, not for quaffing, wines over which you linger, observing them develop in the glass. Some logic in the serving of wine may also help: go from dry to sweet, lighter bodied to fuller bodied and from younger to older.

Drinking wine is above all about pleasure; it is about friendship, love, conviviality, good conversation, wit and humour, the enjoyment of some of the finer things of life. A bottle of wine will enhance a simple meal, turning it into a special occasion. There are people who remember an incident by the wine associated with it, coining the quip 'I can't remember her name, but the wine was Chambertin'. There is no doubt that particularly memorable bottles have associations with enjoyable occasions, and that drinking wine constitutes part of a fundamental enjoyment of life, alongside music, art and literature.

# 2 The history of wine as a medicine

*P. A. Norrie*

## Introduction

Viticulture, or grape-growing, began in Georgia (which lies on the eastern shore of the Black Sea, near the Caucasus Mountains) some 9000 years ago (Johnson 1989, p. 17). From here it spread to all the great wine-loving cultures of the Middle East via the Tigris and Euphrates rivers to Mesopotamia, and then on to Persia, and each of these cultures has its own myth about the origins of wine. For example, in Persia (today's Iran) wine is supposed to have been discovered by the mistress of King Jamsheed, who was so fond of grapes that he had them stored in jars so that he could eat them all year round. One particular year, the grapes in one jar had fermented and were no longer sweet, so he assumed the new liquid in the bottom of the jar was poisonous and marked the jar accordingly. His mistress, so the story goes, had a bad headache and wanted to die, so she drank the liquid in the 'poisonous' jar. The wine made her feel better, greatly easing her pain and letting her fall asleep. Upon hearing of this miraculous cure, King Jamsheed tested the 'poison' himself and enjoyed the wine so much that the wonderful tonic was named the 'Royal Medicine' (Johnson 1989, p. 23). Wine was consequently held in the highest esteem by the Persians because of its fame as a cure, and this may have given rise to the oldest desert proverb: 'He that hath health hath hope, and he that hath hope hath everything'. Thus wine was the choice of the privileged.

In the ancient world life was short and harsh, so when something came along, such as wine, that had the power to make one feel good and even to preview paradise in the hereafter, it was adopted with enthusiasm. Wine was stronger than ale, the alcoholic beverage of poorer people from about 6000 years ago; it could be kept longer than ale; it improved with keeping and it tasted better; thus it was valued more highly than ale and continues to be so today.

Later it was realized that wine aided the digestion of meals that may have been bland and normally hard to digest, given the cruder cooking techniques of the past, and thus wine-drinkers became better nourished, stronger, healthier, more confident and more capable than others. 'It is no wonder that in many early societies the ruling classes decided that only they were worthy of such benefits and kept wine to themselves' (Johnson 1989, p. 12).

## Mesopotamia

Armenian wine sellers to the south of Georgia spread the knowledge of wine down the River Euphrates to the cities of Uruk, Ur and later Babylon and Kisk (today's Iraq) where the Sumerian culture began around 5000 BC. Here, in part of the fertile crescent – an arc of land from the Zagros Mountains in Iran in the east to southern Israel and Jordan in the west – domestication of animals and plants began, thus allowing people to settle in the one place permanently and hence develop cities. Grapevines were imported from Armenia, Syria and Lebanon into Sumer. The Sumerians had the earliest form of writing that we know of, consisting of stylized pictures called pictograms drawn using a stylus on a moist clay tablet. The oldest known medical handbook, a Sumerian pharmacopoeia written on a clay tablet dated approximately to 2200 BC to 2100 BC and excavated at Nippur in 1910, recommended the use of wine with various drugs as treatments for various ailments, such as sweet wine with honey to treat a cough (Bang 1973). '*Tabatu*' was a Babylonian medical drink made from water and small amounts of fermented fruit juice or wine. This makes wine man's oldest documented medicine (Burke 1984).

The Babylonian pharmacopoeia was extraordinarily extensive, using 250 medicinal plants, 120 mineral substances and 180 other drugs as well as solvents or vehicles for the actual medicinal substances – called menstruums – such as various kinds of milk, honey, oil and wine (Ciba Symposia 1940).

The Sumerians also provided the first reference in literature to a vineyard in the Babylonian 'Epic of Gilgamesh', written about the time of King Hammurabi in the eighteenth century BC (Skovenborg 1990, p. 4), plus the first representation of wine-drinking in the Standard of Ur. This is a wooden panel inlaid with semi-precious stones housed in the British Museum in London. It is 5000 years old and depicts seated courtiers raising their wine cups to their ruler (Johnson 1989, p. 26).

## Ancient Egypt

Wine was a gift to the Egyptians from their god Osiris, the God of Wine and of the six Nile floods. Ptah-Hotep was a nobleman who lived at Memphis in North Egypt about 2400 BC. In his tomb are the oldest inscriptions depicting winemaking (Burke 1984, p. 191). Egyptian papyri dating from 2000 BC record the medicinal use of wine. The medicinal wines used in Egypt were mainly from grapes but also from dates and palm sap.

Egyptian wines were stored in sealed clay jars that were imprinted with the seal of the owner, the name of the vineyard, the type of wine and the vintage, just like on a modern wine label. By the time of the fifth Dynasty

*Table 2.1* All known and translated Egyptian medical papyri

| *Name* | *Text* |
|---|---|
| Kahun | Gynaecological papyrus written in 1900 BC |
| Edwin Smith | Papyrus written in 1650–1550 BC but the original author was an experienced physician from 2200 BC |
| Ebers | Papyrus written in 1500 BC |
| Hearst | Papyrus written in 1500 BC |
| London | Papyrus written in 1350–1100 BC |
| Berlin | Papyrus written in 1350–1100 BC |
| Brugsch Minor | Papyrus written 1350–1100 BC |

(2494–2345 BC), six different vineyard appellations were recorded in Egypt. This all goes to show that viticulture and oenology were quite advanced and taken seriously in ancient Egypt. The funeral offerings of King Unas at the end of the Fifty Dynasty include five kinds of wine (Nunn 1996, p. 18).

The papyri show that wine was used (along with beer) as a solvent for mixing other medicines – whether they be plant, animal or mineral. Other mixing media included water, honey, milk and oil. But wine was also used as an integral part of a prescription. Table 2.1 lists all the known and translated Egyptian medical papyri.

From these papyri, an extensive knowledge of Egyptian medical practice can be gained, including hundreds of specific prescriptions and combinations of medicines (polypharmacy). These papyri also show a consistency of prescribing and 'contain evidence of having been copied in part from earlier medical treatises dating back to 2550 or even 3400 BC', thus making them Man's oldest wine prescriptions (Lucia 1963, p. 161).

Wine (*irep* in Ancient Egyptian) made from grapes would have enough alcohol in it to extract alkaloids such as in Ebers' prescription 287, which gives a remedy to cause the heart to receive bread (i.e. restore one's appetite like a tonic). The tonic remedy required that wine and wheat grouts 'spend the night' (i.e. get mixed and time to be dissolved) before being drunk (Lucia 1963, p. 140). The wine would have also made disagreeable components of a prescription more palatable and the 'mild intoxication would have eased the burden of many complaints' (Lucia 1963, p. 159). Ebers 9 and 12 show other examples of wine as a vehicle for dissolving materia medica (Lucia 1963, p. 159).

Ebers 804 gives a prescription 'to release a child from the belly of a woman' (Lucia 1963, p. 195) using wine, and wine was also one of the drugs used for the treatment of coughing (*seryt* in Ancient Egyptian). Most remedies for cough are in Ebers 190, 305–25, Berlin 29, 31–4, 36–47 and Hearst 61 (Lucia 1963, p. 161).

Grape wine was used extensively in prescriptions for anorexia (Skovenborg 1990, p. 7) because it would stimulate the appetite and be a good source of nutrition.

Six of the 10 prescriptions in Ebers 326–35 contained wine to drive out the 'great weakness', i.e. act as an uplifting agent (McGovern *et al.* 2000, p. 229). Wine was also used in salves, in enemas and in bandaging to prevent infection. Wine lees (the grape skins remaining after pressing) were also used in these ways, as well as being applied externally to bring down swelling in limbs and reduce fevers, as recommended in Ebers 162–3 (McGovern *et al.* 2000, p. 229).

Other Ebers' prescriptions follow 'To eradicate asthma: honey 1 ro (a mouthful, used as a measure in Ancient Egypt), beer 8 ro, wine 5 ro, are strained and taken in one day. To cause purgation: 6 senna (pods) (which are like beans from Crete) and fruit of . . . colocynth are ground fine, put in honey and eaten by the man and swallowed with sweet wine 5 ro. To cause the stomach to receive bread: fat flesh 2 ro, wine 5 ro, raisin 2 ro, figs 2 ro, celery 2 ro, sweet beer 25 ro, are boiled, strained and taken for four days. To expel epilepsy in a man: testicles of an ass are ground fine, put in wine and drunk by the man; it (i.e. the epilepsy) will cease immediately. To treat jaundice: leaves of lotus 4 ro, wine 20 ro, powder of zizyphus 4 ro, figs 4 ro, milk 2 ro, fruit of juniperus 2 ro, frankincense 1/2 ro, sweet beer 20 ro, [it] remains during the night in the dew, is strained, and taken for four days. Remedy for dejection: colocynth 4 ro, honey 4 ro, are mixed together, eaten and swallowed with beer ro ro or wine 5 ro' (Lucia 1963, p. 12).

Thus wine and wine lees, used internally and externally, were an integral part of Ancient Egyptian medicine.

## Ancient India

Medicine in India appeared around 2000 BC and developed independently of Mesopotamia, Egypt and Greece. During the Vedic period of Indian history (2500 to 200 BC), based originally around the Indus River system, wine was worshipped as the liquid god Soma because of its medicinal attributes. The Hindus' most ancient sacred text, called the Vedas, credited Soma with great medicinal powers. Another Hindu sacred text is the Rig-Veda, which contained hymns praising Soma such as: 'This is Soma, who flows wine, who is strength giving . . .' (Bose 1922, p. 6); and 'the god Soma heals whatever is sick . . . makes the blind see and the lame walk' (Sarma 1939).

Soma is originally thought to have been the fermented juice of an east Indian leafless vine (*Asclepias acida*) and other wild indigenous grapevines. Later, the normal cultivated European grapevine *Vitis vinifera* was introduced, originally from Persia during the early Christian era, then later from Europe in general as trade between India and Europe developed.

The Vedas included a life science and medical text called Ayur-Veda, part of which, the famous Charaka Samhita, deals extensively with the use of wine as a medicine. The Charaka Samhita states that wine is the 'invigoration of mind and body, antidote to sleeplessness, sorrow and fatigue . . . producer of hunger, happiness, and digestion . . . if taken as medicine, and not for intoxication, it acts as Amrita (Soma), it cures the natural flow of internal fluids of the body . . . Wine is natural food but taken indiscriminately produces disease, but when taken properly, it is like Amrita, the immortal drink' (Bose 1922, p. 35).

The Charaka Samhita also contained an ode to wine: '[Wine] who is worshipped with the gods, invoked in Sautra-moni Yajna, who is Amrita to the gods . . . Soma juice to the Brahmans . . . the destroyer of sorrow, fear and anxiety . . . who is pleasure, happiness and nourishment [to men]' (Bose 1922, pp. 35–6).

The Tantras are part of the Hindu Shastras or scriptures and refer to wine as the god-beverage with the power of Soma 'the supreme being in liquid form . . . the medicine of humanity . . . the cause of great joy . . . the mother of enjoyment and liberation' (Bose 1922, pp. 23–4). Wine was an important and indispensable part of Tantric worship, so with the rise of Buddhism and its use of Tantric rituals, wine became part of medical practice in Bengal, Nepal and Kashmir.

The early Hindus were advanced surgeons who also had an extensive materia medica including 760 medicinal plants that were prescribed in diets, baths, gargles, enemas and inhalations. They were among the first to record the use of wine as an anaesthetic as the following quote from an ancient Sanskrit medical text testifies: '. . . the patient should be given to eat what he wishes and wine to drink before the operation, so that he may not faint and may not feel the knife' (Jolly 1951).

Alexander the Great of Greece invaded India in 327 BC and took back to Greece knowledge from Hindu surgeons and physicians, so the Indians may have influenced Greek medicine. When the Muslims conquered India in the seventh and eighth centuries AD, Indian medicine and the use of wine in medicine, religion and sacrifice declined.

Today Ayurvedic medicine is still practised in India, and the text *Fundamentals of Ayurvedic Medicine* lists wine (*madya*) as one of the 12 ingredient groups for food and drinks (Dash 1980, p. 117). The text also has a list of the best drugs, diets and regimens for certain conditions, with wine appearing as number three on the list 'Wine is the best for dispelling fatigue, and it is exhilarating' (Dash 1980, p. 134). At number 18 the text states: 'Wine and milk as well as meat of goat are exceedingly useful in treatment of *sosa*' (emaciation caused by tuberculosis) (Dash 1980, p. 139).

## Ancient China

The Chinese have been using alcoholic beverages as menstruums for more than 5000 years. The wine (*chiu*) used could have been made from grains as well as grapes. Chinese materia medica not only mixed plants and minerals with wine but also selected parts of various animals that were thought to have special virtues. Opium (not available to European medicine for many centuries) was also commonly mixed with wine.

Examples of prescriptions include one where animal parts were mixed in wine to procure an abortion: rub a mixture of lizard's liver, skin of the cicada locust and wine on to the navel. Or the flesh of a pit viper was prepared by placing the snake in a gallon of wine then burying the sealed jar under a horse's stall for one year. The resultant liquid was a cure for apoplexy, fistula, stomach pain, heart pain, colic, haemorrhoids, worms, flatulence and bleeding from the bowel. Alcoholism could be cured by donkey's placenta mixed in wine, the liver of a black cat in wine for malaria, and to cure a bad cold an owl was smothered to death, plucked and boiled, its bones charred and taken with wine (Read 1931–7).

The German Franz Hubotter spent 25 years studying in China and Tibet, then wrote a book called *Chinesische-Tibetische Pharmakologie und Rezeptur*, which was published in 1957 (Hubotter 1957, p. 144). In his book, 19 of the 87 prescriptions listed had wine in them.

Hubotter also states in his book that the wines were from the European cultivated grape *Vitis vinifera* and not from grain or wild indigenous grapes. *Vitis vinifera* was introduced into China by Chang Ch'ien during the second century BC after he had learnt winemaking in Persia. Wines must have been made from indigenous grapes prior to this, however, because during the Chou Dynasty (1000 BC) red wine, which could only be made from red grapes and not from grain, was used in sacrifices because its colour was associated with blood. The wine was mixed with human blood and bone marrow and then drunk (Ackerman 1945, pp. 75, 98, 100).

## Ancient Greece

The vine then spread north to Greece. By about 2000 BC Dionysus became the Greek god of wine. The Greeks adopted wine as part of their daily nutritional needs along with bread and meat, believing it strengthened them as is amply described throughout Homer's *Odyssey* and *Iliad*, where wine was not only the medicine most frequently mentioned but characters such as Achilles and Ulysses recognized its ability to sustain the body. Normally, as a medicinal remedy, wine had been prescribed diluted three to five times by water. The Greek physicians were the first to prescribe wine undiluted and it was one of their main medicines.

Hippocrates (450–370 BC) was one of the leading physicians of the ancient world. He lived on the island of Kos and is recognized as the father of modern Western medicine because he was the first to say that illness was due not to the wrath of the gods but to poor nutrition or disease. Hippocrates believed in assisting the forces of nature to restore harmony in the body and thus promote recovery by using dietary treatments along with fresh air and exercise. This he called his Regimen. Hippocrates believed that if there was any deficiency in either food or exercise then the body would fall sick. A nourishing pottage called *kykeon* was made of barley with wine and milk as a nutrient (Phillips 1973, p. 77).

He used wine extensively as a wound dressing, as a nourishing dietary beverage, as a cooling agent for fevers, as a purgative and as a diuretic. He made distinctions among the various types of wine, described their different effects, directed their uses for specific conditions and advised when they should be diluted with water. In addition, he stated when wine should be avoided. In his essay on wounds, Hippocrates said: 'No wound should be moistened with anything except wine, unless the wound is in a joint' (Burke 1984, p. 193). He taught that the wound should be thoroughly cleansed with wine, that all the blood should be removed and a clean piece of linen soaked in wine should be applied directly to the wound before bandaging. Alternatively, a sponge soaked in wine and kept moist with wine from a vessel above the sponge could be applied to the wound. This was good medicine as infection was one of the greatest causes of death in the ancient world and the polyphenols and alcohol in wine are potent antiseptics. Old blood left in a wound is also a good culture medium for bacteria and thus a source of further infection. So, instead of poking around inside a wound that had not been washed with wine, and with unclean hands, as ancient physicians did, 'Hippocratic physicians, by contrast, used surgical probes which had been disinfected with wine or vinegar' (von Staden 1989, p. 15) in a wound also disinfected with wine.

Regarding the therapeutic uses of wine, Hippocrates noted that the yeast and unaltered sugar of new wines were irritants to the gastrointestinal tract; white, thin and acid wines are the more diuretic; wines rich in tannin are antidiarrhoeic. These observations are described well in the following passage from Lucia (1963).

> The therapeutics of Hippocrates were based on rational observations of the responses of patients to treatment, and on strict hygienic rules. He made no extravagant claims for wine, but incorporated it into the regimen for almost all acute and chronic diseases, and especially during the period of convalescence. Although he advised against its use in illnesses involving the central nervous system, particularly in meningitis, he suggested that even in this disorder, if fever were absent, enough wine should be added to the water to ensure an adequate intake and exchange of fluid. By varying the proportion of water, he tempered the dose of wine to the requirements of the illness and the needs of the patient.
>
> Hippocrates described water as 'cooling and moist,' and wine he characterized as 'hot and dry' and containing 'something purgative from its original substance'. Dark and harsh wines, however, were said to be 'more dry', and to '. . . pass well neither by stool nor by urine, nor by spittle. They dry by reason of their heat, consuming the moisture out of the body'. The latter constitutes the earliest recorded observation of the biophysiological effects of wines with an excessive tannin content – an agent that retards the motility and mobility of the bowel, decreases

> the production of urine, and suppresses the flow of salivary and other glandular secretions. Of other wines used therapeutically, he observed: 'Soft dark wines are moister; they are flatulent and pass better by stool. The sweet dark wines are moister and weaker; they cause flatulence because they produce moisture. Harsh white wines heat without drying, and they pass better by urine than by stool. New wines pass by stool better than other wines because they are nearer the must, and more nourishing; of wines of the same age, those with bouquet pass better by stool than those without, because they are riper, and the thicker wines better than the thin. Thin wines pass better by urine. White wines and thin sweet wines pass better by urine than by stool; they cool, attenuate and moisten the body, but make the blood weak, increasing in the body that which is opposed to the blood. Must causes wine, disturbs the bowels and empties them. It causes wind because it heats; it empties the body because it purges; it disturbs by fermenting in the bowels and passing by stool. Acid wines cool, moisten and attenuate; they cool and attenuate by emptying the body of its moisture; they moisten from the water that enters with the wine. Vinegar (sour wine) is refreshing, because it dissolves and consumes the moisture in the body; it is binding rather than laxative because it affords no nourishment and is sharp.'
>
> This passage epitomizes the logic of a mastermind in its observations of human physiology and of the chemical changes upon which physiological reactions are dependent. The yeast and unaltered sugar of new wines are irritants to the gastrointestinal tract; white, thin and acid wines are the more diuretic; wines rich in tannin are antidiarrhoeic. Thus, in terse phrases, the mechanisms for acceleration and retardation of bowel movement and urinary flow and for hydration and dehydration of the body in relation to the ingestion of grape extractives, acids, tannin and alcohol were established for the ensuing centuries.
>
> (Lucia 1963, pp. 37–9)

Hippocrates also had the following to say about wine as a medicine: 'Wine is fit for Man in a wonderful way provided that it is taken with good sense by the sick as well as the healthy' (Norrie 2000, p. 14) in accordance with the circumstances of each individual person.

The following are other writings about the use of wine as a medicine by Hippocrates.

> Infants should be bathed for long periods in warm water and given their wine diluted and not at all cold. The wine should be of a kind which is least likely to cause distension of the stomach and wind. This should be done to prevent occurrence of convulsions and to make the children grow and get good complexions.
>
> (McGovern *et al.* 2000, p. 3)

> The main points in favour of . . . white strong wine . . . It passes more easily to the bladder than the other kind and is diuretic and purgative, it is always beneficial in acute diseases . . . These are good points to note about the beneficial and harmful properties of wine; they are unknown to my predecessors.
>
> (McGovern *et al.* 2000, p. 3)

> His jaws are fixed, and he is unable to open his mouth . . . Grind wormwood (*Artemisia absinthium*), bay leaves, or henbane seed with frankincense; soak this in white wine, and pour it into a new pot; add an amount of oil equal to the wine, warm and anoint the patient's body copiously with the warm fluid, and also his head . . . Also give him a very sweet white wine to drink in large quantities.
>
> (McGovern *et al.* 2000, p. 3)

For an obstinate ulcer, sweet wine and a lot of patience should be enough (Skovenborg 1990, p. 9). Wine removes the sensation of hunger (Coar 1822, p. 27). Pains of the eyes are cured by wine, by the bath, by formentation, by bleeding, or by purging (Coar 1822, p. 173). In pains of the eyes, after having administered pure wine, and free ablution with warm water, a vein must be opened (Coar 1822, p. 211). Anxiety, yawning and rigor are removed by drinking equal parts of wine and water (Coar 1822, p. 217). If the wound is in a good state but the adjacent parts are inflamed, a cataplasm, composed of the flower of lentils boiled in wine, will be found serviceable; but if you want to close and heal, you must employ the leaves of the blackberry bush, nasturtium, park leaves or allum, macerated in wine or vinegar.

For the wounds of the head and ears, whether recent or old, Hippocrates recommended unripe grapes, myrrh and honey, with a small proportion of nitre, and a still smaller one of flower of brass, boiled together in wine, for at least three days (Riollay 1783, pp. 58–9). During the whole course of the disorder, it is useful to give honey and water, and now and then wine (Riollay 1783, p. 135).

Sweet red wine is more powerful than the white for promoting expectoration (Riollay 1783, p. 136). White (wine) is best for exciting a flow of urine: this diuretic quality renders it very serviceable in acute complaints (Riollay 1783, p. 136).

When a violent headache or a delirium supervenes, wine must be entirely laid aside, and water substituted in its place; or, at most, a watery sort of white wine: observing to give some water after it (Riollay 1783, p. 136).

Hippocrates was also one of the first of the ancient physicians to attribute feelings of joy, sadness, grief and sorrow to the brain. Prior to then the heart was considered the house of the soul and mind, while the ancient Egyptians thought one's personality lived in the liver. Mental illness was seen as a possession by evil spirits, whereas Hippocrates saw mental illness as a physical disease that would respond to his physical treatments. Aristotle also believed mental illness was physical, but resulting from an excess of 'black life', and advocated that the melancholia be treated with wine, aphrodisiacs and music (Horsley 1998, p. 15) – surely a combination that would work even today. The great Greek philosophers Socrates, Plato and Aristotle, contemporaries of Hippocrates, were all great oenophiles.

After Hippocrates, the father of Greek medicine, came many more physicians. Theophrastus of Eresus (372–287 BC) described many medicinal plants that were mixed with wine (Burke 1984, p. 192). He was a pupil of Aristotle and wrote many books including *Inquiry into Plants*, comprising nine books, and *Growth of Plants*, comprising six books. He had a great knowledge of plants that could be used as medicines with wine.

Mnesitheus (320–290 BC) was a famous Hippocratic physician practising in Athens who wrote a treatise called *Diet and Drink* in which he claimed: 'In medicine it is most beneficial; it can be mixed with liquid drugs and it brings aid to the wounded . . . While dark wine is most favourable to bodily growth, white wine is thinnest and most diuretic; yellow wine is dry, and better adapted to digesting foods'. (Lucia 1963, p. 12). Thus he observed that red wines contained more vitamins 2200 years before Morgan reported on vitamins in wine in 1939.

Athanaeus (AD 170–230), a Greco-Egyptian physician from Naucratis, wrote about the use of wine as a medicine: 'in medicine it is most beneficial; it can be mixed with soluble drugs and it brings aid to the wounded' (Skovenborg 1990, p. 4). He also commented on wine from Mareo, noting its diuretic effect: 'it is white and pleasant, fragrant, easily assimilated, thin, does not go to the head and is a diuretic' (Skovenborg 1990, p. 8). Athanaeus also quoted the original writings, which no longer exist, of Diocles of Carystus (*c.* 375 BC) and his pupil Prascagoras, both of whom wrote about the therapeutic uses of wine.

Between 300 BC and 50 BC the centre of Greek medicine moved to Alexandria where Erasistratus (300–260 BC) founded a school for progressive physicians known as the Erasistrateans who favoured therapies involving mild laxatives, barley-water and wine in small doses. In the first century BC, followers of Erasistratus founded the medical school at Smyrna to advance his work. Hikesios led this group and wrote a treatise on the preparation of wine called *De Conditura Vini* which advised the use of wine as a medicine. Apollonius of Citium (*c.* 81–58 BC) was a contemporary of Hikesios who also wrote a treatise on wine as a medicine. Cleophantus was another famous Alexandrian physician who tried to simplify treatments and taught the use of wine and cold water in therapy, especially in dealing with fevers such as malaria, to reduce the fever and to sedate the patient.

The Greeks also developed their theriacs and alexipharmics – antidote medicines using wine as part of the therapeutic agent. The term theriaca comes from *therian*, a wild beast which later became a venomous serpent; thus, theriaca was about the symptoms and treatment of venomous bites and animal stings. Alexipharmaca is from the Greek *alexein* meaning to ward off; thus alexipharmaca was about antidotes to poisons in food and drink. Both terms were first used by Nicander (190–130 BC), a poet and physician.

The next great advocate of these antidote medicines was King Mithradates the Great (132–63 BC). Mithradates developed his 'true medicines' (Burke 1984, p. 194), or mithradatium – antidotes and prophylactic medicines – on an empirical basis by giving guinea-pigs and human prisoners a certain poison or bite and seeing which of his medicines worked and which didn't, much to the disadvantage of the subject being tested, but such was the absolute power of being king. These mithradatium were later used by Roman pharmacies, Arab physicians, doctors in Medieval Europe as a cure for plague, and by English doctors as a cure-all well into the eighteenth century.

The Hippocratic physicians broadened the art and science of therapeutics and championed the use of wine as a medicine. They ushered in the Greco-Roman period of medicine where wine became the most important therapeutic agent of the time.

## Ancient Rome

After the destruction by the Romans of Corinth in Greece in 146 BC, Greek medicine moved to Rome. The Etruscans had introduced viticulture into northern Italy, mainly around Tuscany, and the Romans developed their own god of wine, Bacchus.

After the ascent of Rome, the Romans mistrusted Greek physicians, believing them to be possible poisoners or assassins. It took Asclepiades (124–40 BC), physician to Cicero and pupil of Cleophantus in Alexandria, to gain Roman acceptance of Greek medicine. He recommended restriction of diet, use of wine, music and exercise in the open air to treat illness. He later wrote an essay describing the virtues of various Greek and Roman wines called *Concerning the Dosage of Wine*. Asclepiades was popular with the Roman nobility because of his therapeutic slogan '*cito, tuto, iucunde*', meaning 'swiftly, safely, sweetly' (Skovenborg 1990, p. 14), and because he invented the shower bath for its hygienic value. He was given the nickname '*Oinodotes*' meaning 'giver of wine'.

Menecrates of Tralles was the physician to Emperor Tiberius, a sign of true Roman acceptance of Greek physicians. He was called *Physikosoinodotes*, or 'natural philosopher who advises the use of wine'.

One Roman who was against this Greek influence was Marcus Porcius Cato (234–149 BC). Also referred to as Cato the Elder or Cato the Censor, he was a great Roman orator and statesman who championed Roman ways and culture. He was a prolific writer and included medicine as one of his subjects, prescribing cabbage and wine mixed in special formulae as remedies against disease (Skovenborg 1990, p. 14).

Aurelius Cornelius Celsus (25 BC–AD 37) wrote *De Re Medicina*. Comprising eight books, it was a vast text on medicine. He was one of the leaders of

Roman medicine and wrote much about the therapeutic uses of wine, discussing the medicinal values of the various wines from different regions of Italy, Sicily and Greece, and which diseases they should be prescribed for.

For indigestion, for example, Celsus recommended: 'Those who have a slow digestion and for that reason get a distended abdomen, or because of some kind of fever feel thirst during the night, they should, before going to bed, drink three or four cups of wine through a thin straw' (Skovenborg 1990, p. 14).

Sextius Niger (*c.* AD 40) was a disciple of Asclepiades. He went against the usual Roman philosophy, which was very superstitious, attributing specific diseases to different gods. He advocated the extensive medicinal uses of what he called natural wine.

Pliny the Elder (AD 23–79) was a famous Roman scholar, statesman and physician who wrote *Naturalis Historia* (Natural History) – a unique encyclopaedia about plants and medicine in which he noted that 'wines have a remarkable property of drawing into themselves the flavour of some other plant' and listed 60 kinds of 'artificial wines being used for medicinal purposes' as antidotes to snake bite and poisonous mushrooms (Skovenborg 1990, p. 14). Twelve books in his natural history were devoted exclusively to medicine, where he listed 200 grape varieties, 50 Roman wines, 38 foreign wines, seven kinds of salted wines and 18 varieties of sweet wines. Pliny the Elder also advocated taking wine *with* meals, and disapproved of the custom of taking wine *before* meals, which was made popular by the Emperor Tiberius. He also wrote: 'There are two liquids that are especially agreeable to the human body – wine inside and oil outside' (Skovenborg 1990, p. 14), and advocated the use of herbs and spices administered in wine.

Dioscorides (AD 40–90) was a Greek army surgeon under Nero. He wrote *De Universa Medicina* in approximately AD 77. While accompanying Roman armies on their expeditions, especially in eastern Mediterranean countries, he gathered information for his writings, describing various substances – animal products, plants, spices, salves, oils, minerals and wines – and detailing their dietetic and therapeutic values. Hence he became the founder of 'materia medica' or the study of medical substances as an applied science. He used wine in many conditions and a particular type was always specified. 'In general wine warms the body, it is digestible, increases the appetite, helps the sleep and has reviving properties' (Skovenborg 1990, p. 15). *De Universa Medicina* comprised five books listing more than 1000 drugs, and was a leading medical text for the next sixteen centuries.

Columella (*c.* 4 BC–AD 65) was a Roman agricultural writer who was a contemporary of Pliny and Dioscorides. He also wrote about the virtues of wine as a medicine, explaining the different effects of different wines from various grape varieties.

Galen (AD 131–201) was another famous Greek physician, second only to Hippocrates. He was a physician to the gladiators and as such treated countless wounds – lacerations, stab wounds, amputations and evisceration (where the abdominal cavity has been punctured and the abdominal contents of bowel and organs are exposed). Here Galen, like Hippocrates, favoured the use of wine to prevent infection. He even went to the extreme of soaking the exposed abdominal contents in wine before putting them back into the abdominal cavity in the cases of evisceration.

For fistulous abscesses Galen recommended: 'Before applying the agglutinant, I am in the habit of cleaning the sinus with wine alone, sometimes with honeyed wine. This wine should be neither sweet nor astringent' (Burke 1984, p. 195). He also insisted that any putrid wound should be washed with wine or that a sponge or a piece of wool soaked in wine be applied to the wound. Galen wrote a complex list of drugs made from vegetables, most mixed with wine, which were called Galenicals. He wrote a catalogue of wines from different areas noting their chemical characteristics and physiological effects. He advocated using wine as a suitable treatment for the diseases of the aged in his book *De Sanitate Tuenda* (Galen's Hygiene). He wrote: 'Wine . . . for old men it is most useful' (Green 1951, p. 204). He prescribed matured Palernian wine, the Emperor's wine, because it was made in the vineyards near Rome. Galen's thoughts and his Galenicals dominated European medicine until the Middle Ages and, especially after he had praised them, mithradatium and other theriacs were also popular up until the eighteenth century.

Other pro-wine physicians in Galen's time included Athenaeus of Attalia who founded the 'pneumatic' school of medicine, which stated that wine aroused the pneuma – the vital spirit or breath in a person – and hence was good as a tonic or restorative. Athanaeus's teachings were continued by Archigenes of Apamea and later by the great physician Aretaeus of Cappadocia. Aretaeus wrote an extensive medical text called *Therapeutics of Chronic Diseases*, which was noted for its accurate descriptions of diseases and for highly recommending Italian wines as medicines.

With the expansion of the Roman Empire along the valleys of rivers such as the Rhône and the Rhine, viticulture spread until most climatically favourable areas of Europe grew grapes.

During this expansion of the empire, Roman generals such as Julius Caesar recommended that soldiers drink wine to preserve good health, to give them strength and to help prevent dysentery. The two major medical conditions harming soldiers, besides being killed in combat, was infection of wounds and dysentery from unclean water in new lands – both of which were prevented with wine.

## *Byzantium*

In AD 330, Emperor Constantine transferred his capital to Byzantium in Asia Minor (later to become Constantinople, today's Istanbul). During this Byzantium era, only a few medical writers were prominent but they were vital because they continued the influence of Greco-Roman medicine during the Dark Ages when intellectual stagnation gripped Europe after the fall of the Roman Empire with the sacking of Rome in AD 410 and the destruction of Alexandria in AD 640.

Oribasius (AD 325–403) was greatly influenced by Galen and wrote an encyclopaedia of medicine called the *Synagoge*. He used papyrus leaves soaked in diluted wine, for example, to stop infection and to stop bleeding. Paul of Aegina (AD 625–690) continued the pro-wine Greco-Roman tradition, as did Aetius of Amida (AD 502–575), who was the first notable Christian physician. He wrote a text called *Tetrabiblion*, which reflected his medical studies at Alexandria. He advocated red, slightly astringent wines 'for persons in good health, and those who are convalescent from diseases' (Ricci 1950, pp. 214–15), and for nausea in pregnant women 'old, tawny, fragrant wine which is a little tart' (Ricci 1950, pp. 21–2).

Alexander of Tralles (AD 525–605) was the fourth great pro-wine Byzantium physician. He described intestinal worms, gout and insanity. He recommended narcotics, bleeding, warm baths and wine for certain mental diseases. For dandruff he recommended rubbing wine with salves and washing with salt water.

## Biblical/Jewish world

Wine, *yayin* in Hebrew, was – and still is – an integral part of Jewish culture, religion and medicine. Wine accompanied meals, was drunk on religious occasions and was an important part of celebrations.

In the parable of the Good Samaritan (Luke 10: 30–7), Jesus mentions the medicinal use of wine. The Good Samaritan bound up the wounds of the assaulted traveller and poured on olive oil and wine to prevent infection. In the Middle East today a mixture of olive oil and wine known as Samaritan balm is still available as an antiseptic for skin wounds. St Luke was a Greek (Hippocratic) physician for Antioch.

Jews used wine to rinse the wound after circumcision to prevent infection, and in the Old Testament Proverbs 31: 6–7 states: 'Let him drink, and forget his poverty, and remember his misery no more', suggesting the use of wine as a sedative. In the New Testament, St Paul advised St Timothy: 'No longer drink only water, but use a little wine for the sake of your stomach and your frequent ailments' (Timothy 5: 2–3). St Paul and St Luke lived in Rome at the same time as Pliny, Dioscorides and Columella and would thus have been influenced by their pro-wine teachings.

The Jewish Talmud, written between 536 BC and AD 427, states the following: 'Wine taken in moderation induces appetite and is beneficial to health . . . Wine is the greatest of medicines. Where wine is lacking, drugs are necessary' (McGovern *et al.* 2000, p. 5).

Moses Ben Maimon of Cordoba, better known as Maimonides (1135–1204), was a very famous Jewish physician and philosopher. He was a great advocate for the use of wine as a medicine, as can be seen in his book *De*

*Regimine Sanitatis* and in the following quotes. He was the leading Jewish medical authority of the Middle Ages. 'Wine is a nutrient . . . It is a very good nutrient . . . It generates praiseworthy blood . . . [it] will generate flatus, and possibly tremor . . . nevertheless if mixed and left for 12 hours or more and then drunk, it is very good . . . and the temperament improves' (McGovern *et al.* 2000, p. 5). '(Mad dog) . . . (if done before onset of hydrophobia, otherwise patients always die) . . . flour of vetch kneaded in wine and applied as a poultice' (McGovern *et al.* 2000, p. 5).

## Arabic period

The Koran presented Arab doctors with a dilemma. According to its teaching, wine was a 'device of the devil' and was therefore forbidden. But the Koran also states: 'Of the fruits of the date palm and grapes, whence ye derive strong drink and good nourishment, is healing for mankind'. The main influence on Islamic medicine was Greek, with its extensive therapeutic use of wine. So Islamic doctors used wine as a medicine only and did not prescribe it for 'social' reasons.

Rhazes (AD 860–932) was a great Arabic doctor and was the first to describe smallpox and measles in the literature. He may have been the first outside China to distil alcohol from wine and use it to prevent infection in wounds. He used compresses soaked in warm wine to compress the intestines back into the abdominal cavity in the case of abdominal evisceration.

Avicenna (980–1037), known as the 'Prince of Physicians', wrote the main medical textbook for Western and Eastern medicine. Called the *Canon of Medicine*, it was used until 1650. He recommended wine for dressings and observed: 'Wine is also very efficient in causing the products of digestion to become disseminated through the body' (Burke 1984, p. 196). Avicenna was philosophical about the use of wine: 'Is wine to blame that it raises the wise to heaven, but plunges the fool into darkness?' (Skovenborg 1990, p. 4).

In his *Canon of Medicine*, he devotes a whole section to wine, covering recommendations 800–814 inclusive. It starts with the 'Virtues of Wine': 'As to the advantages that be in wine – it strengtheneth the viscera and banisheth care, and moveth to generosity and preserveth health and digestion; it conserveth the body, expelleth disease from the joints, purifieth the frame of corrupt humours, engendereth cheerfulness, gladdeneth the heart of man and keepeth up the natural heat; it enforceth the liver and removeth obstructions, reddeneth the cheeks, cleareth the brain and deferreth grey hairs' (Gruner 1930, p. 409). And ends with 'Anaesthetics': 'If it is desirable to get a person unconscious quickly, without his being harmed, add sweet smelling moss to the wine, or lignum aloes'.

If it is desirable to procure a deeply unconscious state, he continues, so as to enable the pain that is involved in painful applications to a member to be borne, place darnel-water into the wine, or administer fumitory, opium or hyoscyamus (half-dram dose of each), or nutmeg or crude aloes-wood

(4 grains of each). Add this to the wine, and take as much as is necessary for the purpose. Or, boil black hyoscyamus in water with mandragore bark until it becomes red. Add this to the wine (Gruner 1930, p. 413).

Albucasis (936–1013) was more interested in surgery. His was the first complete book of surgery with illustrations on techniques and instruments. His *Treatise on Surgery* was translated from Arabic into Latin in the late twelfth century, then into English in 1778. He recommended treating wounds with cotton wool soaked in rose oil or rose oil mixed with astringent wine and preventing the wound from being exposed to the air. He was thus a pioneer of aseptic surgery. He used tepid astringent black wine in compound abdominal wounds (evisceration), irrigated infected sinuses with honey and dry wine, washed venesection (blood-letting) sites with old wine and recommended that 'men of frigid constitution should also take perfumed raisin wine, not too old and not too new' (Burke 1984, p. 197).

Haly ben Abbas, a famous Arabic physician of the tenth century, described Arabic medical practices of the time in his medical encyclopaedia called *Almaleki* (Royal Book), which contained a section on the action of natural and artificial wines. Arabic medicine's major contribution was the separation of pharmacy, then known as the arts of apothecary and alchemy, from medicine. The first apothecary shop was opened in Baghdad in AD 754. The great tenth-century Persian pharmacologist Mansur the Great wrote his extensive *Book of the Foundations of the True Properties of the Remedies* in which he combined Greek, Syrian, Arabic and Hindu materia medica. From this background he would naturally be pro-wine and advised, for example, that if you gargled with 'wine in which plum leaves have been boiled, he alleviates all complaints of catarrh which exist in the throat, neck and chest' (Reed 1942, p. 53).

## Medieval medicine

During the Byzantine period (fourth to seventh centuries AD), physicians preserved Greek medicine during the Dark Ages to be used during the Arabic period, which in turn preserved these medical traditions and organized them into proper coherent texts to reintroduce them later into Western Europe via Spain after the Muslim conquests by the Moors from north-west Africa starting in AD 711. Via this complex circuitous route, Greek medicine had thus gone to Rome, then to Byzantium, then to Arabia, finally to re-emerge in post-Moorish Spain, whence it was re-introduced into Western (Christian) Europe, some 2000 years later.

Healing in medieval times was undertaken mainly by monks using medicines based chiefly on herbs and secondarily on animal products and mixed in wine. Minerals were rarely used until Paracelsus's time. After the thirteenth century the profession of physician/surgeon/pharmacist was separated under Arabic influence into the professions of pharmacist on one hand, and physician/surgeon on the other. Medicine was based on the classic Greco-Roman works with some Islamic input. Blood-letting by bleeding, cupping or leeching was popular.

Wine was used as a medicine by itself or mixed with other compounds to make a palatable concoction out of foul-tasting substances. Monks preserved medical knowledge from the past in their libraries and advanced science and viticulture in the protection of their monasteries, which also housed the hospitals and the pharmacies. Monks did not practise surgery (but used Galenicals instead) because it was thought to be unholy and was prohibited in 1162 by an edict of the Council of Tours. Many of the liqueurs used today owe their pedigree to medicines used in the Dark Ages by monks. Hence different monasteries developed their own famous liqueurs, such as D.O.M. Benedictine which came from the Dominican Order of Monks (D.O.M).

Wine in medieval times suffered from oxidation. The art of sterile wine making and the making of airtight containers were both lost after Roman times, resulting in secondary fermentation in wine barrels and goat skins, which turned the wine into vinegar, and no ageing in cellars.

Monasteries eventually developed into medical schools and one of the finest was at Salerno in southern Italy. The Salerno medical school was founded on the site of a ninth-century Benedictine hospital and had both clerical and lay medical practitioners as teachers. Salernitan medicine eventually spread throughout Europe, and the use of wine as a medicine was an integral part of its teachings. In fact, wine was the most frequently mentioned therapeutic agent in the *Regimen Sanitatis Salernitanum*, its code of health. Wine was prescribed as a nutrient, as a tonic, as an antiseptic and as the universal menstruum for other medicinal substances.

Women students were taught at Salerno. The most famous female student was Trotula or Mother Trot of Salerno, who in the eleventh century wrote about obstetrics and advised 'hot wine in which butter has been boiled' (Trotula 1940, p. 28) be applied to treat prolapse of the uterus after childbirth.

Constantine the African (*c.* 1020–87) brought many Arabic medical manuscripts to Salerno, which later ended up as an important compilation of medical recipes called the *Antidotarium Nicolai*, named after the author Nicholas Praepositus who was the director of the Salerno medical school at the time. His final text listed 150 Galenicals previously lost to Christian European medicine, many of which contained wine. Thus the Salerno medical school helped bridge the gap between Arabic medicine, based on Greek medicine, and the later Middle Ages.

Monasteries developed vineyards to make their wines for religious and medicinal purposes. They also developed herb gardens, or herbularia, where they grew the specific herbs needed for their medicines. During this period the belief that diseases were caused by miasmas or bad odours developed, so spices started to be used, in addition to plants and herbs, as medicines to ward off disease. Even in recent times it was common practice to hang cubes of camphor around children's necks in winter to ward off illness, just as it is common today to rub Vicks VapoRub on to children's chests when they have a cold. During the Middle Ages spices were more important as medicines than as supplements to cooking. They were added to wines as well to make an even stronger medicine. This practice continues today as vermouths, bitters, aperitifs, liqueurs and cordials.

Monks also introduced distillation from the Arabs into Europe in the fourteenth century, and made brandy known as 'aqua ardens'. This new spirit of wine became the 'super wine' or head medicine that was the elixir of life and therefore later called 'aqua vitae'.

After Salerno other medical schools were established in Naples, Palermo, Montpellier and Bologna. The founder of the Bologna medical school was Thaddeus of Florence (AD 1223–1310), who compiled a text of prescriptions called 'De virtute aqua vitae, quae etiam dicitur aqua ardens' or 'On the virtues of the water of life, which is also called fiery water' (Lucia 1963, p. 99), i.e. brandy. These medical schools and the monasteries preserved Greek and Arabic medicine for future generations.

Arnauld de Villeneuve (AD 1235–1311) (also known as Arnald of Villanova) wrote *Liber de Vinis*. He was a great advocate for the use of wine as a tonic, as part of a poultice, as an antiseptic, especially in wound dressing, and for sterilizing polluted water. His book established wine as part of recognized therapy throughout Europe during the late Middle Ages.

The following extract from *Liber de Vinis* sums up what Arnald thought about wine as a medicine.

> If wine is taken in right measure it suits every age, every time and every region. It is becoming to the old because it opposes their dryness. To the young it is a food, because the nature of wine is the same as that of young people. But to children it is also a food because it increases their natural heat. It is a medicine to them because it dries out the moisture they have drawn from their mother's body. No physician blames the use of wine by healthy people unless he blames the quantity or the admixture of water . . . Hence it comes that men experienced in the art of healing have chosen the wine and have written many chapters about it and have declared it to be a useful embodiment or combination of all things for common usage. It truly is most friendly to human nature.
>
> (von Hirnkofen 1943, pp. 24–5)

Arnald wrote 123 books and treatises, was one of the first men in the Middle Ages to pursue original independent medical investigation, invented tinctures by extracting herb essences with alcohol, and pioneered disease classification. But his greatest contribution to medicine was indoctrinating European doctors in the therapeutic uses of wine for centuries to come.

During the Middle Ages wine was mainly used internally. Not everyone used wine externally to prevent infection of wounds. Theodoric (Teodorico Borgognoni) (AD 1205–1296) was an advocate of the use of wine for washing wounds, however, along with the removal of all foreign matter, this being contrary to the popular contemporary belief in 'laudable pus', which theory held that the best way to treat a wound was to promote suppuration, or pus-formation, by keeping the wound open. 'No error can be greater than this. Such a practice is indeed to hinder nature, to prolong the disease and to prevent the conglutination and consolidation of the wound,' he wrote (Albutt 1905, p. 30).

Theodoric was the son and pupil of Hugh of Lucca (Ugo Borgognoni) who was the first surgeon in the Middle Ages to question the doctrine of laudable pus while treating Christian casualties during the Fifth Crusade. His theories were supported in 1252 by Bruno da Longoburgo of the University of Padua who expressed his ideas in his book *Chirurgia Magna*, in which he advocated wound antisepsis with wine.

These antiseptic practices were continued by William of Salicet (Guglielmo Salicetti AD 1210–77) city physician of Verona and pupil of Bruno da Longoburgo as well as Salicet's pupil Lanfranc (Guido Lanfranchi) who practised in Paris and founded the French school of surgery at the Collège de Saint Côme. Unfortunately, Lanfranc later lapsed back into the practice of suppuration or laudable pus. But one of Lanfranc's pupils, Henri de Mondeville (AD 1260–1320), a lecturer at the University of Montpellier, followed Hippocrates and wrote that pus 'is not a stage of healing (laudable pus) but a complication' (Albutt 1905, p. 40). He also advocated the use of wine as a tonic or 'wound drink' to help strengthen his patients. John of Arderne (AD 1307–77) was the champion of antiseptic surgery in England during this period and used wine as a menstruum.

During the fourteenth and fifteenth centuries, gunpowder made from bird-dung nitrates and gunshot wounds from lead projectiles covered in gunpowder complicated battlefield surgery due to the increased risk of infection and because the wounds were more extensive.

The use of wine as an antiseptic was not as extensive and universal as one might wish and it was not until Lord Lister introduced aseptic surgery with the use of carbolic acid in the nineteenth century, after Louis Pasteur proved bacteria cause infections, that asepsis was universally adopted.

Hieronymus Brunschwig (AD 1450–1533) was a surgeon in the Alsation army whose forte was the treatment of gunshot wounds and the distillation of alcohol. He promoted a mixture of strong Gascony wine, brandy and herbs called '*aqua vite composita*' for cleansing wounds and also used it to 'cure palsy, putteth away ring worms, expel poison and it was most wholesome for the stomach, heart and liver. It nourisheth blood' (Burke 1984, p. 199). His essay *The Vertuose Boke of Distyllacyon of the Waters of All Manner of Herbs – for the help and profit of surgeons, physicians, apothecaries and all manner of people* introduced distillation into England in 1525. The spirit subsequently distilled was originally used as a medicine but later became a source for alcoholism and poisoning when gin was distilled from sawdust and wood shavings. Wine did not have this effect as it was made from pure, healthy grapes and unadulterated.

The word alcohol was first used by Paracelsus in the sixteenth century. The physicians of ancient Egypt used the brittle metallic element antimony as a medicine. The Arabs called powdered antimony 'al-kohl'. Finally, Paracelsus applied this Arab term to the spirit in wine, calling it 'alcohol', presumably because of its healing qualities. Theophrastus Bombastus Von Hohenheim, or Paracelsus (AD 1493–1541), was a medical teacher in Switzerland who had a great interest in alchemy, astrology and the occult. He emphasized the revolutionary ideas of observation and experience and that humans were chemical machines. He popularized chemical medicine, or the use of minerals as therapeutic agents, as against the accepted followers of Galen who used plant medicines, and he thus earned the title of 'father of chemical pharmacology'. He is also famous for stating: 'Whether wine is a nourishment, medicine or poison is a matter of dosage', and was a great believer in aseptic surgery using wine.

Other famous surgeons who used wine as an antiseptic during this period include Ambroise Paré (AD 1510–90) in France and Richard Wiseman (AD 1622–76) in England.

## Post-medieval Europe

After the Middle Ages, wine was prescribed constantly: 'the astringent red wines for diarrhoea, the white wines as diuretics, port in acute fevers and for anaemia, claret and burgundy for anorexia, champagne for nausea and catarrhal conditions and port, sherry and madeira in convalescence' (Burke 1984, p. 200).

Theriacs, first compounded by the Greeks Nicander and Mithradates, started to be questioned as medicinal potage or fraudulent mixtures of polypharmacy that had no scientific basis. When the eminent English physician William Heberden (1710–1801) investigated theriacs and published his negative findings in 1745 in an essay entitled *Antitheriaka*, the days of the theriacs were numbered and eventually the College of Physicians eliminated them from the *London Pharmacopoeia*.

Patients were given alcoholic drinks for another reason also – the alternative drinks were very suspect. Water was polluted or infected (often with typhoid or cholera) in the cities, and milk had a very real risk of containing tuberculosis bacilli. The only safe, infection-free drink was alcoholic and the most therapeutic of the alcoholic drinks was wine, not spirits. Hence the famous microbiologist Louis Pasteur described wine as 'the most healthful and hygienic of all beverages'. Even as late as 1892 Professor Alois Pick of the Vienna Institute of Hygiene recommended adding wine to water to sterilize the water in the cholera epidemic of Hamburg. Research today has shown that the reason wine is so much more effective as an antiseptic than pure alcohol alone is because wine contains sterilizing compounds other than alcohol and it is the polyphenols such as malvoside (the principal pigment in red wine) that have the major antibacterial effect.

Hospitals used wine as a medicine all the time in the Middle Ages. The single biggest expenditure of Leicester Hospital, England, in 1773, for example, was for wine for the patients. In Germany, at the Alice Hospital in Darmstadt, between October 1870 and early April 1871, i.e. in less than six months, 755 patients used 4633 bottles of white wine, 6332 bottles of red wine, 60 bottles of champagne and 360 bottles of port besides some superior white wines and some Bordeaux (Banckes 1941, pp. 50, 52–3, 70).

The post-medieval period was also a time for the regulation of drugs and wine. The theriac frauds eventually forced governments to pass laws regulating the practices of apothecaries and vintners. During the fourteenth century, Italian states passed laws to discourage the counterfeiting of drugs and the watering down and adulteration of wine. Next came standard formulae for drugs with official authority (initially by local legislatures then by national governments) which became known as pharmacopoeias. The first such pharmacopoeia was produced in 1535 by Valerius Cordus of Erfurt, a young Prussian doctor who got his *Pharmacorum Conficiendorum Ratio, Vulgo Vocart Dispensatorium* ratified by the Nuremberg High Senate. They ordered its printing in 1546 and instructed all pharmacists to prepare their medicines according to Valerius Cordus's *Dispensatorium.* The original pharmacopoeias contained good medicine as well as elements of magic, witchcraft and mysticism. In England these collections of recipes were called leech books and contained herbs mixed in wine according to early Saxon tradition.

The physician to King Edward II of England was John of Gaddesden (1280–1361). He prepared a pharmacopocia called *Rosa Anglica* in which he used fennel and parsley in wine to cure blindness, using the Saxon belief that fennel was one of the nine sacred herbs. While in England '. . . for a cold stomach that is feeble of digestion and for the liver, give him wine that nutmegs is boiled in. Also good for the same, boil nutmegs and mastic in wine and drink it' (Druitt 1873, p. 110)

Paracelsus made the use of iron and antimony in wine popular. The *vinum ferri* was used as a remedy for anaemia until the late nineteenth century, while his wine of antimony continues to be used today as an expectorant and emetic.

The first *Pharmacopoeia of London* was published in 1618 and contained three medicated wines and ten medicated vinegars. Other European cities and countries then followed suit with their own pharmacopoeias, namely Amsterdam in 1636, Paris in 1639, Spain in 1651, Brussels in 1671 and Russia in 1778, for example. With the demise of the theriacs after 1745, preparations became simpler and those containing wine or alcohol grew rapidly. In 1818 Thomson's *London Dispensatory* contained an extensive chapter on wines and their medicinal properties and uses, listing 10 formulae for medicated wines such as wine of ipecac for coughs and dysentery and wine of opium for inflamed eyes. The *Pharmacopoeia Universalis of Heidelberg* listed 170 wines in its 1835 edition, while the *Pharmacopée Universelle of Paris* in 1840 listed 164 wines, and the first *Pharmacopoeia of the United States*, published in 1820, listed nine 'vina medicata'. *Vinum* was specified as the wine produce of *Vitis vinifera* or the European cultivated wine grape. Later, *vinum porterse* (port) and *vinum xericum* (sherry) were added to the standard pharmacopoeias.

Eventually the anti-alcohol Temperance Movement, or prohibitionist lobby, gained momentum in the USA and the UK in the nineteenth century and all wines were removed from the US pharmacopoeia in 1916 and from the British one in 1932. The pharmacopoeias of continental Europe continued to list wine, however. For example, in France (seven wines in the 1960 French *Codex*), in Italy (six wines in *Farmacopea Ufficinale*) and in Germany (41 wines in *1958 Hagers Handbuch der Pharmazeutischen Praxis*).

During the Great Plague of London, which flared up in May 1665 after beginning at the end of 1664, the doctors who remained in London to help treat the victims used wine all the time. Dr Heinrico Sayer, for example, fortified himself with a good strong wine before entering his patients' houses (Quintner 1999). One of the most famous doctors at the time of the Great Plague was Dr Nathaniel Hodges, who swore by sack or sherry 'to warm the stomach, refresh the spirits and dissipate any beginning lodgement of the infection [plague]' (Quintner 1999). He drank his sack before dinner and upon retiring to bed. He used it as his main antidote to the plague and credited it with saving his life on two occasions when he became ill.

One of the most famous images of the plague in Europe during the Middle Ages was provided by the 'beak doctors', who wore an overcoat, gloves, boots, hat and a mask that completely covered the face and neck and had a beak protruding from it that was lined with antidotes (wines) to prevent infection. In 1672 Dr Hodges wrote a poem about this strange outfit entitled *The Beak Doctors*.

As may be seen on picture here,
In Rome the doctors do appear,
When to their patients they are called,
In places by the plague appalled,
Their hats and cloaks of fashion new,
Are made of oilcloth, dark of hue,
Their caps with glasses are designed,
Their bills with antidotes are lined,
That foulsome air may do no harm,
Nor cause the doctor man alarm
The staff in hand must serve to show
Their noble trade where'er they go.
(Quintner 1999)

In 1724 a doctor, an unnamed 'Fellow of the Colleges' wrote a book in which he described wine as 'the Grand Preserver of Health and Restorer in most Diseases' (A Fellow of the Colleges 1724, pp. 10, 38), and in 1775 Sir Edward Barry FRCP, FRS wrote a book called *Observations Historical, Critical and*

*Medical on the Wines of the Ancients and the Analogy between them and Modern Wines*, in which he described the use of wine by Hippocrates, Artaeus, Galen, Celsus, Dioscorides and others. He also talked about Dr Sydenham, 'the English Hippocrates', because he most closely pursued Hippocrates' rules in the use of wine.

Dr William Sandford a surgeon at the Worcester Infirmary published *A Few Practical Remarks on the Medicinal Effects of Wine and Spirits* in 1799.

> Wine . . . is undoubtedly one of those real blessings with which a kind Providence has favoured us; and its true uses and effects have long been known, and considered, by medical writers of very high eminence and authority . . . with regard to the uses of wine, and its good effects on the human body in certain states of indisposition, especially, where the persons have not been in the habit of daily using it: to such it proves particularly beneficial when taken in moderate quantity, as its tendency is to increase the circulation of the fluids, and to stimulate all the functions of the mind and body . . . And this was probably the principal reason that wine, when first introduced medicinally as a cordial into this kingdom, was sold only by the apothecaries, which we are well assured it was about the year 1300 . . . Wine quickens the pulse, raises the spirits, and gives more than common animation for the time; but no sooner has the intoxicating delirium ceased than the patient becomes weak, enervated, and depressed in mind and body: here we distinctly see both the stimulant and sedative powers of wine . . .
>
> (Sandford 1799, pp. IV, V, 8–9, 20)

One Dr Alexander Henderson, a London physician, caused a storm in 1824 when he published a history of wine in which he adopted a temperance stand and wrote profusely about 'the deplorable effects of the abuse of wine' (Henderson 1824, pp. 346–8). However, Dr Henderson later wrote: 'Temperately used, it acts as a cordial and stimulant; quickening the action of the heart and arteries, diffusing an agreeable warmth over the body, promoting the different secretions, communicating a sense of increased muscular force, exalting the nervous energy, and banishing all unpleasant feelings from the mind' (Lucia 1963, p. 160).

Between 1863 and 1865 a series of articles by Dr Robert Druitt about the medicinal virtues of foreign wines appeared in the *Medical Times Gazette*. The series was later published as a book in 1873 entitled *A Report on the Cheap Wines from France, Germany, Italy, Austria, Greece, Hungary and Australia – their Use in Diet and Medicine*. Among his recommendations, Druitt suggested old sherry to stimulate the heart, clarets for gout and for measles in children as well as champagne for neuralgia and influenza. He concluded: 'The medical practitioner should know the virtues of wine as an article of diet for the healthy and should prescribe what, when, and how much should be taken by the sick' (Druitt 1873, p. 110).

But the most comprehensive set of directions for the medical prescription of wine came, in the 1870s, from Dr Francis Anstie, the editor of the *Practitioner* journal and a physician at Westminster Hospital in London. Dr Anstie's recommendations also appeared as a series of articles, this time in the *Practitioner*, and, like Druitt's, were later published as a book in 1877 called *On the Uses of Wine in Health and Disease*. In his book, Dr Anstie argued against the opponents of wine, stating that its medical use was 'established . . . by widespread custom' and was not therefore subject to discussions of 'lawfulness or . . . advisability' (Anstie 1877).

He wanted to standardize the medical use of wine and criticized the haphazard way in which different wines were being prescribed by doctors for the same disease, stating: 'It is common to meet with invalids and others who have received diametrically opposite directions as to the choice of beverages from different practitioners of equal standing'.

Anstie divided wines into strong wines, such as the fortified wines port, sherry, madeira and marsala, which were good 'as a dietetic aid in debility of old age', and light wines such as table wines with no more that 10% alcohol. He recommended the light wines on a daily basis with lunch and dinner for healthy people. In fact half his book was devoted to the use of wine by healthy people as a prophylactic against disease and as a dietetic aid, while the other half considered the use of wine as a medicine in acute and chronic disease.

In his book *Wine and Health – How to Enjoy Both*, published in 1909, Dr Yorke-Davis, a member of the Royal College of Surgeons, shared and reenforced Dr Anstie's views. For example, for anaemia he recommended wines containing iron, such as Burgundy.

During the nineteenth century, when the Temperance Movement was gaining a following in the UK and the USA, doctors in continental Europe were not compelled to defend their use of wine, and so its use as a medicine flourished. Dr Loebenstein-Lobel of Strasbourg wrote his extensive text *Traité sur l'Usage et les Effects des Vins* in 1817 and administered wines by the spoonful, the goblet and even as an enema.

Armand Trousseau was a professor of medicine in Paris. In 1861 he published *Clinique Médicale de l'Hôtel Dieu*, in which he recommended old red *vin ordinaire* during convalescence after typhoid fever and a mixture of white wine, juniper berries, squill, digitalis and potassium acetate for heart disease.

Dr Jean-Martin Charcot (1825–93) was a French neurologist of world renown whose name has been given to many medical signs and diseases such as Charcot's joints, Charcot's cirrhosis, Charcot's crystals, Charcot's disease, Charcot's fever, Charcot's gout, Charcot's pain, Charcot's posterior root-zone, Charcot's sensory crossway, Charcot's sign, Charcot's syndrome, and Charcot's zones, as well as many other diseases and syndromes named in conjunction with other doctors, such as Charcot-Marie-Tooth syndrome, Charcot-Guinon's disease, Charcot-Leyden crystals, Charcot-Marie's symptom, Charcot-Marie type of progressive muscular atrophy, Charcot-Neumann crystals, Charcot-Robin's crystals and Charcot-Vigourouse's sign. Such was the fame of this man in the medical world.

Between 1899 and 1905 his 10-volume *Traité de Médecine* was published in Paris. It was an encyclopaedia of the current European treatments and included wine diluted in water or with Cognac added to it for pernicious anaemia, diets for aortic aneurysms, scurvy, convalescence after diphtheria and bronchopneumonia, in tuberculosis, stomatitis and endocarditis, to name just some indications. An endorsement of wine by one so famous as Charcot was not to be taken lightly.

The Spanish doctors Alexandre and Aparici published a book called *Valor Terapeutico del Vino de Jerez* in 1903. It praised sherry as an aid for convalescence, for the aged, in pneumonia, and for the overworked.

Thus, by the end of the nineteenth century wine was established as a medicine in continental Europe, but the Temperance Movement had decreased its popularity in the UK and the USA. In the coming years it was to face the test of all the new 'wonder drugs' produced by the pharmaceutical industry to replace its role, such as aspirin for pain and fever, barbiturates for sedation, vitamins for deficiency diseases and tonics, sulphurs and other antibiotics for infections, tranquillizers for anxiety, and so on.

## Australia

During the period 1787 to 1868, the British Government transported about 160 000 convicts to its penal colonies in Australia. Their survival during the long sea voyage depended to a great extent on the surgeon-superintendents of the transport ships, who were instrumental in effecting a 10-fold reduction in mortality during the first 35 years of transportation (Pearn 1996). After the notorious voyage of *HMS Surrey* in 1814, during which more than one-quarter of the ship's company died, including convicts, the captain and the first and second mates, Dr William Redfern was appointed to ensure the safety of future voyages. Redfern, himself a former transported convict, recommended a daily ration of a quarter of a pint of wine with added lime juice. Wine played a major preventive role in maintaining the health of the convicts and was used throughout the voyage to counter malnutrition and illness, including waterborne diseases such as dysentery. Convict mortality fell from one in three for those embarked on the second fleet in 1790, to one in 20 in 1833, and to zero in 1859.

Redfern was later to establish Australia's first 'wine doctor' vineyard to the south-west of Sydney in 1818, although Governor Philip had planted the very first vineyard 30 years earlier at Sydney Cove. The tradition of Australian physicians establishing vineyards has continued to this day, with many of the largest and most prestigious companies – Lindeman, Penfold,

*Table 2.2* Summary of the medicinal uses of wine by the ancients

- Antiseptic: of wounds, of water, and preoperatively
- Tranquillizer/sedative
- Hypnotic
- Anaesthetic
- Antinauseant
- Appetite stimulant
- Tonic/restorative during convalescence
- Treatment of anaemia
- Diuretic
- Purgative (some types of wine)
- Antidiarrhoeal (other types of wine)
- Cooling agents
- Poultices
- Mixing medium for other medicines, sometimes with honey

Hardy, Houghton, Angove, Stanley, Minchinbury – being founded by doctors. Two-thirds of any vintage in modern times is made by companies founded by Australia's 180 wine doctors.

## Conclusion

Thus, by the end of the nineteenth century, wine was established as a medicine in continental Europe and Australia, although the Temperance Movement had eroded its position in the UK and the USA. Its many medicinal uses over the ages are summarized in Table 2.2, and the chronology of wine as a medicine is shown in Table 2.3. Since then, the scientific evidence on wine as a medicine has multiplied enormously, especially during the past decade. The medicinally active ingredients have been mostly identified, their pharmacological actions have been largely characterized, and the role of wine in the epidemiology of various disorders from cancer to cardiovascular disease is a hot topic. Rather than reinventing the wheel we are only now beginning to understand the mysteries of an ancient medicine.

*Table 2.3* Chronology of wine as a medicine (modified from Lucia 1963)

| *Event* | *Date** |
|---|---|
| First wine produced in Georgia | 7000 BC* |
| *Mesopotamian culture (5000–1400 BC*)* | |
| Introduction of wine | 4000 BC* |
| Invention of writing as pictograms | 3300 BC* |
| First representation of wine-drinking in Standard of Ur | 3000 BC* |
| Use of wine as medicine as illustrated by a Sumerian pharmacopoeia inscribed on a clay tablet at Nippur in cuneiform script | 2100 BC* |
| *Egyptian civilization (3000–332 BC*)* | |
| Ancient Egyptian medical papyri, forebears to the current known medical papyri | 3000 BC* |
| Earliest depiction of winemaking in the pictographs of the tomb of Ptah-Hotep at Thebes | 2400 BC* |
| Medical papyri: | |
| Kahun | 1900 BC* |
| Edwin Smith | 1650 BC* |
| Ebers | 1500 BC* |
| Hearst | 1500 BC* |
| London | 1350 BC* |
| Berlin | 1350 BC* |
| Brugsch | 1350 BC* |
| *Biblical times (1220 BC–AD 70*)* | |
| Use of wines as sedatives, antiseptics and vehicles for other medicines is illustrated in the Sacred Writings: | |
| Talmud, written after | 536 BC |
| Old Testament, written before | 400 BC |
| New Testament, first recorded | 1st century AD* |
| *Ancient India (200BC–AD1000*)* | |
| *Vedic period*: Soma, the supreme deity of healing, was conceived as a being in liquid form; in the Vedas the healing potential of wine was made equal to the power of Soma | 2000–200 BC* |
| *Brahmanic period* | 200 BC* |
| Use of wine in medicine illustrated in the Charaka Samhita | AD 1000 |
| *Ancient China (1800 BC–AD 220*)* | |
| Wines were incorporated in the materia medica and appeared as menstruums in ancient Chinese writings | |
| Wine used in libational ritual in the Chang dynasty | 1766–1122 BC |
| Wine used in sacrificial rituals in the Chou Dynasty | 1122–222 BC |
| *Early Greek medicine (900–100 BC*)* | |
| *Homeric times* | 900–500 BC* |
| In the *Odyssey* and the *Iliad*, wine was described as an antiseptic, a sedative, and as a staple food | 850 BC* |
| Hesiod described wine as a nutrient and tonic | 8th century BC |

*Table 2.3 (cont'd)*

| *Event* | *Date** |
|---|---|
| *Hippocratic times* | 450–300 BC |
| Hippocrates used wine as an antiseptic, diuretic, sedative and menstruum as described in his medical text *Regiment* | 460–370 BC* |
| Diocles of Carystus wrote on the use of sweet wines in medicine | 375 BC |
| Theophrastus of Eresus described plant-embellished wines | 372–287 BC |
| Mnesitheus wrote of wine in *Diet and Drink* | 320–290 BC |
| *The Alexandrians* | 300–50 BC* |
| The centre of medicine moved to Alexandria | |
| The judicious use of wines in therapeusis was stressed in the teachings of the medical school founded by Erasistratus | 300–260 BC |
| Nicander used wine as a menstruum for his theriacs and alexipharmics | 190–130 BC |
| Mithradates, King of Pontus, used wine as the menstruum for his antidote, mithradatium | 132–63 BC |
| Hikesios wrote a treatise and commentary on wine, *De Conditura Vini* | 1st century BC* |
| Apollonius of Citium wrote on the medicinal value of European wines in a letter to Ptolemies | 81–58 BC* |
| *Greek Medicine in Rome (100 BC–AD 100*)* | |
| With the establishment of the Greek physicians in Rome, the therapeutic use of wine became a vital question; physicians who adopted the medical use of wine were known as *physikos oinodotes* | |
| Cato the Elder described wine as a medicine | 234–149 BC |
| Asclepiades, leader of the wine-prescribing physicians | 124–40 BC |
| Zopyrus used wine as the menstruum for a mithradatium called ambrosia | 80 BC* |
| Menecrates of Tralles used wine clinically | 1st century BC* |
| Celsus wrote on wine as a medicine in *De Re Medicina* | 25 BC–AD 37 |
| Pliny the Elder described the therapeutic uses of wine in *Naturalis Historia* | AD 23–79 |
| Columella emphasized wine as a medicine | 4 BC–AD 65 |
| Sextius Niger advocated the use of natural wine in medicine | AD 40* |
| Dioscorides recommended wine as materia medica for many diseases | AD 40–90 |
| Discorides published *De Universa Medicina* | AD 77* |
| *The era of Galen (AD 100–400*)* | |
| After the death of Asclepiades, independent medical schools were established | |
| The School of Eclecticism: Athenaeus of Attalia taught that wine in small doses rouses the pneuma and restores vitality | AD 41–54* |
| Galen used wine-based mixtures called Galenicals and wrote about wine as a medicine in *De Sanitate Tuenda* (Galen's Hygiene) | AD 131–201* |
| Aretaeus of Cappadocia recommended Italian wines | 2nd to 3rd centuries AD |

*Table 2.3 (cont'd)*

| *Event* | *Date** |
|---|---|
| Athenaeus of Naucratis, the encyclopaedist, recorded valuable information on the medicinal uses of wine in *The Deipnosophists* | 3rd century AD |
| Oribasius recommended wine as a medicine | AD 325–403 |
| Roman generals such as Julius Caesar recommended wine for their soldiers to increase their strength, preserve good health and prevent dysentery | |
| *Byzantine era (AD 400–700*)* | |
| Following the transfer of the Roman capital to Byzantium, the centre of learning became displaced but the teachings of Galen prevailed | AD 330 |
| Aetius of Amida detailed the medical uses of wine in the *Tetrabiblion* | AD 502–575 |
| Alexander of Tralles followed the tradition of the wine-prescribing physicians | AD 525–605 |
| Paul of Aegina recognized as the link between Greek and Arabic medicine | AD 625–690 |
| *Arabic period (AD 600–1300*)* | |
| Arabic culture influenced Western thought for many centuries after the death of Mohammed | AD 632 |
| Conquest of Alexandria | AD 641 |
| First apothecary shop established in Baghdad | AD 745 |
| The precepts of Galen prevailed and the use of wine in medicine continued: Rhazes wrote on the washing of wounds with wine | AD 860–932 |
| Haly ben Abbas discussed wine as a medicine in *Almaleki* | 10th century AD* |
| Avicenna promulgated rules for the proper use of wine in the *Canon of Medicine* | AD 980–1037 |
| Mansur the Great discussed wine as a pharmacological menstruum | 10th century AD* |
| Avenzoar adhered to the emphasized Hippocratic teaching | AD 1162 |
| Maimonides elaborated on the medicinal value of wine in *De Regimine Sanitatis* | AD 1135–1204 |
| Averroes applied Aristotelian teaching to medicine | AD 1198 |
| Albucasis recommended wine as an antiseptic in his treatise on surgery | AD 936–1013 |
| *The School of Salerno (AD 1050–1300*)* | |
| The first lay medical school in Europe established at Salerno | 10th century AD |
| Arabic medical manuscripts brought to Salerno by Constantine the African | AD 1027–1087 |
| The *Regimen Sanitatis Salernitanum* illustrated the therapeutic uses of wine | 11th century AD* |
| Ugo Borgognoni used wine as an antiseptic, died | AD 1258 |
| Teodorico Borgogoni advocated the use of wine as an antiseptic | AD 1205–1296 |
| Salicet used strong wine as an antiseptic in surgery | AD 1210–1277* |
| Bruno da Longoburgo achieved wound antisepsis with wine | AD 1300* |

*Table 2.3* (*cont'd*)

| *Event* | *Date** |
|---|---|
| Lanfranc lapsed back into the practice of suppuration (laudable pus) | AD 1306* |
| *Late Middle Ages (AD 1300–1543*)* | |
| The physicians of the period began to realize the importance of the treatment of disease based on clinical experience | |
| Arnald of Villanova established the therapeutic use of wine in *Liber de Vinis* and popularized *aqua vitae* | AD 1235–1311* |
| Henri de Mondeville advocated the use of wine as a 'wound drink' | AD 1260–1320 |
| Guy de Chauliac used wine in the treatment of wounds and as a mouthwash | AD 1300–1368 |
| John of Arderne employed wine as a menstruum | AD 1307–1377 |
| Hieronymus Brunschwig ascribed miraculous healing powers to '*aqua vite composita*' | AD 1450–1533* |
| The *Antidotarium Nicolai* printed | AD 1471 |
| Paracelsus, known as the father of modern pharmacology, stressed the tonic value of wine and invented the word alcohol | AD 1493–1541 |
| *Beginnings of modern medicine (AD 1543–1850*)* | |
| The publication of *De Corporis Humani Fabrica* by Andreas Vesalius marked the beginning of an important era in medicine, an era that witnessed many departures from tradition and in which the foundations for the scientific age were laid | AD 1543 |
| Ambroise Paré used wine as a tonic and to dress wounds | AD 1510–1590 |
| Richard Wiseman wrote on the medicinal uses of wine in his textbook of surgery | AD 1622–1676 |
| Sir John Haryngton published the first English translation of the *Regimen Sanitatis Salernitanum* | AD 1607 |
| Era of dispensatories and pharmacopoeias established by Valerius Cordus | AD 1546 |
| Wine as an official therapeutic agent depicted in: | |
| *The Pharmacopoeia of London* | AD 1618 |
| *The Pharmacopoeia of Amsterdam* | AD 1636 |
| *The Pharmacopoeia of Paris* | AD 1639 |
| *The Pharmacopoeia of Spain* | AD 1651 |
| *The Pharmacopoeia of Brussels* | AD 1671 |
| *The Complete English Dispensatory* | AD 1741 |
| *The Pharmacopoeia of Russia* | AD 1778 |
| *Codex Medicamentarius of France* | AD 1819 |
| *The Pharmacopoeia of the United States* | AD 1820 |
| The inclusion of many of the theriacs in the dispensatories and pharmacopoeias led to a polemic that resulted in the final demise of the theriacs | |
| De Diemerbroeck published a defence of the theriac | AD 1646 |
| Dr Hodges recommended sherry-sack as a preventive against the plague | AD 1665 |

*Table 2.3* (*cont'd*)

| *Event* | *Date** |
|---|---|
| Wine remains an important therapeutic agent: an anonymous author (a Fellow of the Colleges) published an essay on the preference of wine to water | AD 1724 |
| Heberden gave the final blow to the theriacs in *Antitheriaka* | AD 1745 |
| Loebenstein-Lobel published a treatise on the uses and effects of wine | AD 1817 |
| Henderson published *A History of Ancient and Modern Wines* | AD 1824 |
| Charcot discussed the clinical uses of wine in *Traité de Médecine* | AD 1825 |
| *The Pharmacopoeia Universalis of Heidelberg* listed 175 wines | AD 1835 |
| *The London Pharmacopoeia* included a description of wines and their medicinal uses | AD 1835 |
| The *Pharmacopée Universelle* of Paris listed 164 wines | AD 1840 |
| The new edition of the US pharmacopoeia added port and sherry | AD 1850 |
| *The British Pharmacopoeia*, revised edition, included sherry and other medicated wines | AD 1851 |
| McMullen published *A Handbook of Wines* | AD 1852 |
| Mulder published a chemical analysis of the constituents of wine | AD 1857 |
| *Australian era (1787–to present)* | |
| Surgeon White uses wine as main medicine for convicts in the First Fleet to Australia | AD 1787 |
| Dr Redfern's letter to Governor Macquarie about use of wine as medicine for convicts | AD 1814 |
| Wine used in convict ships and later in migrant ships bound for Australia | AD 1815 |
| Dr Redfern becomes first Australian medical vigneron or wine doctor by planting Campbellfields vineyard | AD 1818 |
| Dr Lindeman founded Lindeman Wines | AD 1841 |
| Dr Penfold founded Penfold Wines | AD 1842 |
| Dr Kelly founded Hardy's Wines | AD 1843 |
| Dr Angove founded Angoves | AD 1889 |
| Lunatic asylum vineyards in Australia began | AD 1870 |
| Dr Lindeman's letter to *New South Wales Medical Journal*: 'Wine as a therapeutic agent and why it should become our national beverage' | AD 1871 |
| Dr William Cleland's speech 'Some remarks upon wine as a food and its production' | AD 1880 |
| Dr Thomas Fiaschi lecture 'The various wines used in sickness and convalescence' | AD 1906 |
| *The modern epoch (1850–present*)* | |
| The experimental method in physiology introduced by Claude Bernard | AD 1813–1878 |
| Scientists became absorbed in the study of alcohol and alcoholic beverages, including detailed studies of wine: Claude Bernard studied the effect of pure alcohol on digestion | AD 1857 |

*Table 2.3 (cont'd)*

| *Event* | *Date** |
|---|---|
| Pasteur described fermentation | AD 1857* |
| Trousseau discussed the medical uses of wine in *Clinique Médicale* | AD 1861 |
| Dr Anstie published his comprehensive work on the therapeutic uses of wine | AD 1870 |
| Parkes and Wollowicz published the first study on the physiological effects of wine | AD 1870–71 |
| Carles investigated the iron content of wines | AD 1880 |
| Buchner published the first comparative study detailing the effects of wine, beer and alcohol on the stomach | AD 1882 |
| Alois Pick published his findings on the bactericidal effects of wines | AD 1892 |
| Krautwig and Vogel published a study on the physiological effects of various alcoholic beverages on respiration | AD 1893–1897 |
| Chittenden and co-workers investigated the effect of wines and spirits on the alimentary tract | AD 1898 |
| Wendelstadt published his findings on the effect of wines on respiration | AD 1899 |
| Benedict and Torok investigated the role of wine in diabetic diets | AD 1906 |
| Neubauer published findings on the use of wine in diabetes | AD 1906 |
| Kast reported on gastric digestion and the effect of wine and alcohol on the diet | AD 1906 |
| Sabrazes and Marcandier published their results on the bactericidal properties of wine | AD 1907 |
| Pavlov demonstrated the appetite-stimulation effect of wine | AD 1910 |
| Carles reported on the diuretic action of wines | AD 1911 |
| Carlson published his findings on the effects of wine on hunger | AD 1916 |
| Sir Edward Mellanby published findings on the physiological and dietetic effects of alcohol and alcoholic beverages | AD 1919 |
| Koutetaladze isolated an amine, a coronary stimulant, from wine | AD 1919 |
| Haneborg investigated the effect of alcoholic beverages on digestion | AD 1921 |
| Pearl first to prove in *Alcohol and Longevity* that moderate drinkers live longer | AD 1926 |
| Lucille Randoin published findings on the vitamin content of wines | AD 1928 |
| Loeper and co-workers reported on the effects of wine on the liver | AD 1929 |
| Winsor and Strongin reported on the effects of wine on salivary digestion | AD 1933 |
| Soula and Baisset investigated the effect of wine on blood sugar levels | AD 1934 |
| Fessler and Mrak reported on the effects of wine on urinary acidity | AD 1936 |
| *US Dispensatory* deleted all wines | AD 1937 |
| Remlinger and Bailly reported on the bactericidal effect of wines | AD 1938 |

*Table 2.3 (cont'd)*

| *Event* | *Date** |
|---|---|
| Flavier demonstrated nutritionally important amounts of vitamin B in wines | AD 1939 |
| Morgan reported on vitamins in wine | AD 1939 |
| Newman published findings on the absorption of wine | AD 1942 |
| Ogden studied the influence of wine on gastric acidity | AD 1946 |
| Goetzl and co-workers reported on wine as an appetite stimulant | AD 1950–53 |
| Flanzy published a study of the comparative physiological effects of wine and alcohol | AD 1953 |
| Lolli and co-workers reported on the relation between wine in the diet and the carbohydrate intake | AD 1952 |
| Castor reported on B vitamins in wines | AD 1952 |
| Gardner presented findings on the bactericidal property of wines | AD 1953 |
| Hall and co-workers reported on the effect of wine on cholesterol metabolism | AD 1957 |
| Engleman published findings on the relationship between wine and gout | AD 1957 |
| Macquelier and Jensen reported on the bactericidal activity of red wines | AD 1960 |
| Pratt and co-workers published findings on the grape anthocyanins | AD 1960 |
| Althausen and co-workers reported on the effect of wine on vitamin A absorption | AD 1960 |
| French *Codex* listed 7 wines | AD 1960 |
| Balboni discussed the role of wine in obesity | AD 1961 |
| Carbone reported on the relation of wine to cirrhosis of the liver | AD 1961 |
| Masquelier published findings on the polyphenols of red wine as a cholesterol-reducing agent | AD 1961 |
| Henneckens and Stamfer showed that moderate alcohol consumption reduces coronary disease and stroke in women | AD 1988 |
| Rimm showed inverse relationship between alcohol consumption and coronary disease | AD 1991 |
| Renaud published *The French Paradox* | AD 1992 |
| Doll published study on British doctors' hearts | AD 1994 |
| Groenbaek published Copenhagen Study, the first to compare the health effects of beer, wine and spirits | AD 1995 |
| Orgogozo showed moderate wine consumption reduced dementia | AD 1997 |
| Heart study by Doll showed that the beneficial effects of alcohol outweigh its harmful effects when taken in moderation | AD 1998 |
| Bertelli showed resveratrol stimulates MAP-kinase (mitogen-activated protein kinase), thus preventing neurodegenerative disease | AD 1999 |
| Pezzuto and Renaud show wine reduces cancer | AD 1999 |

* *circa*.

## References


Ackerman, P. (1945) *Ritual Bronzes of Ancient China*. Dryden Press, New York.

A Fellow of the Colleges (1724) *The Juice of the Grape – or Wine Preferrable to Water*. Printed for W. Lewis, under Tom's Coffee House, Covent Garden, London.

Albutt, T. C. (1905) *The Historical Relations of Medicine and Surgery to the End of the Sixeenth Century*. Macmillan, London.

Anstie, F. E. (1877) *On the Uses of Wine in Health and Disease*. Macmillan, London.

Banckes, R. (1941) *An Herball* (edited and translated into modern English and introduced by S. V. Sarkey and T. Tyles). Scholars' Facsimiles and Reprints, New York, pp. 52–3, 50, 70.

Bang, J. (1973) Vinen traek af dews historie ogdews medicinske virkning. *Archiv for Pharmaciog Chemi*: **83**, 1145–56.

Bose, D. K. (1922) *Wine in Ancient India*. Connor, Calcutta.

Burke, P. (1984) Wine as a medicine. *Medico-Legal Society of Victoria Proceedings*: April.

*Ciba Symposia Babylonian Medicine* (1940) December (2)9, 680.

Coar, T. (1822) *The Aphorisms of Hippocrates*. Valpy, London.

Dash, V. B. (1980) *Fundamentals of Ayurvedic Medicine*. Barsal, New Delhi.

Druitt, R. (1873) *Report on the Cheap Wines from France, Germany, Italy, Austria, Greece, Hungary and Australia – their use in Diet and Medicine*. Henry Renshaw, London.

Green, R. M. (1951) *A Translation of Galen's Hygiene*. Charles C. Thomas, Springfield, USA.

Gruner, O. C. (1930) *A Treatise on the Canon of Medicine of Avicenna Incorporating a Translation of the First Book*. Luzac, London.

Henderson, A. (1824) *The History of Ancient and Modern Wines*. Baldwin, Cradock and Joy, London.

Horsley, K. (1998) Hippocrates and his legacy. *Veteran's Health*, Department of Veteran's Affairs: 62, July.

Hubotter, F. (1957) *Chinesische-Tibetische Pharmakologie und Rezeptur*. Hakg, Ulm, p. 144.

Johnson, H. (1989) *The Story of Wine*. Mitchell Beazley, London.

Jolly, J. (1951) *Indian Medicine*. Translated from the German, supplemented and published by C.K. Kashikar, Poona.

Lucia, S. (1963) *A History of Wine as Therapy*. Lippincott, Philadelphia.

McGovern, P., Fleming, S., Katz, S. (2000) *The Origins and Ancient History of Wine*. Gordon and Breach, Amsterdam.

Norrie, P. (2000) *Dr Norrie's Advice on Wine and Health*. Apollo Books, Sydney, p. 14.

Nunn, J. F. (1996) *Ancient Egyptian Medicine*. British Museum Press, London.

Pearn, J. (1996) Surgeon-superintendents on convict ships. *Australian and New Zealand Journal of Surgery*: **66**, 254.

Phillips, E. D. (1973) *Greek Medicine*. Thames and Hudson, London, p. 77.

Quintner, J. (1999) Plague doctors in the face of death. *Medical Observer*, December: 59.

Read, B. E. (1931–7) Chinese Materia Medica, Peiping. *Peking Natural History Bulletin Series*: **I–IX**.

Reed, H. S. (1942) *A Short History of the Plant Sciences*. Chronica Botanica Co., Waltham, MA.

Ricci, J. V. (trans.) (1950) *Actios of Amida* (translated from the Latin edition of Cornarius 1542). Blakiston, Philadelphia, pp. 214–15.

Riollay, F. (1783) *Doctrines and Practice of Hippocrates in Surgery and Physic*. Cadell, London.

Sandford, W. (1799) *A Few Practical Remarks on the Medicinal Effects of Wine and Spirits*. J. Tymbs, London, pp. IV, V, 8–9, 20.

Sarma, P. J. (1939) The art of healing. *Rigveda Annals of Medical History*, Third Series: 1, 538.

Skovenborg, E. (1990) In vino sanitas. *Saertryk Fra Bibliotek for Laeger*: **182**, 4.

Trotula (1940) *The Diseases of Women* (translated by Elizabeth Mason-Hohl). Ward Ritchie Press.

von Hirnkofen, W. (1943) *Arnold of Villanova's Book on Wine* (translated by H. E. Sigerist from the German). Schuman's, New York.

von Staden, H. (1989) *Herophilus, The Art of Medicine in Early Alexandria*. Cambridge University Press, Cambridge.

Yorke-Davis, P. (1909) Wine and Health – How to Enjoy Both. Chatto & Windus, London.

# 3 Archaeology and the origins of wine production

*J. M. Renfrew*

Winemaking is essentially a natural process whereby the juice of ripe grapes comes into contact with the natural yeast present in the bloom on grape skins. One has only to observe the behaviour of birds in a vineyard when feasting on overripe grapes to see that the fermentation process occurs in nature without human intervention. It is, however, facilitated by the storage of gathered grapes, or their juice, in a waterproof container that retains the juice during fermentation and can later be sealed for long-term storage. It is perfectly possible to make wine from the grapes of wild vines; cultivation of the grapevine is not a prerequisite for winemaking. Human intervention is more as facilitator and refiner of this natural process. For humans, it is the fact that this pleasant alcoholic beverage can be stored in sealed containers for long periods, and is a desirable, tradeable product that led to its influential role in the development of early civilizations in the Old World.

Archaeologists have to use all the evidence they can find to reconstruct the past: in the case of winemaking the physical remains of wine sediments in pottery vessels give the clearest evidence. Added to that are finds of the pips, skins, fruit stalks, leaves (or leaf impressions on pottery), charcoal derived from vine stems, and even the root holes left by vines in early vineyards. Sometimes these types of evidence are found associated with vessels used as presses and spouted containers, which do suggest that they were involved with winemaking. In some cases, vineyards and winemaking procedures and the consumption of wine are depicted in art. Elaborate mixing and drinking vessels as well as the wineskins, barrels and amphorae used to transport and store wine also give clues to its importance in early civilizations.

Clearly grapes could have been consumed fresh or dried as raisins or currants. Not all finds of grape pips necessarily indicate that they were the residues of winemaking. Having said that, it interesting to note that some of the earliest finds of grape pips associated with humans belong to the palaeolithic period in Europe. Pips of wild grapes, *Vitis vinifera* ssp. *sylvestris*, have been found associated with the palaeolithic hut shelter at Terra Amata, Nice, southern France, dating back some 350 000 years; at the Grotta del' Uzzo in Sicily, in the Franchthi Cave in Greece, at Tell Abu Hureyra in Syria, at Tell Aswad and at Jericho, all in levels dating between the 12th and 9th millenium

BC, so that the merits of grapes, if not wine, were appreciated well before the beginnings of agriculture.

Winemaking, as an industry, requires the cultivation and domestication of the wild grapevine. In neolithic times (9000–4000 BC) when farming was established in the Near East and the craft of making pots became well understood, the conditions for domesticating the grape were in place. Wild grapes are to be found around the Mediterranean basin and in the area between the Black and Caspian Seas, and also in eastern Anatolia extending as far east as the northern Zagros Mountains, although in the damper climates that prevailed at this time they may well have been more widely distributed. It seems likely that the domestication of the grapevine took place in the region between the Caucasus and Armenia, where the greatest diversity of wild vines are found today. The process of domestication was quite a complex one, rather different from the domestication of annual cereal and pulse crops on which the early farmers depended. The wild vine is dioecious, with male and female plants occurring separately; only occasionally are hermaphrodite forms found in the wild. The domesticated grapes are all hermaphrodite, and may well have arisen from taking cuttings from the hermaphrodite wild forms, or by grafting hermaphrodite forms on to wild stock. (A number of vine cuttings enveloped in close-fitting silver sleeves, which even detail the buds on the enclosed twigs, form a unique find from a burial site at Traleti, Georgia dating to 3000 BC). Whatever the mechanism, the result was that self-pollinating forms arose that were selected for yielding larger, juicy fruit with fewer seeds. The wild plants are rampant climbers, often reaching 30 metres into the branches of host trees. Cultivation of the domesticated grapes included a strict regime of pruning to ensure that the plants' energies were concentrated on fruit production and to make the bunches of fruit more accessible for harvesting. The establishment of vineyards was a labour-intensive activity that would not yield instant results, but that with careful management would last for several generations.

It seems likely that the domestication of the grapevine took place somewhere in the northern part of the Near East around 6000 BC, and that by 4000 BC it had been brought as far south as the Jordan valley. By a millennium it later had become established in the Nile valley. The earliest physical traces of wine have been identified in two pottery vessels excavated at the site of Haji Firuz Tepe, south of Lake Urmia, dating to 5400–5000 BC. Sediments found in the bottom of these vessels were in one case pale yellowish and in the other reddish in colour. They were analysed in the laboratory at MASCA (Museum Applied Science Centre for Archaeology), at the University Museum of Philadelphia, using infrared, liquid chromatography and other chemical analyses; they clearly showed the presence of tartaric acid, calcium tartrate and terebinth resin. These vessels had been embedded with another four similar ones in an earthen floor along one wall of a kitchen. Clay stoppers were found nearby showing how they had been sealed originally. It was concluded that these vessels had contained resinated wine. If all six vessels found had contained wine, then about 14 gallons (64 litres) were being stored in that kitchen; this suggests that wine production was well underway here at that date, and maybe that both white and red wines were being made.

Once wine production had been established, a trade in wine soon arose since once it is sealed in containers it can easily be transported and keeps well for quite long periods. The earliest indications of a trade in wine come from the analysis of residues found in the bottom of a large amphora from Godin Tepe, in modern Iran, dating to 3500 BC. This analysis was also carried out at MASCA by Patrick McGovern, and again showed traces of tannins and tartaric acid. By the late fourth millennium BC resinated wine appears to have been traded south and east to Uruk and Tello in southern Iraq and Susa in Iran, as is shown by the analyses of residues in piriform and spouted vessels. The trade routes probably followed the major rivers, and for part of the way at least the wine was probably transported by water. Mesopotamia was not the ideal place for growing vines since it has such a hot, dry climate, and beer appears to have been the favourite alcoholic drink of the ancient Sumerians, being drunk through straws from large jars. Some cylinder seals, however, depict another special drink being consumed from hand-held cups and goblets, and this was probably wine. Finds of charcoal from grapevines at Tepe Malayan as well as grape pips, identified by Naomi Miller at MASCA, suggest that by the mid-third millennium BC some grapes were being cultivated in the southern Zagros region of Iran.

Wine, however, remained a rare and expensive commodity in Babylonia from the third to the first millennium BC and was the prerogative of the gods and the rich; poorer people had to content themselves with drinking beer. Although there was limited wine production in the Middle Euphratess as mentioned in the cuneiform tablets from the royal palace at Mari, *c.* 1800 BC, most of it was imported from regions to the north and west and transported down the Euphrates by boat. With the establishment of the Assyrian Empire in the first millennium BC, wine production seems to have expanded in the mountainous areas stretching from western Iran through southern Turkey to Syria and the Mediterranean. That it was plentiful is indicated by the record that 10 000 wineskins were used in the feast given by Ashurnasirpal II (883–859 BC) to celebrate the inauguration of his new residence at Nimrud. The oldest visual representations of wineskins are on the bronze gates of Shalmaneser from Balawat, now in the British Museum. They seem to be a feature of the Syrian wine trade, being less breakable than jars for long distance transport. Wine rations were given to members of the royal household and to the soldiers, according to eighth century texts found at Nimrud. There is a delightful relief of Assurbanipal (668–627 BC) and his queen from the palace at Nineveh: it shows them feasting and drinking wine from shallow bowls in a garden under an arbour of fruit-laden vines. Evidence from cuneiform tablets shows that Nebuchadresser (605–562 BC) imported wines from eight foreign countries to offer to the gods, these included areas of northern Iraq, Turkey and Syria. Esarhaddon (689–668 BC) mentions in his texts a mixture of wine with beer made from barley malt with additives of emmer wheat and grape syrup. Recent analyses of residues found in vessels in King Midas's tomb at Gordion in central Turkey dating to *c*. 700 BC, undertaken by the MASCA laboratory in Philadelphia, have shown that the guests at this funeral feast drank a similar potent beverage made of wine, barley beer and honey mead to wash down a tasty lamb stew.

The invention of viticulture is attributed in the Book of Genesis to Noah: 'And Noah began to be a husbandman, and he planted a vineyard. And he drank of the wine and was drunken'. Noah's drunkenness shocked his son, Ham, and later generations of Jews, and preoccupied early Christian commentators on the Bible, although Genesis did not condemn Noah for it. In fact, the vine and wine are the most frequently mentioned plant and plant product in the Bible. There are numerous references to grapes, wine, wine presses and the vintage in the Old Testament. For Moses' followers, the first glimpse they had of the fertility of the Promised Land was huge bunch of grapes borne on a staff between two of the spies returning from Canaan and reporting on the land flowing with milk and honey (Numbers 13: 23). Isaiah contains advice on how to plant a vineyard, and the books of Micah, Amos, Joel, Jeremiah, Ezekiel, Zachariah and Nehemiah all use the vine as an indicator of a happy state. When these prophets threaten doom they say that the Lord will withold the products of the vintage. As Psalm 104: 15 says, 'wine maketh glad the heart of man'. New wine is used to evoke images of joyful abandon and drunken revelry among the Israelites: 'Awake ye drunkards, and weep, and howl, all ye drinkers of wine, because of the new wine, for it is cut off from your mouth' (Joel 1: 5). But God's mercy is shown when 'the mountains shall drop down new wine (Joel 3: 18). In the New Testament, Luke (5: 37–9) not only explains that new wine should not be put in old bottles, but also goes on to explain: 'No man also, having drunk old wine, straightway desireth new; for he saith the old is better'. The first miracle that Christ performed was when he turned water into wine at the marriage at Cana, when the wine provided for the guests ran out. The governor of the feast protests to the bridegroom: 'Everyman at the beginning doth set forth good wine and, when men have well drunk, that which is worse, but thou hast kept the good wine until now' (John 2: 10). Christ uses the image of the vine in his teaching: 'I am the true vine, and my Father is the vinedresser. Every branch of mine that bears no fruit he takes away, and every branch that does bear fruit he prunes, that it may bear more fruit' (John 15: 1–2). In the Eucharist, the drinking of wine symbolizes the drinking of Christ's blood, and recalls for Christians the Last Supper Christ enjoyed with his disciples before His crucifixion. The medicinal benefits of wine are mentioned in the first letter to Timothy (1 Timothy 5: 23) where he is encouraged to give up drinking water in favour of using 'a little wine for thy stomach's sake and thy other infirmities'.

This translation of a Hebrew song supposedly sung by a royal mother to her princely son sums up the ancient Hebrew attitude to wine, as also reflected in the Book of Proverbs:

> It is not for the king to drink wine
> Nor for rulers to drink strong drink,
> Lest, drinking they forget the Law,
> And disregard the rights of the suffering.
>
> Give strong drink to him who is perishing,
> Wine to him who is in bitter distress;
> That drinking he may forget his poverty,
> And think of his poverty no more.
>
> (Hyams 1987)

A remarkable find indicating the early and extensive trade in wine in the Near East comes from the analyses of jars stacked in three rooms of the tomb of Scorpion I, one of Egypt's first kings, at Abydos in Middle Egypt, dating to around 3150 BC. It appears that this tomb held about 700 wine jars, and, if they were full, this would represent about 1200 gallons (5455 litres) of wine. Some of the jars contained grape pips, some contained whole grapes, while others contained slices of figs that had been perforated and strung together to be suspended in the wine, probably for flavouring and sweetening it. The jars had been sealed with clay over cloth or leather covers tied over the mouths of the vessels to prevent the wine from turning to vinegar. The wine had long since evaporated leaving yellowish residues in the jars. It was these that Patrick McGovern analysed at MASCA using diffuse-reflectance Fourier-transform infrared spectrometry (DRIFTS), which is based on the principle that different chemical compounds absorb infrared light at different wavelengths. The presence of tartaric acid, which occurs naturally in large amounts only in grapes, indicated that these were wine residues. The use of the complementary technique high performance liquid chromatography (HPLC) on these residues suggested the presence of aromatic hydrocarbons derived from the resin of the terebinth tree. This may well have been added to the wine to prevent it from turning to vinegar. The wine was not produced in Egypt: analysis of the clay from which the wine jars were made, using neutron activation, suggests that they were imported from the southern Levant. Once a market for wine had developed in Egypt it appears that vineyards soon became established in the eastern delta area.

Although the standard funerary offerings in the Old Kingdom consisted of bread, beer, oxen, geese cloth and natron, by the Sixth Dynasty the officials of Teti, the first king of this dynasty (*c.* 2323–2291), were listing five different types of wine as approved funerary offerings. They came from different parts of the Nile delta.

Pictorial evidence for the production of wine features on the tomb walls from the Old Kingdom, e.g. from the tomb of Ptahhotep, and slightly more frequently in Middle Kingdom tombs, e.g. the tomb chapels at Beni Hasan, especially those of Baquet III (no. 15) and Khety II (no. 17) of the late Eleventh Dynasty (2050–2000 BC). The processes depicted include picking the grapes, treading them, several men pressing the residues through a cloth by twisting it using poles to get more purchase, bottling it, and sealing the bottles. Sometimes a scribe is shown checking off the vessels as they are filled. Many more depictions are known from the New Kingdom, for example from the Eighteenth Dynasty Theban tombs of Khaemwaset (no. 261) showing the watering of vines and the harvesting of bunches of grapes; and the tomb of Nakt (no. 52), which also shows the treading of the grapes in a spouted vat, and the storage of wine in sealed wine jars. A lively depiction of many of the processes involved in wine production can be seen on the walls of the tomb of the Royal Herald Intef (no. 155) together with a text commentary. Although both have suffered damage, they do bring the whole process alive. Many garden scenes show the cultivation of vines around garden pools in walled enclosures, for example the Theban tomb of Ken-Amun (no. 93). The walled garden of the temple of Amun shown on the walls of Theban tomb no. 96 may well depict one of the temple's estates in the delta: it is entered through a pyloned gateway which opens on to a vine arbour surrounded by pools, orchards and storerooms.

One of the most romantic finds of labelled wine jars must be that from the annexe to Tutankhamun's tomb. Here 26 wine jars were found, carefully labelled and with clay stopper sealings, their contents reduced to dried residues. The wine labels contained a great deal of information: the location of the vineyard, the year of the vintage expressed as the regnal year of the king, the ownership of the vineyard was indicated by the stamp on the sealing of the jar, the labels also indicated the chief winemaker who was responsible for its contents. Tutankhamun died at the early age of 19. The majority of the wines in his tomb dated from the fourth, fifth (the most popular with 12 jars) and ninth years of his reign (1345 BC, 1344 and 1340 BC, respectively). Most of the vineyards represented were located in the western delta of the Nile, but one came from the north-eastern delta, and one from the southern El Kharga oasis. Two of the 15 vintners named on the labels had Syrian names: Aperershop made the wine in jar 1 and Khay the contents of jars 4, 5, 14, 15, 16 and 17. The bottle dated to year 31 could be from Amenhotep III's reign (*c.* 1372 BC) and could have been a very special wine, or it may have been a case of reusing the jar without changing the original label. Only four jars were labelled 'sweet wine', and three others were described as '*šdh* wine', probably pomegranate wine.

Wine jar labels are also known from other New Kingdom sites. At Tell el Amarna, Akhenaton's capital city and religious centre, some of the wine labels give additional information apart from the date, estate, vineyard and vintner; they describe the wine as 'good wine', or 'very good wine', or 'genuine', or 'sweet', or 'for merry-making'. Some 26 different vineyards are named here, the majority being in the western delta of the Nile, and the wine was shipped the 300 miles upstream by boat. A few bottles came from the El Kharga oasis to the south.

At El Malqata, the palace city of Amenhotep III in western Thebes, the wine labels sometimes refer to the specific occasion for which the wine was made: 'wine for offerings', 'wine for taxes', 'wine for a happy return'. Some of these wines are described as being 'blended'. On the walls of Theban tomb no. 113 is a scene showing the blending of wines from different jars using siphons.

At the end of the New Kingdom, the pharoah Ramses III (1184–1153 BC) gave huge quantities of wine in offerings to the gods in their temples, as is recorded on the Papyrus Harris (I, 7, 10 ff.); in the section addressed to the god Amun, Rameses says:

> I made vineyards without limit for you in the southern oasis and the northern oasis as well, and others in great number in the southern region. I multiplied them in the north in the hundreds of thousands. I equipped them with vintners, with captives of foreign lands and with canals from my digging, that were supplied with lotuses and with *šdh* wine and wine like water for presentation to you in victorious Thebes.

He also claimed to have presented Amon with 59 588 jars of wine (Papyrus Harris I, 15a, 13 and I, 18a, 11).

The ancient Egyptians used wine for various medical purposes, as described in the Ebers medical papyrus. One section (50.21–51.14) details its uses in restoring appetite; in another section (55.1 ff.) it is used for 'driving out great weakness'. Wine was also used in salves, in enemas and in bandaging and wine lees were used for these purposes too, as well as for reducing swelling in the limbs (33.13–19).

Grape pips of the wild form are found on prehistoric sites in Greece from pre-agricultural times onwards, but it is not until the early Bronze Age that clear evidence of winemaking can be identified. Finds of pressed grape skins, pips and stalks were recovered from inside a spouted vessel in Early Minoan II levels at Myrtos, Crete, dating to *c.* 3000 BC, and give the earliest direct evidence for winemaking. The series of grape pips recovered from the stratified sites of Sitagroi, Dikili Tash and Dimitra in the north of Greece show that domesticated vines were present there from *c.* 2500 BC, whereas the earlier levels had only pips of wild grapes. That the manufacture and trade in wine and olive oil was of great importance to the development of civilization in the Aegean is shown by the number of huge pithoi in the storerooms of the Minoan palaces in second millennium Crete, and by the occasional finds of presses on villa sites such as at Vathypetro or the most recent find at Akrotiri on Thera. An ideogram for wine has been identified in Linear A, and artistic representations suggest the use of wine in religious libations.

The close links between the Minoans and the Egyptians, and the finds of Minoan pots in sites like Tell el Amarna, may indicate a trade in wine between these areas and possibly other parts of the Near East. The accumulation of wealth by the Minoans may have been partly due to trade in wine.

The Mycenaeans of mainland Greece (*c.* 1600–1200 BC) attached great importance to wine. Mycenaean palaces at Mycenae, Tiryns and Sparta have yielded physical remains of grapes, and many sites have large storage vessels suitable for the storage of wine. At Pylos a complete cellar was found containing 35 such large storage jars with some of them labelled in Linear B script as containing wine. Many of the Linear B tablets refer to 'wine', 'vineyard', and 'wine merchant'. It is clear from the widespread finds of Mycenaean pottery in the eastern Mediterranean that they were exporting wine and olive oil to Syria, Palestine, Egypt, Cyprus and even as far west as Sicily and southern Italy. The discovery of a few small Canaanite jars at Mycenae suggests that the wine trade was not only one way. Wine appears to have been consumed using elaborate drinking vessels made of special materials, such as the two-handled gold cup of Nestor from Mycenae, and the silver drinking service from Dendra. In both Minoan and Mycenaean society, wine seems to have been used originally with ceremonies of initiation and pouring libations, and then to have been used as an indicator of genteel, courtly, aristocratic behaviour. Drinking wine may also have taken place during business transactions: at Pylos, Blegan found two rooms that he thought might have been some sort of reception rooms for the palace. In one were two storage jars suitable for holding wine, and in the inner room he found the remains of between five and six hundred *kylikes* (drinking cups) smashed on the floor. He felt that this was the remains of some sort of ceremonial drinking possibly connected to conducting business in the palace.

Wine features largely in Homer, and was clearly quite commonly drunk in his day. It is the drink of mortals in his poems, whereas the gods drink nectar. It was used by the Greeks and Trojans in their feasts, and also for rituals in sacrifice, prayers and funerals. He speaks of two specific wines – Pramnian and Ismarian – and his contemporary Hesiod also refers to Bibline wine. Hesiod in his *Works and Days* describes some of the activities of the vineyard, and describes the drying of grapes before their use for making wine.

In classical times the Greeks not only cultivated vines throughout their country but also spread the practice of viticulture to Sicily, southern Italy (which they called 'Oenotria', the land of trained vines), to southern France and to their colonies on the Black Sea as far as the Crimea. Most of the vineyards were small, but some were more extensive: up to 74 acres (30 ha) on the island of Thasos in the fifth century BC, and about 30 acres (12 ha) in Attica in the fourth century BC.

Greek wine was widely traded. Within Greece, Athens provided the major market. Most of the wine for the export trade was produced on the

Aegean islands. Chios was perhaps the most famous: its amphorae stamped with the emblems of a sphinx, an amphora and a bunch of grapes are found in eastern Russia, in Scythian tombs on the Dnieper, at Naucratis in the Nile delta, in Cyrenaica, at Massilia (Marseille) in southern France, in Spain, in Tuscany and in Bulgaria, that is in most of the places that the Greeks had trade links with from the seventh century BC onwards. Another famous and widely traded wine was produced on the island of Lemnos, and this may have been the source of the famous Pramnian wine (although it is also claimed by Smyrna and Ikaria in the Dodecanese). The wines of Thasos were lighter in quality and flavoured with apples. The wine trade in Thasos was particularly well regulated. Their amphorae were of standard sizes and were sealed with the name of the annual magistrate to prevent fraud. The islanders were forbidden to import wines from elsewhere. The vine's social and economic importance is emphasized by the use of bunches of grapes and drinking cups on the coins and medals of towns and regions famous for their vineyards. These include the Greek colonies of Trapezus on the Black Sea, Soli near Tarsus, and Sicily. The Greek islands of Naxos and Tenos also showed grapes on their coins, as did Mende in mainland Greece. The most frequently praised wines in Athens were those of Thasos, Lesbos, Mende and Chios (especially Ariousian), and wines from Ismaros in Thrace, Naxos, Peparethos (Skopelos), Acanthos and Kos were also admired, according to the lyric poets.

Soon the Greek colonies were producing their own wines. Amphorae from Massilia are found along the southern coast of France and up the Rhône valley. In the Crimea it appears that they established extensive vineyards in which were planted indigenous varieties of grapes that they domesticated locally, rather than importing stock from elsewhere, and they protected them from the prevailing winds by constructing low walls.

Various flavourings were added to the wines of classical Greece to make them more attractive. Brine or sea water was added to wine from Kos in the fourth century BC, which may also have had preservative qualities. Wines for special occasions on Thasos were flavoured with dough and honey. In other places aromatic herbs were added to make a sort of vermouth. Some blending of wines was done and Theophrastus describes the mixing of the aromatic wine of Heraclea with the soft, salted Erythrean wine that had no bouquet. Wines were described as being sweet, honeyed, ripe, or soft. Some wines, made from unripe grapes, must have been dry. Wine was almost always drunk diluted with water.

Wines were usually transported in amphorae, of characteristic shapes, that were lined with pitch or resin to reduce their porosity (which gave an additional flavour to their contents). They held up to 75 litres each and were used commercially for trading. Stamps on the amphorae handles indicated where the wine had originated from. Of the thousands of amphorae discovered in the excavations of the Athenian Agora (the market place of ancient Athens), some 62% came from Knidos and 23% from Rhodes). Smaller painted amphorae were used for serving wine. Wineskins, made from the skins of sheep or goats, appear to have been used for transporting small amounts of wine over short distances, presumably for rapid consumption.

Wine was drunk after dinner at a symposium, with men reclining on couches, propped up on an elbow, sipping wine from a shallow two-handled cup. Women, if present, sat on the edge of a couch or on a chair. The symposium had a chairman whose job it was to stimulate conversation. Sometimes those present at a symposium were entertained by girls playing flutes or harps, or dancing. A light-hearted drinking game called kattabos was developed about 600 BC, in which drinkers at a symposium competed with each other to extinguish a lamp on the top of a tall stand by throwing the dregs of their wine cups at it. Eventually a special bronze disc called a plastinax replaced the lamp, and the aim was then to dislodge it and cause it to fall with a clatter on to a lower, larger one called the manes. This game was extremely popular for about 300 years.

Hippocrates, the father of medicine, was born on the island of Kos about 460 BC. He recommended that wine be used in almost all his remedies – for cooling fevers, as a diuretic, a general antiseptic and to help convalescence. He was precise in his instructions, always recommending a particular wine for a particular case, and occasionally advising against the consumption of any wine at all. Eubulius, writing about 375 BC summarizes Greek wisdom on wine drinking thus:

> Three bowls do I mix for the temperate: one to health, which they empty first, the second to love and pleasure, the third to sleep. When this bowl is drunk up wise guests go home. The fourth bowl is ours no longer but belongs to violence; the fifth to uproar, the sixth to drunken revel, the seventh to black eyes, the eighth is the policeman's, the ninth belongs to biliousness, and the tenth to madness and hurling the furniture.

Wine was associated with the god Dionysus, and his festivals in Athens in the fifth century BC were concerned with wine. The most important was the Anthesteria, held in early spring to mark the opening of the first jars of the new vintage. The vintage celebration in autumn was marked by the festival of Oschophoria, during which a procession was led by young men carrying branches bearing bunches of grapes.

Socrates summed up the virtues of wine in a civilized society thus:

> Wine moistens and tempers the spirits and lulls the cares of the mind to rest . . . it revives our joys, and is oil to the dying flame of life. If we drink temperately, and small draughts at a time, the wine distills into our lungs like the sweetest morning dew . . . It is then the wine commits no rape upon our reason, but pleasantly invites us to agreeable mirth.

Viticulture was almost certainly introduced into Sicily and southern Italy by the Mycenaean Greeks, and it was certainly well established when the Greek colonies were founded here around 800 BC. Further north, in Tuscany, the Etruscans were also cultivating vines and making their own wines from the eighth century BC. These they traded into France, reaching Burgundy even before the Greeks had established their wine trade in this region. It is not clear where the Etruscans had obtained their cultivated vines; they imported fine Greek drinking cups and amphorae and made their own copies of them for use in their symposia.

By the time Hannibal invaded Italy in the late third century BC, vines were being grown throughout the peninsula of Italy, mainly on smallholdings as part of the family farm. Greek wines continued to be imported and to dominate the market. We know little about special vintages, however, until 121 BC when Opimius was consul and Opimian wine was accepted by Pliny as belonging to one of the greatest vintages. It was so prized that some of it was kept and drunk when it was 125 years old.

The most famous Roman wines were made in the first centuries BC and AD and correspond to the establishment of vineyards on an industrial scale, owned by financiers, managed by a steward and run by slaves. They arose to meet the considerable demands from Rome and other large urban centres. (Outside the walls of Rome, close to one of the gates used for the wine trade, is a mound 115 feet (35 metres) high and 1000 paces in circumference called Monte Testaccio, which is entirely formed from broken shards of discarded amphorae.) Most of the famous wines come from Latium and Campagnia. Among the most famous were Caecuban (produced in the marshes around the Lago di Fondi), Falerian (from the slopes of Monte Massico), and Gauranum (from Monte Barbaro overlooking the northern shore of the Bay of Naples). There were also notable vineyards at Pompeii (where excavations have revealed the layout of the vines by finding holes left by their roots in the volcanic ash that overwhelmed the town), and at Sorrento, which produced a light, dry white wine. The wines of Alba, to the south of Rome were also highly rated. Pliny and others rated the following wines highly: Mamertine from Messina in Sicily, Praetutian from Ancona on the Adriatic, Rhaetic from Verona, Hadrianum from Atri on the Adriatic, Luna from Tuscany, and Genoa from Liguria. Of the Campanian wines, Pliny notes specially the Trebellian from Naples and the Cauline from Capua.

The streets of towns like Pompeii, Herculaneum and Ostia were well provided with bars and 'thermopolia', where snacks and wine could be purchased. (There were at least 200 in Pompeii alone.) The wine was usually served mixed with water, and in cold weather the mixture would be served hot as a kind of mulled wine. Their marble counters have huge basins let into them for serving wine, and racks behind the counters held amphorae to replenish the supply. One particularly successful Pompeian wine merchant, Marcus Porcius, exported wine widely in the western Roman world but more especially to southern Gaul, and numerous amphorae bearing his stamp have been found from Narbonne to Toulouse and Bordeaux.

After the destruction of Pompeii and its surrounding vineyards by the eruption of Vesuvius in AD 79, there followed an unseemly scramble to plant new vineyards on any piece of cultivatable land in the vicinity of Rome. Cornfields became vineyards overnight at the expense of normal agricultural production. It was the effect of this that probably prompted Domitian to issue his edict in AD 92 banning the planting of any new vineyards and ordering the destruction of half the vines in the overseas provinces of the Roman Empire. In another edict he banned the planting of small vineyards within towns in Italy. Although they were never fully enforced, they did have the effect of raising the price of wine. They were eventually repealed by Probus in AD 280, and this was followed by a burst of vineyard planting throughout the empire, including in Roman Britain.

The Roman wine trade was well organized by associations of merchants who had their own headquarters on the Tiber, and their own ships and lighters specially designed to carry 3000 amphorae, which were replaced in the second and third centuries AD by the use of wooden barrels and casks.

The earliest evidence that wine-drinking had reached the Celtic world of eastern France, southern Germany and Switzerland appears in the archaeological record in the sixth century BC. Griffon-headed cauldrons, huge craters for mixing wine, jugs, strainers and finely painted Greek drinking cups were being imported, chiefly through Massilia, and transported inland along the navigable rivers using an elaborate system of gift exchange between the chieftains and the Greek and Etruscan merchants from the south. The large ceramic amphorae in which the wine was transported were made in southern France, suggesting that some of the wine they contained may also have been produced there. Sites like Mount Lassois, where the rich female burial at Vix contained a magnificent bronze crater some 1.64 metres tall and manufactured in Greece, and the Heueneberg in southern Germany, were clearly the courts of the elite in the period 530–520 BC, but had ceased to exist by the end of the century.

The next period of remarkable imports occurs at the end of the third century, when Rome was growing rapidly in power having defeated the Carthaginians in Spain. The Carthaginians had inherited wine-making skills from the Phoenicians, and Spanish wine was soon being traded to Rome. Pompeii traded with Tarragona, both buying and selling wine, and the rich Marcus Porcius had a wine estate here. Enormous quantities of wine were also exported to Rome from the province of Baetica in southern Spain. The Phoenicians had penetrated up the river valleys and the Romans followed them up the Ebro to Rioja. Remains of a remarkable winery have been found at Funes, which indicate that it had the capacity to produce and store as much as 75 000 litres of wine.

This was a period that saw a good deal of traffic between Italy and Spain through the south of France, and Roman entrepreneurs were not slow in opening up new markets for their considerable surplus of wine. In the Celtic world, elaborate feasts were used to establish the status of their chieftains, the more exotic the goods offered, the greater the prestige of the host. Wine from the Mediterranean was in considerable demand and was eagerly consumed, not diluted with water.

Provence and Languedoc were taken into the Roman Empire in the second century BC, and at this time there seems to be little evidence of local wines being drunk. On the contrary, an increasing number of shipwrecks off the southern French coast seem to be filled with Roman amphorae. Once offloaded at Massilia and Narbo (Narbonne), the wine was taken inland by road or river. At two places, one outside Toulouse and the other at Chalon-sur-Saône, large quantities of discarded Italian amphorae fragments have been found. It is suggested that these were major transhipment points and that the wine was transferred from pottery amphorae to wooden barrels for the onward journey by boat or cart into the interior. Not all wines were decanted, however, since considerable numbers of amphorae were recovered from the Celtic towns of Montmerlhe, Essalois, Joeuvres and Mont Beuvrais, between 20 and 100 km beyond the transhipment points. It is difficult to estimate how much wine was being imported into Gaul at this time but it may well have been about 2.6 million gallons (11.8 million litres) a year. Italian wines were reaching all parts of France by the end of the second century BC. Mostly it was distributed by river, but some still in amphorae appears to have been loaded on to ships in the Gironde estuary to be transported along the Atlantic coast to Brittany. Most of it was consumed there, but a small quantity was transported via Guernsey to Hengistbury Head overlooking Christchurch Harbour, the first wine to reach Britain.

The first vineyards in the south of France were probably established by the Greeks around Marseilles about 600 BC, followed by more in Languedoc around 200 BC. The vineyards of Bordeaux and Vienne appear to have originated about AD 50. Pliny says that the resinated wines of the Rhône valley and Vienne were a source of pride and commanded high prices. Viticulture had reached Bourgogne by AD 150 and the Loire valley by AD 250. It was established in the Rhine and Moselle valleys by AD 300 and reached Champagne by AD 350. The only part of France where vineyards had not been established during Roman times was Alsace.

## Reference

Hyams, E. (1987) *Dionysus, A Social History of Wine*. Sidgwick & Jackson, London, p. 59.

## Further reading

Darby, W. J., Ghaliounguis P. and Grivetti, L. (1977) *Food: the Gift of Osiris*. Academic Press, London, New York, San Francisco.

Hyams, E. (1965) *Dionysus. A Social History of the Wine Vine*. Sidgwick & Jackson, London.

Johnson, H. (1971) *World Atlas of Wine*. Mitchell Beazley, London.

Johnson, H. (1989) *The Story of Wine*. Mitchell Beazley, London.

Lesko, L. (1977) *King Tut's Wine Cellar*. B.C. Scribe Publications, Berkeley, California.

Lutz, H. F. (1922) *Viticulture and Brewing in the Ancient Orient*. J.C. Heinrichs, Leipzig.

McGovern, P. E. (1998) Wine for eternity. *Archaeology*: **51**(4), 28–32.

McGovern, P. E. (1998) Wine's prehistory. *Archaeology:* **51**(4), 32–4.

McGovern, P. E., Fleming, S. J., Katz, S. H. (eds) (1995) *The Origins and Ancient History of Wine*. Volume 11 in *Food and Nutrition in History and Anthropology Series*. Gordon and Breach, Luxembourg.

Murray, O. and Tecusan, M. (eds) (1995) *In Vino Veritas*. British School at Rome, London.

Renfrew, J. M. (1973) *Palaeoethnobotany. The Prehistoric Food Plants of the Near East and Europe*. Methuen, London.

Robinson, J. (ed.) (1999) *The Oxford Companion to Wine*, 2nd edn, Oxford University Press, Oxford.

Theophrastus (1961) *Enquiry into Plants* (trans. Sir A. Hort), Loeb edn, William Heinemann, London.

van Zeist, W., Wasilykowa, K. and Behre, K.-E. (1991) *Progress in Old World Palaeoethnobotany*. A. A. Balkema, Rotterdam.

Zohary, D. and Hopf, M. (2000) *Domestication of Plants in the Old World: the Origin and Spread of Cultivated Plants in West Asia, Europe, and the Nile Valley*, 3rd edn. Oxford University Press, Oxford.

# 4 Saving the vine from *Phylloxera*: a never-ending battle

*G. Gale*

## Introduction

During 1866, a 5 hectare (*c.* 13 acre) patch of lower Rhône grapevines sickened and died from unknown causes. Over the next year the patch extended considerably, with the area of sick and dying vines moving out in all directions from the original center of infection. By July 1868 the situation had become so grave that growers in the region appealed for help to Montpellier's highly respected Société Centrale d'Agriculture de l'Hérault (SCAH). SCAH appointed a three-member commission to inspect stricken vineyards on the west bank of the Rhône. Members of the commission were Georges Bazille, President of the SCAH, Jules-Émile Planchon, a physician and professor of pharmacy and botany at Montpellier, and Felix Sahut, a well-regarded winegrower. The commission began its work on 15 July, spent three days at the task, and immediately reported their results in all available media. At first the commission had focused upon examination of the dead vines; nothing exceptional was found. But when, by happy accident, an apparently healthy vine was uprooted and inspected, something extremely exceptional was found. As Planchon relates:

> Loupes were trained with care upon the roots of uprooted vines: but there was no rot, no trace of cryptogams; but suddenly under the magnifying lens of the instrument appeared an insect, a plant louse of yellowish color, tight on the wood, sucking the sap. One looked more attentively; it is not one, it is not ten, but hundreds, thousands of the lice that one perceived, all in various stages of development. They are everywhere . . .
> (Cazalis 1869a, p. 237)

In this fashion was discovered that worst of all scourges of the vine: *Phylloxera*. From this small patch of an unimportant Rhône vineyard came the destroyer of the world's traditional winegrowing: within several years *Phylloxera* had spread to all the major European winegrowing areas. Within a decade it had begun devastating California. Winegrowers in Australia, South America, Algeria and South Aftrica prepared their defences in anticipation of the bug's arrival on their shores. In the end, the entire winegrowing world was not only affected but transformed – agriculturally, economically and socially – by the plague. Effects of the disaster rippled out from the vineyards into their embedding cultures at large, invoking such large-scale consequences as rural depopulation and massive emigration. Even recent troubles in places like Algeria and the former Yugoslavia have their roots in the *Phylloxera* disaster.

What is worse is that the disaster is not really over. The vicious and terribly expensive *Phylloxera* outbreak in California during the 1980s and 1990s revealed again the truth of the French conclusion that *Phylloxera* will never go away, it is a fact of life that must be lived with.

Battling the *Phylloxera* to a standstill took 30 years of all-out warfare, from 1870–1900. French scientists, joined at crucial junctures by their American colleagues, tried anything and everything they knew, gradually coming to an understanding of their enemy, finally learning enough to take back their devastated vineyards. Understanding how the battle was fought reveals much about the way in which science and scientists interact with each other, with nature, and with their cultural environments. And, perhaps just as importantly, understanding how the battle was fought will serve permanently to alert us to the dangers of complacency in the face of this devastator of vines.

## Identifying the cause

It took nearly seven years after the initial discovery of the bug for scientific consensus to be reached about the actual cause of the vine sickness. Although Planchon and his two colleagues were immediately convinced that the sickness was caused by the bug (Cazalis 1869a, p. 238), they initially formed a small, little-regarded minority. Arrayed against them was most of the French scientific and professional establishment, which waged a loud and widespread campaign to discredit the views of the Montpellierians. A number of factors, sociopolitical as well as scientific, weighed against Planchon and his colleagues. In the first place, even though Montpellier housed an ancient and highly regarded university, plus one of the three national medical schools, as well as the École de Gaillard, France's major agricultural school, it was far from Paris, it was southern, and, most importantly, it was of an ethnically, culturally and politically different tradition from that of the capital. Second, Montpellier was capital of the Midi wine-growing region or *vignoble*,[1] known for its only ordinary (and lower) quality bulk wines. Midi wines were generally disrespected; transferrence of this attitude to Midi wine scientists and producers could only be expected. Third, although the three men had hypothesized an entomological aetiology for the new malady, none of them were entomologists. For whatever reason, the national entomological establishment immediately took the offensive against the Montpellierian theory that the insects themselves caused the disease. As always, the ever-outspoken large estate owner Duchesse Fitz-James gets it exactly right: Planchon, Bazille and Sahut are mockingly derided as 'the Hérault entomologists' because: 'not one of them is an entomologist by profession, and the number of ignoramuses who think they know more defies arithmetic' (Fitz-James 1881, p. 689). Chief among the professional entomological opponents was Signoret, president of the Entomological Society of France, who published a strongly polemical monograph against Planchon and his theory about 'the alleged cause of the current malady of the vine' (Signoret 1870). Although professional jealousy is clearly involved in this attack upon the Hérault (non)entomologists, it also must be noted that the professional insect scientists had never before seen an insect do such severe damage; hence, the *Phylloxera* as cause was beyond their experience.

But the major reason behind the opposition to the theory of Planchon *et al.* was deeper, more pervasive, and fundamentally scientific: their hypothesis that the insect was the cause of the malady contradicted the disease paradigm then prevailing in French medicine and plant pathology. From shortly after the French Revolution, French medicine had increasingly renounced the *ontological* model of disease in favor of the *physiological* model. Finally, in 1827, the French Academy of Medicine officially adopted the physiological approach to disease (Ackerknecht 1948, p. 573). According to the physiological approach, disease is due to some sort of disequilibrium in the animal or plant body's internal function, due, perhaps, to predisposition, diathesis or degeneration. Galenic medical practice exemplifies this tradition. Opposed to this model is the ontological approach, which looks for the cause of disease in some external agent, such as a contagion or a germ. As Cohen notes: 'the two notions varying a little in content and occasionally overlapping have persisted' since antiquity; 'the dominance of the one or the other at different epochs reflecting either the philosophy of the time or the influence and teaching of outstanding personalities' (Cohen 1961, p. 160). During the epoch in question, French medicine, and thereby plant pathology – most of whose practitioners were physicians (Whetzel 1918, p. 33) – was dominated by physiological thinking.

According to these plant pathologists, attacks by parasites, including funguses and insects, occurred only *after* the physiological imbalances had rendered the diseased plant defenceless against them (Whetzel 1918, p. 28; Ainsworth 1981, p. 34). In the case of the *Phylloxera*, the original cause of the malady was attributed variously to meteorological conditions (Cazalis 1869a, p. 234), the state of the soil and or the vines (Barral 1868), the processes or structures of the plant itself (Falières 1874, p. 16) or, finally, 'general circumstances' (Cazalis 1869b, p. 29). Included in the plant science establishment espousing this disease paradigm were Guérin-Méneville, an entomologist and academician who had solved the silkworm pebrine problem; Naudin, a botanist, academician, director at Paris's Jardin-des-Plantes, the national botanical gardens; Trimoulet, a leading entomologist in Bordeaux; and, of course, Signoret. They were a formidable opposition.

Planchon most often characterized the opposition's view as *Phylloxera*-effect, since they believed that the insects' arrival was only an effect following the plant's original illness. His own view was, naturally, called *Phylloxera*-cause, since the *Phylloxera* itself was taken to be the 'original and unique cause' of the vine malady (Planchon 1874, p. 554).

Battle between the two viewpoints was fierce. Very early in the fray a rivalry grew up between Montpellier and the high-quality wine region of Bordeaux, where a separate and distinct disease focus was discovered shortly after that in the Rhône. Confrontation between the two regions began in summer 1869, when the Viticulture Section of the Société de Agriculteurs de France set up a national commission to survey the extent of the damage and report on its findings. A large committee of scientists and viticulturalists, most taken from the membership of the SCAH, was formed and sent on its way throughout the Midi, and, at sudden notice and unbeknownst to local officials, the Gironde region, most especially Bordeaux. The commission's subsequent report, which unanimously took a strong and explicit *Phylloxera*-cause position, was published in the autumn and given wide circulation (Vialla 1869). Shortly after its publication, a series of increasingly polemical written attacks against the views of Planchon *et al.* were launched by the Linnean Society of Bordeaux; the series ended six years later with Trimoulet's vituperative Fifth Report (Société Linnéenne de Bordeaux 1869a, 1869b; Trimoulet 1875). The battle was well and truly joined.

## Some progress is made

The years 1869–71 marked both the worst conflict in the controversy over the cause of the malady, and some first inklings of how progress in fighting the insect might begin. Unfortunately, these years also marked *Phylloxera* salients into départements to both the north and the east of the Midi, and north-east out of Bordeaux. Worried officials in the newly threatened regions commissioned investigations into the disease's progress, typically instructing the investigators to examine both sides of the aetiological controversy (Barral 1869a).

At the École, Planchon found a strong ally in Camille Saintpierre, who had entered the Montpellier faculty in 1860 after completing a medical dissertation called *On Fermentation and Putrefaction* (Saintpierre 1860). The dissertation was not only strictly Pasteurian, it was in fact dedicated to Pasteur 'as a modest tribute in recognition to the savant whose work and ingenious discoveries have offered a signposted path to our researches' (Saintpierre 1860). Ten years later, in 1870, Saintpierre was to ascend to the directorship of the École, there to carry out his earlier promise regarding Pasteur: 'This École will totally follow the path he has so surely traced'. It is clear that Saintpierre's Pasteurian thinking, and resolve to realign Montpellier along the lines of the ontological disease paradigm is the major philosophical-cum-theoretical reason why Planchon *et al.* were so immediately ready to hypothesize that the insect was the cause of the malady.

In a more practical vein, Planchon succeeded in finding answers to several specific questions that he had posed, in particular: 'Where did it come from? Had it been described? What were its closest relatives?' (Planchon 1874), p. 547). His investigations soon revealed that the bug was a close relative of the *Phylloxera quercus*, which lived on oaks. It had indeed been described, both in the USA (described by New York state entomologist Asa Fitch in 1854 as the so-called *Pemphigus* aphid) and most recently in England. The question of its origin, however, was difficult to work out. In this Planchon had some help from the USA.

British-born C. V. Riley became the state entomologist of Missouri in 1868. His seven successive *Annual Reports on the Noxious, Beneficial and Other Insects of the State of Missouri* became almost instantly famous for their unique blend of humour, quotes from classical literature, technical advice and, above all, enormous practical wisdom. Each of the seven sold out immediately it was published. Riley got involved in the French crisis in mid-1870. As soon as he saw Planchon's published description of the French vine devastator, Riley wrote both to him and to Signoret, announcing his suspicions that Planchon's *Phylloxera* and Fitch's *Pemphigus* were the same beast. Unfortunately, Riley's hypothesis failed in one important regard: the US bug lived solely in galls on the leaves of the various American vine species, while the French bug had been observed only on the roots of the European species. Tipped off by Riley, Planchon looked for the bug on the leaves of American vines in Europe and found them (Planchon 1874, p. 548). Several days later, Leo Laliman, Bordeaux's brilliant but irascible vine collector and grower, replicated Planchon's observation. Planchon, with assistance from his brother-in-law Jules Lichtenstein, soon demonstrated *in vitro* that leaf-gall bugs would rapidly infest clean European vine roots (Lichtenstein and Planchon 1870). It was clear that the investigators were seeing two phases of the same insect's life-cycle. The origin of the bug was now clear: it had come from the USA. How and why it got to France was not difficult to discover.

During the 1850s there had been a viticultural crisis caused by powdery mildew, an import from the USA. It was widely noted during the crisis that the US vine specimens on exhibition in botanical gardens and private collections were immune to the mildew's depredations. A small but significant number of French vineyardists decided to investigate the wide-range of American species to see if any of them were both disease-resistant and of interest for winemaking. Hundreds of US vines were imported, transported on the new, fast steamships. So fast, in fact, that the *Phylloxera* on the US vines didn't die during the trip, and lived to flourish in their new country. Two of the largest collections of US vine varieties were near the sites of the original *Phylloxera* outbreaks in Bordeaux and the Rhône. Once Planchon's experiment solidified the connection between the US bug and the French bug, evidently settling the question of the pest's origin, anger against the US vines and those who had imported them grew markedly. This attitude would harm the practical campaign against the disease for years, as is noted below.

Planchon's work had another important result as well. Within the next few months, as word of his experiment circulated around France, converts were made for the phylloxera-cause side. Of course, not all were convinced – it was during this period that Trimoulet's first attack was published in Bordeaux, followed at the end of the year by Signoret's notorious polemic. But in the USA, Riley was convinced that Planchon and Lichtenstein had it right. In his *Third Annual Report* he attacked Signoret, and came out solidly in favor of the identity of the US and French insect (Riley 1871). Late the next year, in one of the first of several instances of direct international cooperation in the war against *Phylloxera*, Riley himself visited Planchon in Montpellier, observed the bug, and took specimens back to the USA with him (Legros and Argeles 1994, p. 214). In his next report, Riley concluded: 'That the two are identical there can no longer be any shadow of a doubt' (Riley 1872a, p. 2).

Identification of the origin of the bug led some workers to suspect that, if the devastation came from the USA, then salvation would ultimately come from there as well. As many have noted, beginning with Planchon, this is the great paradox of the phylloxera, that the source of the problem might also be the solution to the problem (Pouget 1990, p. 51). Somehow, through grafting or growing them as wine grapes, or special breeding, US vines held the secret to solving the phylloxera problem.

Riley's Darwinian views supported the paradox. According to Riley's understanding of Darwinian theory, prey and predators evolve together, each becoming 'adjusted' to the other over long eons of adaptation (Riley 1872b, p. 623). Accordingly, in the case of the US vine + phylloxera combination, the vine would have evolved defences to sustain it under attack from the insect. European vine species, which had evolved without exposure to *Phylloxera*, had no such defences and consequently perished. Europeans, including Planchon and Laliman, although they were not staunch Darwinians, took Riley's word for it (Planchon 1877; Laliman 1872). For his part, Laliman had earlier observed that the US vines in his collection did not suffer from the insects' attacks and had suggested that perhaps US vines could be used to produce wine directly (Barral 1869b, pp. 666–7). His suggestion had been taken by many, and experimentation began immediately. By the Winter of 1872–3, enough success had been demonstrated that over 400 000 dormant US vine cuttings had been sent to winegrowers around Montpellier by the St Louis firm Bush and Sons and Meissner (Morrow 1972, Chapter 4).

Two other trials also spoke to the efficacy of the *Phylloxera*-cause position. In late spring 1869, Louis Faucon, a commited *anti-Phylloxeriste* – as *phylloxera*-effect proponents were frequently called – with a small vineyard on the east bank of a tributary of the Rhône, had the apparent misfortune of losing his vineyard to a severe flood. It must have seemed like overkill to Faucon: his vineyard was already doomed by a bad case of the malady, and the insects had flocked to the roots. But lo and behold, when the flood receded a month later, the *phylloxera* was gone, drowned by the flood. Faucon's vines immediately perked up, and he harvested a decent crop at the end of the year instead of having a dead vineyard. Needless to say, Faucon dropped his former beliefs and became a thorough-going *Phylloxeriste*, joining Planchon *et al.* (Vialla 1869, p. 358). At the Beaune agricultural congress in November, Faucon distributed information about his observations, and provided some directions about methods for flooding. Within the next year, his method was proven; from that time forward, flooding was accomplished wherever conditions – and finances, since it was an expensive scheme – allowed (Faucon 1870; Borde de Tempest 1873). A small number of estates continue to use the method today.

Water was not the only protective medium. It had been early noticed that sandy soils seemed to protect vines from the disease. *Phylloxera*-cause explanations of the result hypothesized some lethal effect of the sand upon the insect. Some test vineyards were planted in the sand dunes around the medieval walled city of Aigues-Mortes, near the eastern bank of the Rhône's mouth. Although no right-thinking winegrower would ever before have planted vines in these dunes, these were desperate times. Nicely enough, the vines succeeded. Moreover, although the vines in the nearest soil-based vineyard were completely *Phylloxerated*, the vines in the sand around Aigues-Mortes flourished, remaining clean and clear from the infestation.

The success of these two methods provided practical support for the efficacy of the *Phylloxera*-cause theory, evidence that was sorely missing from the purely theoretical research side of the position. Theoreticians from both camps were still at it, vigorously bombarding each other with journal articles and pamphlets throughout the time from 1869 until late in 1872. But the tide began to turn during early 1873. Most importantly, the government had finally awakened from the torpor induced by the Prussian war of 1871, suddenly noticing the disaster building in the southern vineyards. Two projects were funded by a national *Phylloxera* commission that had been set up under the directorship of the renowned chemist Dumas (Pouget 1990, p. 30). The first was a three-month visit by Planchon during autumn 1873 to the vinegrowing areas of the USA in order to examine all the potential anti-*Phylloxera*-weapon candidates present in US viticulture. Riley, of course, hosted Planchon, and shepherded him through all the most important US winegrowing districts. Many valuable things were learned, especially during an extended stay in and around the Missouri vineyards, St Louis and the Bush nursery. Plants, techniques and cultural data for US vines all returned to France with Planchon (Planchon 1875).

A second project also focused on the US vines. At that time, the systematic taxonomy of the US vines was a great unknown. Nomenclature varied wildly, and no one could ever be sure what vines they were dealing with. In early 1874, Alexis Millardet, professor at the University of Bordeaux and France's greatest botanist, was commissioned to produce order out of the chaos of US vine classification. He began work immediately, systematically and, ultimately, successfully.

Seeing the government funding as a conclusive tilt toward *Phylloxeriste* thinking, Guérin-Méneville, one of Planchon's strongest *Phylloxera*-effect opponents, made a plea for renewed funding of physiological research (de Ceris 1873, p. 674). But it was not to be. Guérin-Méneville died shortly afterwards, leaving the fight to the increasingly contentious Trimoulet. By late 1874 the game was up. Planchon and the other ontologist, *Phylloxera*-cause, *Phylloxeriste* workers had won. Dumas, in a letter to Falières, chides the latter for not being tough enough on the physiologists in his recent book:

> You have been too indulgent to the promoters of the dangerous idea of the *Phylloxera*-effect; this idea has caused the ruin of a great number of vines and reduced to powerlessness the most authoritative among those who wish to devote themselves to this so grave question.
>
> (Falières 1874, p. 1)

Unfortunately, after having committed themselves to *Phylloxera* as the unique cause of the malady, Dumas and the government then resisted accepting the promise of the US vines. Thus began *la défense* – the attempt to defend the old vines and the old ways against the devastating US invader.

### *La défense*

Once the bug had been accepted as the unique cause of the malady, thought immediately turned to stopping it, preferably by stopping it dead in its tracks. Since the goal was to defend France against the foreign invader, this phase of the battle with the insect soon came to be called *la défense*. Especially in official circles in Paris, the explicit goal was to defend and preserve the traditional wine varieties, traditional growing practices and, wherever possible, traditional winegrowing regions. Coupled with the widespread anger against them, this goal ruled out any use of the US vines, whether as graftstocks or as direct producers of wine. Animosity against US vines as the carriers of the disease was not only prevalent in the viticultural regions, it reached all the way to the top. The Academy of Sciences, which had been charged by the government with directing the battle against the insect, 'at first was not at all favorable to use of the American vines; its prejudices were prolonged for a long time afterwards' (Convert 1900, p. 513). Cornu and Dumas, speaking for the Academy of Science's national *Phylloxera* commission, set it out thus:

> The commission considers the *Phylloxera* as the cause of the maladie of the vine. It proposes for a precise goal the conservation of the French vines, their principal types being the product of secular practice, it is important above all to save them.
>
> (Cornu and Dumas 1876, p. 1)

Given their goal to conserve a completely *French* viticulture, and their animosity toward the US vines as disease 'carriers', the commission ruled out from the first any serious official experimentation based on the US vines. With this attitude, they 'incontestably held back' the eventual long-term solution (Pouget 1990, p. 89).

From the start, *la défense* relied on physical and chemical strategies rather than biological ones. Efforts to plant vines in sand and apply submersion techniques accelerated during the early 1870s, reaching a peak use by the turn of the decade. Although both techniques were clearly efficacious, their use was severely limited. Submersion required an enormous infrastruture of dikes, gates, pipes, steam-powered pumps and, most clearly, an abundant supply of water. Although plans were made early on to build a huge canal between the Midi and Bordeaux's Gironde estuary precisely for the purpose of supplying water for immersion, the government dithered for nearly 20 years before even beginning construction. At no time did the area immersed total more than 40 000 hectares (Garrier 1989, p. 73).

Planting in the sand was equally fraught with difficulties. Since sand is biologically inert, all the vine's nutritional requirements had to be supplied by fertilization. Traditional fertilization methods involved pumping rich river-bottom silt on to the field. This technique was immediately discovered to be no longer an option: as soon as the sand was diluted with the silt, the *Phylloxera* colonized the planting. Chemical fertilizers were the only possibility, and they were expensive. Another problem came from the sea. Since almost all the useable sandy regions were shorelines along the Mediterranean Sea and Atlantic Ocean, wave action frequently washed out the plantings, uprooting the vines and carrying them off to sea. And coastal wind was a problem, too. 'The strong coastal breezes and especially the mistral are the great enemies; they uplift the sand and uncover the vine roots' (Gachassin-Lafite 1882, p. 139). To prevent this, a mulch of reeds was planted between the rows. Once the reeds were established, the sand was considerably stabilized. Vernette, of Béziers, designed and developed a special machine that disced open four furrows into which small rooted pieces of reed could be dropped as the machine was dragged over the dune (Ordish 1972, p. 95).

But perhaps the greatest problem with planting in sand was that the wines no longer had their traditional character, those qualities that grew out of the local conditions of soil, terrain and climate. Nonetheless, the technique worked, it produced a drinkable beverage, and winegrowers could profit from the enterprise. (Even today, one sees small vineyards along the beach roads near Aigues-Mortes, complete with signs advertising 'Wines from vines grown on their own roots!' as if this compensated for the wine's totally ordinary quality!)

Although the method was successful, total sand plantings topped out at about 20 000 hectares (Garrier 1989, p. 73). Taken together, immersed and sand-planted vines never accounted for more than 4% of the overall French vineyard surface. But then, *la défense*'s main focus never was these physical methods; rather, from the start it put its hope in chemistry and the search for an effective chemical insecticide.

The search for insecticides counted some strong allies. Dumas, Cornu and Mouillefert, the 'most ardent and zealous' among '*les chimistes*', 'figured among the most influential members of the Superior *Phylloxera* Commission' (Pouget 1990). From their position, they were able to do some real harm, especially hindering attempts to experiment with the American vines:

> Persuaded that only insecticidal treatment, to the exclusion of all other procedures, was capable of battling effectively against the *Phylloxera* and making it definitively disappear, they adopted an intransigent attitude and proclaimed obstinately their hostility toward the American vines and their partisans, the 'Americanists'.
>
> (Pouget 1990, p. 41)

Serious trials of insecticides were mandated by the national commission. Montpellier's Professors Durand and Jeannemot first set up a trial field at a nearby vineyard called Mas des Sorres in 1872, and by the Summer of 1873 had established the first fully scientific, replicated and controlled agricultural field experiment ever attempted. In 1874, the national commission set a prize of FFr 300 000 for anyone who could come up with a practical insecticidal remedy; suggestions were invited, vetted and then sent to Montpellier for trial. During the following three years, a huge number of suggestions for methods to try were received by the academy; 696 were forwarded from the academy to the Montpellier professors; out of the proposals forwarded, 317 were actually tried (Commission Départmentale 1877). Given the eccentricity of some of the suggestions deemed sensible enough to forward for trial (goat's urine, shrimp bouillion, garlic peels), one wonders about the suggestions the academy deemed *unworthy* of forwarding on to Montpellier! Planchon was scathing in his denunciation of the whole prize-seeking, public suggestion process: 'The remains of this dossier of stupidities reveals a sad day for the state of mind of the great public in terms of its scientific instruction' (Pouget 1990, p. 31). In the end, nothing worked and the vines all died.

Earlier insecticidal efforts had been made by Baron Paul Thenard, son of the pioneering chemist Baron Louis-Jacques Thenard. After trying a long list of things, Thenard the son settled upon carbon disulphide ($CS_2$), a volatile chemical solvent just then becoming available in French industry. Thenard and colleagues worked out a mechanical system to inject the chemical around the plant, then proceded to treat a large parcel. Sure enough, the pests died. Unfortunately, so did the vines (Laliman 1872, p. 20). Thenard retreated to reconsider. His next trial a year later had more favourable results; after treatment, the insects perished, but the vines did not; indeed, they slowly began to recover from the bugs' effects. $CS_2$ and its salts became the weapon of choice among the chemists.

$CS_2$ is an oily liquid with a nauseating odour – anyone familiar with a school chemistry laboratory is familiar with $CS_2$. Because it is heavier than air, it settles into the soil, setting up an asphyxiating layer that soon kills many bugs, especially *Phylloxera*. But it does not kill *all* the bugs, again especially *Phylloxera*. Thus it is only a palliative in the long run; applications must be renewed at least annually, and, in some locations, two applications a year are required. Finally, and this is the saddest part of all, continued use of the insecticide without taking exquisitely careful steps to strengthen the vine after treatment 'has the effect of weakening the vine in a certain measure, and sometimes even opposing its rapid restoration' (Plumeau *et al.* 1882, p. 78).

But Thenard's work was only a beginning. Reliable general success with $CS_2$ did not happen until after the series of experiments and workshops conducted by the Compagnie des Chemins de Fer de Paris à Lyon & Marseilles, the famous PLM railway. PLM's motive, at least in part, was to make up for revenue lost as a result of decreasing wine shipments, by shipping $CS_2$ to the winegrowers from the new chemical factories along the lower Rhône. In the spring of 1876, PLM instituted a series of experiments on the uses of $CS_2$ and its salts. The experiments were headed by Antoine Marion, Professor of Chemistry at Marseille. Marion's work was well-financed and careful. Within two years, not only had Marion and colleagues worked out an effective protocol for application of the gas, they had also instituted a series of workshops in a travelling 'extension' course, worked with PLM in setting up an efficient distribution system, and written a detailed set of instructions for applying $CS_2$ (Compagnie des Chemins de Fer 1878; Marion 1879).

Although successful, the gas had many problems. It required a skilled workforce to apply it; only the best and most fertile soils were open-textured enough for the gas to work, which left out all but the best estates and holdings from its use; it was only a palliative, and required at least one application annually, but usually more; and finally, it was expensive, far too expensive for a smallholder to afford. A grower with 'only a modest production of table wines' could not support the cost of the insecticide because 'when its cost was added to all the other costs of production, the treatment cost more than the price of the wine' (Lachiver 1988, p. 422). Only the rich estate owners could afford to save their vines. But this then set the smaller growers against the rich landowners, 'and the conflict took on the look of a "species of class warfare"' (Bouhey-Allex, quoted in Laurent 1958, p. 347).

Social strife was made worse by the laws of 15 July 1878 and 2 August 1879, which prohibited importation of US vines into still-healthy areas, and mandated defensive use of $CS_2$ in areas that *Phylloxera* was invading. Resistance to this process, at least initially, occurred nearly everywhere. But it was worst in Burgundy, a region whose vineyards are extremely small and scattered among a vast and wildly diverse number of owners, even by French standards. As Morrow notes: 'In Burgundy the peasants simply hated the features of the act which required chemical treatments if the majority of winegrowers in a commune demanded that they be tried' (Morrow 1972, p. 33). In some places there were riots. In July 1879, at Bouze, demonstrators carried on even in the presence of the Prefect and the gendarmes. It was worse at Chenôve, where 160 growers chased the treatment team out of the area, 'saying that they were scoundrels, more to be feared than the *Phylloxera*' (Laurent 1958, p. 333).

Rather than lose their vines, vignerons took the only action they could: they smuggled US vines – and their attendant *Phylloxera* – into their vineyards to replace the dead or dying French vines. For them, *la défense* was over. Laurent's summary of the situation cannot be bettered:

> Public opinion was profoundly divided about action in face of the plague. The 'sulfureurs', among whom numbered the proprietors of the grand crus, are partisans of forced defense and manifest to excess the most vigorous hostility to the very idea of foreign vines. The 'americanistes', who comprise all the small proprietors of ordinary vines, having abandoned all defenses, or practicing them only half-heartedly in their best regions, desire on the contrary the reconstitution of the vignoble using the American vines.
>
> (Laurent 1958, p. 344)

By the early 1890s, incidental success of *'la reconstitution'* – replanting France with the American vines – had eliminated insecticidal use on all but the finest properties: 'This means of defense, having caused many disappointments in most types of soils, ceased to be applied' (Zacharewicz 1932, p. 279). In the end, everyone agreed that, if French viticulture were to be saved, it wasn't going to be saved by the methods of *la défense*.

### *La reconstitution*

Luckily enough, alternatives to *la défense* were being tried and tested out in the provinces, even as it was being proclaimed official policy in Paris. These alternatives all involved the vines from the USA in one way or another. American vines eventually functioned in three roles: as *direct producers*, giving wine from US plants grown on their own roots; as *graft stock*, US roots supporting French vine tops; and as parents of genetic *hybrids*. Each role had its successes and failures. In the end, decisions about which role to feature in a given region rested upon evaluating a trade-off between aesthetics, economics, and biological viability. What made the decisions so very difficult in each case was the fact that 'success' was a moving target. At first, when nearly every *vigneron* in a region was devastated, success meant simply surviving, making some wine – any wine – a bit for your family to drink, with maybe a little left over to sell. Then, once that goal had been accomplished, and progress had been made in the university laboratories, or at the research station, or in a private concern in the next *arrondissement*, success meant better wine, better production, or even production in pieces of your property that couldn't grow the first wave of vines.

At first, nothing was known about the US vines. This was the period quite rightly called *'l'expérimentation anarchique des vignes américaines'* (Pouget 1990, p. 51). Disasters were commonplace. When Planchon returned from his visit to the USA in 1873, he recommended a number of vine varieties as suitable for direct production, as graftstock, or both. Unfortunately, among his recommendations were several varieties with high percentages of *Labrusca* – an American species from the cool north-eastern woods – with parentage such as *'Concord'* and *'Clinton'*. These vines were rapidly shown to have three crippling faults: they couldn't bear up under the heat of southern France, the 'region of the olive' (Sahut 1888, p. 17); secondly, they were not sufficiently *Phylloxera*-resistant under French conditions; and finally, their wines 'were undrinkable' as Laliman so succinctly put it (Laliman 1872). Unfortunately, well before the first two of these flaws were solidly established, enormous resources had been expended obtaining, planting, and growing the doomed vines. In the two winters of 1872 and 1873, over 700 000 cuttings of these *Labrusca*-based vines were imported from St Louis. Many *vignerons*, devasted once by the *Phylloxera*, were now destroyed again, once and for all, by the 'Concord' disaster: 'These two varieties . . . have thus been the cause of ruin for most of those who imprudently adopted them' (Sahut 1888, p. 17). Years of experience revealed that the best American direct producers were those which, like *'Herbemont'* and *'Jacquez'*, had a large proportion of the southern American species *'Aestivales'* parentage. Unfortunately, even these vines were fastidious about where they were grown, and their wine never achieved better than ordinary quality. Still, they were a reliable source, such as they were.

But grafting on US roots was no less replete with difficulties. Grafting has three parameters of importance: ease of grafting, which reduces the percentage of take; affinity between American rootstock and French scion; and resistance to *Phylloxera* of the roots. Values on all three of these parameters varied wildly. The first attempts, using *Labrusca*-based varieties, failed due to insufficient resistance; the second, using *Aestivales*-based varieties, failed due to unacceptably low takes. As new pure US wild species were discovered – particularly *Riparia* and then *Rupestris* – each started its own roller-coaster fad. First, orders were sent to Missouri and several other states, for cuttings taken from wild vines of the target species. Once in France, the cuttings would be rooted in place, there to be grafted during their second season. Needless to say, take was highly variable, and resistance varied unexpectably as well. Not all wild *Riparia* vines were the same, and nor were *Rupestris*. A second wave of popularity developed around Millardet's idea of raising rootstock plants from pure US species seeds gathered in the wild (Millardet 1877, p. 31). This idea was soon criticized, and rightly so, for the excessive variability in resistance that was found in the seedlings (de Lafitte 1883, p. 535). In the end, Montpellier solved the problems by selecting from among its enormous collection of pure US species only those individual vines that were easy to graft, compatible with most French varieties, and highly resistant to *Phylloxera*. Best among the dozen or so that were eventually propagated and disseminated were '*Riparia Gloire de Montpellier*' and '*Rupestris du Lot*', both of which still see widespread service worldwide. In the end, the campaign was long and hard, but generally successful.

By the 1890s, the original tide of US vines – the direct producers and rootstocks taken directly from US wild species – had ebbed, and the second wave was starting to flow. This new wave consisted of vines created deliberately by cross-breeding various US and European selections. Two distinct roles were played by the newly created vines. First, second-generation *hybrid* rootstocks, designed for a better match with the typical French soil, so different from America's terrain, were being rapidly disseminated from the École in Montpellier and research stations in Bordeaux, the Charentais and Provence. Among these graftstocks were candidates suitably adapted to almost every terrain in France, and compatible with almost all the traditional wine varieties. But competing with the campaign run by Montpellier to graft traditional French vines on to US roots was an alternative programme emphasizing *hybrid direct producers*, an entirely new class of vines whose genetic foundation combined root and fruit elements from both US and French varieties. According to proponents, genetic cross-breeding would eventually produce a superior vine, with US health and French wine quality; in their view, grafting was seen as 'only a transitory cultural process, destined to disappear in the future' (Ganzin 1888, p. 23). Although this campaign's theoretical base was the University of Bordeaux, in practice many of the most valuable additions to the armentarium of available hybrid direct producer (HDP) varieties were developed by private individuals, most of whom were located in the south-east, particularly around the Ardeche.

Since Montpellier did not think that HDPs were a genuine possibility, they elected to stay with promulgating their grafting programme, contenting themselves with frequent sniping at the Bordeaux hybridizers. Differences between the two universities were based on their respective theories about inheritance. Montpellier, from the time of Planchon, had believed in a 'blending' theory of inheritance; thus, crossing a phylloxera-resistant US vine with a French vine could only weaken the resistance of the resulting offspring (Berget 1896, p. 53; Gouy 1903, p. 189). As Paul notes, 'this was the "*thèse classique*" of Foëx', Saintpierre's successor as director at Montpellier (Paul 1996, p. 70). On the other hand, Bordeaux's Millardet held a 'mosaic' theory of inheritance, a major French 'genetic' hypothesis prior to Mendel. On this view, an offspring individual is a mosaic constituted from discrete elements inherited from its parents. One of the more explicit theories of this sort was developed by Naudin based on hybridizing experiments he carried out during the early 1860s (Marza and Cerchez 1967). Millardet was certainly aware of Naudin's hypothesis. As his own work developed in the late 1870s and early 1880s, Millardet came to the conclusion that whole organ systems – e.g.

roots, foliage, fruit – were the heritable elements, the pieces of the mosaic; a vine type existed as a unity of these organic elements, which, during sexual reproduction, would be mixed and matched to produce newly unified individuals in the offspring-generation. Based on this theory, it was entirely plausible to believe that an offspring might combine root systems from its US parent with fruit systems from its French parent. Millardet's proclamation of the brave new hybrid world rallied generations of workers to his side:

> Thus, the year 1887 will mark a date ever remembered in the history of our devastated vineyards, our agonies and our struggles against the formidable plague which has assailed our viticulture since twenty years ago. By grace of the hybridation of our European varieties with diverse American vines, we are, from today onward, absolutely certain to obtain, in first-generation hybrids, either graft-stocks of an assurred resistance, and of an adaptation much easier than those which we have possessed until now, or direct producers, resistant to *Phylloxera* and to the most dangerous plant parasites, which are capable of producing at the same time wine completely correct in flavor.
>
> (Millardet 1888, p. 28)

Noisy conflict between the two university programmes lasted for at least 20 years. But even then it didn't go away; rather, it just got less noisy, even as it spread far beyond France. Indeed, it is quite fair to say that the principals of the two sides are still vigorously competing more than a century later (Galet 1988; Pouget 1990, p. 56; Paul 1996). The issues that divide the two camps are complex, subtle and have deep roots, roots that are firmly embedded in the ever-opposing camps of tradition and modernity. Today, although all but a small number of hybrid direct producers have been outlawed and driven from France, these vines have taken solid hold worldwide and are the foundation upon which viticulture rests in the USA and Canada, except for a narrow tract of land extending inward from the Pacific Ocean, stretching from Vancouver to San Diego.

## Victory in France

By 1900 the battle to save the French *vignoble* from *Phylloxera* had reached an uneasy stand-off. French viticultural interests – from the national councils, to the laboratories and research stations, to the smallest *vigneron* – knew how to live with the devastating insect. There was no possibility of victory if that meant returning to what was before: traditional French viticulture was well and truly finished. Moreover, there was no possibility of victory if that meant having vanquished the bug: every few years, someone would try to grow the old varieties on their own roots, but the attempt would fail within a couple of seasons. The bug was always there, lurking, waiting for some lapse in vigilance. What had been achieved could only be called a truce, with a 'demilitarized zone' established around the US roots of the French vine. No one knew how long the truce might last.

The costs of the battle had been enormous. According to one authority, *Phylloxera* had 'cost us nearly as much as the war-indemnity payments we paid to Germany after the war of 1870' (Convert 1900, p. 337). Although costs varied considerably from region to region, a recent analysis put together from a welter of data taken at the time indicates that a reliable average cost per hectare would be around FFr 3000 (Garrier 1989, p. 131). But costs were far more than merely monetary. Fully one-third of the French *vignoble* had disappeared, never to be replanted. Thousands upon thousands of *vignerons* had left their land, many for the cities – Marseilles doubled in population during the period, even though its *département* lost population overall – while even more emigrated, hoping to start viticulture anew in the *Phylloxera*-free lands of Tunisia and Algeria.

But the most difficult loss was a millennium-old tradition and way of life, with its vast library of accumulated experience. *La nouvelle viticulture* demanded a smart, literate, decently educated *vigneron*, one who could read the literature, read the labels on chemical containers, read the signs of the new diseases on the old vines. Everything had to be learned anew.

Fortunately, not all the effects of the 30-year war were bad. The new viticulture was more productive, grape quality was higher, and the new cultural practices, once mastered, were easier, more efficient, and less-labour intensive. To take just one example: in the new viticulture, vines were planted in straight lines at orderly intervals rather than completely helter-skelter as classical tradition would have had it. Such order was demanded by the necessity to move equipment – sprayers, for example – among the vines without damaging either the plants or the machines. Examples such as this abounded. Put most simply, after *Phylloxera*, the French *vignoble* was rationalized.

In the end, nothing was the same. But as Missouri's Professor George Hussmann is widely reputed to have later observed, it would be difficult to decide whether in the long run the *Phylloxera* disaster was for good or for ill.

## Other places, other times, other battles

*Phylloxera* invaded other European countries only slightly later than France, starting sometimes as original infections from imported US vines, sometimes as secondary infections from imported French vine material. Italy and Spain were definitely hit during the early 1870s, as were Portugal, Germany, and Switzerland. Damage was inevitable, but frequently less devastating than in France. Portugal was probably the worst hit, with Italy next. Even Portugal's offshore *vignobles* in Madeira and the Azores were severely struck. Germany, most likely because of its cooler summers, which limited the bug's reproductive excesses, was least damaged. Italy was the most variable, with Sicily's *vignoble* destroyed by the late 1880s, even while Naples and its regions were spared until at least the 1930s (Morrow 1972, Chapter 5).

During the late 1870s, some French firms planted vines in Dalmatia and Slovenia; Croatia was targeted for a huge build-up (Anonymous 1880, p. 112). Over the following decade, the size of the Yugoslavian wine industry went up by an order of magnitude, mostly in aid of exports back to France. The whole scheme collapsed between 1902 and 1905 as the *Phylloxera* arrived and conquered. A diaspora of ruined wine workers, their families and others dependent on the industry fled the region, with large numbers emigrating to the USA, Canada and Australia.

Countries in the Balkans and Greece were seriously infected shortly after the beginning of the twentieth century. Around the same time, Australia was mildly infected in Victoria and New South Wales, but strict quarantine and *cordon sanitaire* procedures kept other regions, particularly South Australia's important *vignoble*, clean from then until now.

Chile is the only significant wine-growing country that remains free of the insect.

*Phylloxera* was found in California in 1874 near the city of Sonoma. It rapidly spread to the other districts in the Sonoma and Napa valleys. Within 25 years the entire state was infested, and nearly 30 000 acres (12 140 hectares) – a sizeable percentage of the then-planted area – were destroyed (Bioletti 1901, p. 4).

All later-infected areas looked to French experience for guidance, although not quite so slavishly as Montpellier's Prosper Gervais claimed (Gervais 1904). Although the Italian authorities imported French hybrid rootstocks, because of some unique indigeneous terrain, success was not complete. Italian viticulturalists, in particular Paulsen in Sicily, had to develop their own special rootstock varieties. Portugal also bred some of their own, importing the parents directly from the USA. Californian viticulturalists tried the French rootstock varieties, and were generally satisfied with the results: every region had one or more useable stocks. In the end, by roughly the 1930s, worldwide viticulture followed only two alternative French-based paradigms: traditional varieties grafted on to US roots; or hybrid direct producers, genetic crosses between resistant US varieties and traditional wine varieties. And thus things settled down, apparently stable and dependable, until the shocking developments in California in the 1980s.

## A frightening portent?

In 1980, Napa Valley winegrower John Baritelle found four stunted vines in one of his vineyards. Next season, the number had grown to 16, and Baritelle was beginning to be more than puzzled by the problem, especially since he could not find any definite signs of disease. By the 1982 season, the number of diseased vines had multiplied appreciably, and a worried Baritelle asked for, and received, a visit from vine scientists from the nearby Davis Campus of the University of California. In a scene eerily reminiscent of one originally acted out 115 years earlier in a vineyard near the Rhône, Davis's viticulturalist,

Austin Goheen, 'dug up a vine on Baritelle's land and saw at once the telltale yellow colonies of tiny aphid-like insects on the roots. The case was a no-brainer. The vines were succumbing to *Phylloxera*' (Lubow 1993, p. 26). But what could have happened? Since all Californian vines, including Baritelle's, were of necessity grafted on to resistant rootstock, there must have been some mistake at the nursery, leading to a non-resistant rootstock being grafted by mistake, or something like that.

However, when *Phylloxera* started showing up at other vineyards, it became clear that it couldn't be a question of a mistake at the grafting nurseries. Something much more fundamental, much more dangerous, had happened. It took a lot of investigation, a lot of time, to reveal what.

When Californian rootstock trials were initiated at the beginning of the twentieth century, acceptable varieties had eventually been found for each of the diverse terrains of the state (Bioletti 1901). Trials continued uninterruptedly, however, and 'by 1958, the collected data were used to recommend A×R #1 as well adapted to most of California's vineyard soils, climate, water conditions, and scions' (Granett *et al.* 1996, p. 10). A×R #1 is a cross between the traditional variety '*Aramon*', and a wild US *Rupestris* vine, made by Ganzin in the mid-1880s (Ganzin 1888). Although the rootstock enjoyed an initial vogue in France, it was soon discarded because it was 'considered insufficiently resistant to *Phylloxera*' (Galet 1979, p. 200). Yet this rootstock, rejected by the French as insufficiently resistant to phylloxera, was officially recommended by Davis and 'based on this recommendation, was used in 60% to 70% of the Napa and Sonoma county plantings that occurred in the 1960–1980 planting boom' (Granett *et al.* 1996, p. 10).

Just as their French ancestors had a century earlier, Californian growers denied the reality before their eyes, seizing on any possible explanation rather than *Phylloxera* for the growing disaster. Even Davis's scientists appeared to go into denial. After Goheen's discovery in 1982, it took seven long years of argument before Davis's viticulturalists 'like a deadlocked jury, had begun to tilt from the weight of incontrovertible evidence' (Lubow 1993, p. 60). In 1989 the Davis scientists noted that 'it was clear that A×R #1 was not adequately resistant to California *Phylloxera*' (Granett *et al.* 1996, p. 10). The phrasing of this remark suggests a new development in the century-old battle with the beast. It is not that the rootstock is not adequately resistant to *Phylloxera in* California; rather, it is not adequately resistant *to* Californian *Phylloxera*. According to the Davis scientists, the rootstock had not lost its original resistance, but rather the bug populations had evolved in such a way that the rootstock's original resistance was bypassed. In other words, a newly evolved version of the bug – a new *biotype*[2] capable of living on A×R #1 – began emerging as the dominant population of northern California. Thus, the A×R-rooted vines were doomed.

In the end, 50 000 acres (20 234 hectares) in Napa and Sonoma alone need replacing, at an estimated cost of somewhere around US$1 billion (Sullivan 1996, p. 7). Unfortunately this is only the beginning. A×R plantings comprise a vast amount of the total Californian *vignoble*, including vineyards in Lake, Mendocino, San Joaquin, Sacramento, Alameda and Santa Clara counties.

Darwin certainly prepared us for such an outcome. Evolutionary forces never cease, adaptation between prey and predator is a dynamic state, never ever stabilizing into a static relationship. Will the French grafting solution, discovered at the cost of much blood, sweat and tears 100 years ago, always work to keep the *Phylloxera* at bay? An affirmative answer would be foolhardy at present, since we simply don't know how and why the French solution worked. As Granett *et al.* (1996, p. 13) conclude: 'The stability of these rootstocks may not be eternal, and we should be prepared for *Phylloxera* strains that are better adapted and potentially damaging in the future'. Eternal vigilance, as always, is the cost of security.

## Notes

1 Two French technical terms will not be translated in this chapter: '*vignoble*', which not only has a geographic connotation, but also, more importantly, conveys the notion of 'place where the same sorts of vines are subjected to the same sort of climate, terrain, viticultural practices and winemaking styles'. One can speak of something as large as 'the French *vignoble*', thereby distinguishing it from the Spanish *vignoble*; or something as small as 'the Chiroubles *vignoble*' in order to distinguish it from other Beaujolais villages such as Fleurie. The second term is '*vigneron*', which refers to any small-scale winegrower/maker. Typically, the *vigneron* is a freeholder, but not always. The term is a term of art, connoting a mixture of social class, economic level, and political bent.

2 Technically, a specific biotype is associated with a specific genome. But the Davis scientists explicitly assert that their discovery of a new biotype is based 'on the behavior of *Phylloxera*, not necessarily on genetic differences' (Granett *et al.* 1996, p. 10).

## Ackowledgements

A huge number of people have helped me along the way toward this paper – none of whom, of course, are the least bit responsible for any of my mistakes.

UMKC Office of Research Administration funded my visits to the University of Montpellier and the University of California, Davis. At the University of Montpellier's library I was particularly well advised by Mme B. Bye and M. J. Argeles; the Department of Viticulture chair A. Carbonneau was especially helpful and hospitable. At Davis, J. Skarstad in Special Collections, J. Granett in Entomology, and P. Teller and K. Whelan in H.P.S. were very helpful. Pittsburgh's Center for Philosophy of Science supported me famously for a semester; in particular, R. Olby, J. Lennox and P. Machamer gave me exciting things to think about.

Earlier versions of this research have been presented at HSS-Atlanta, HSS-San Diego, and HSS-Vancouver; at colloquia in Davis, Virginia Tech, and

East Tennessee State; plus at the 4th International Pitt Center Fellows Conference, Bariloche, Argentina, June 2000.

Finally, I must thank Ann Mylott who originally pointed me in the direction of medical models of disease.

## References

Ackerknecht, E. H. (1948) Anticontagionism between 1821 and 1867. *Bulletin of the History of Medicine*: 22, 562–93.

Ainsworth, G. C. (1981) *Introduction to the History of Plant Pathology*. Cambridge University Press, Cambridge.

Anonymous (1880) Les vignes de Croatie. *La vigne française*: 1, 112.

Barral, J. E. (ed.) (1868) Editorial. *Journal de Agriculture*: 20 September, 2, 725.

Barral, J. E. (ed.) (1869a) Committee notice. *Journal de Agriculture*: 20 February, 3, 455.

Barral, J. E. (ed.) (1869b) Congrès viticole de Beaune. *Journal de Agriculture*: 5 December, 4.

Berget, A. (1896) *La Viticulture Nouvelle*. Félix Alcan, Paris.

Bioletti, F. T. (1901) *The Phylloxera of the Vine*. Bulletin No. 131, Agricultural Experiment Station, State Printing, Sacramento.

Borde de Tempest, E. (1873) *Phylloxéra: sa Destruction Certaine*. Imprimerie Centrale du Midi, Montpellier.

Cazalis, F. (1869a) *Le Messager Agricole* 9. Imprimerie Typographique de Gras, Montpellier.

Cazalis, F. (1869b) De la maladie de la vigne causée par le Phylloxéra. *L'Insectologie Agricole*: 3, 29–33.

Cohen, H. (1961) The evolution of the concept of disease, in *Concepts of Medicine* (ed. B. Lush). Pergamon, Oxford.

Commission Départmentale de l'Hérault pour l'Étude de la Maladie de la Vigne (1877) *Résultats Pratiques de l'Application des Divers Procédés*. Grollier, Montpellier.

Compagnie des Chemins de Fer de Paris à Lyon et à la Méditerranée (1878) *Instructions pour le traitement des vignes par le sulfure de carbone*. Paul Dupont, Paris.

Convert, F. (1900) La viticulture après 1870: La crise phylloxérique. Parts II–IV. *Revue de Viticulture*: 14, 337–9; 449–52; 512–17.

Cornu, M. and Dumas, J. P. (1876) *Instruction pratique. Commission du Phylloxéra, séance du 17 jan. 1876*. Institut de France, Academie des Sciences. Gauthier-Villars, Paris.

de Ceris, A. (1873) Chronique Agricole. *Journal d'Agriculture Pratique*: 37, 673–6.

de Lafitte, P. (1883) Les vignes américaines obtenues de semis, in *Quatre ans de luttes pour nos vignes et nos vins de France*. G. Masson, Paris, pp. 535–49.

Falières, E. (1874) *Du Phylloxéra et d'un nouveau mode d'emploi des insecticides*. Imprimerie Nouvelle A. Bellier, Bordeaux.

Faucon, L. (1870) Nouvelle maladie de la vigne. *Journal d'Agriculture Pratique*: 34, 512–18.

Fitz-James, M. (1881) Les vignes américaines. *Revue des Deux Mondes*: 51, 685–94.

Gachassin-Lafite, L. (1882) Rapport de la commission des vignes américaines et des sables, in *Compte-rendu général du Congrès international phylloxérique de Bordeaux du 9 au 16 octobre 1881*. Féret & fils, Bordeaux, pp. 100–43.

Galet, P. (1979) *A Practical Ampelography* (trans./adapted L. T. Morton). Cornell University Press, Ithaca.

Galet, P. (1988) *Cépages et vignobles de France, Tome 1: Les vignes américaines*, 2nd edn. Charles Déhan, Montpellier.

Ganzin, V. (1888) Les Aramon-rupestris: Porte-Greffe hybrides inédits, parties 1 et 2. *La vigne française*: 1(2), 15–16; 23–5.

Garrier, G. (1989) *Le Phylloxéra: Une guerre de trente ans 1870–1900*. Albin Michel, Paris.

Gervais, P. (1904) *La crise phylloxérique et la viticulture européene*. Bureau de la Revue de Viticulture, Paris.

Gouy, P. (1903) L'école de Montpellier et l'hybridation. *Revue des hybrides franco-américains*: 5, 189–90.

Granett, J., Walker, A., De Benedictis, J., Fong, G., Lin, H., Weber, E. (1996) California grape *Phylloxera* more variable than expected. *California Agriculture*: **50**, 9–13.

Lachiver, M. (1988) *Vins, vignes et vignerons: une histore du vignoble français*. Fayard, Paris.

Laliman, L. (1872) *Étude sur les Divers Phylloxéra et Leurs Médications*. Librairie de la Maison Rustique, Paris.

Laurent, R. (1958) *Les vignerons de la 'Côte d'Or' au XIXe Siècle*. Société des Belles Lettres, Publications de l'Université de Dijon XV, Paris.

Legros, J.-P. and Argeles, J. (1994) L'invasion du vignoble par le phylloxéra. *Bulletin de l'Academie des Sciences et des Lettres de Montpellier*: N.S. 24, 205–22.

Lichtenstein, L. and Planchon, J.-É. (1870) On the identity of the two forms of phylloxera. *Journal d'Agriculture Pratique*: 34, 181–2.

Lubow, A. (1993) What's Killing the Grapevines of Napa? *New York Times Magazine*: October 17, 26–63.

Marion, A. F. (1879) *Rapport sur les expériences contre le Phylloxéra et les résultats obtenus, campagne de 1878*. Paul Dupont, Paris.

Marza, V. D. and Cerchez, N. (1967) Charles Naudin, a pioneer of contemporary biology (1815–1899). *Journal d'agriculture tropicale botanique appliqué*: 14, 369–401.

Millardet, A. (1877) *La Question des Vignes américaines au point de vue théorique et pratique*. Féret & fils, Bordeaux.

Millardet, A. (1888) Nouveaux cépages hybrides: Résistant au Phylloxéra et au Mildiou. *La vigne française*: 9(2), 25–9.

Morrow, D. W., Jr (1972) *The Phylloxera Story*, Chapters I–VI. Unpublished manuscript, Manuscripts, Shields Library, University of California, Davis. Used with permission.

Ordish, G. (1972) *The Great Wine Blight*. Charles Scribner's and Sons, New York.

Paul, H. (1996) *Science, Vine, and Wine in Modern France*. Cambridge University Press, Cambridge.

Planchon, J.-É. (1874) Le phylloxéra en Europe et en Amérique. I. *Revue des Deux Mondes*: 44, 544–65.

Planchon, J.-É. (1875) *Les vignes américaines, leur culture, leur résistance au phylloxéra et leur avenir en Europe*. C. Coulet, Montpellier.

Planchon, J.-É. (1877) La question du phylloxéra en 1876. *Revue des Deux Mondes*: 47, 241–7.

Plumeau, M., Gayon, R., Princeteau, R., Falières, E. (1882) Rapport sur le sulfure de carbone et les sulfo-carbonates, in *Compte-rendu général du Congrès international phylloxerique de Bordeaux du 9 au 16 octobre 1881*. Féret & fils, Bordeaux.

Pouget, R. (1990) *Histoire de la Lutte contre le Phylloxera de la Vigne en France*. Institut National de la Recherche Agronomique, Paris.

Riley, C. V. (1871) *Third Annual Report on the Noxious, Beneficial and Other Insects of the State of Missouri*. Public Printer, Jefferson City.

Riley, C. V. (1872a) *Fourth Annual Report on the Noxious, Beneficial and Other Insects of the State of Missouri*. Public Printer, Jefferson City.

Riley, C. V. (1872b) On the cause of deterioration in some of our native grapevines, and one of the probable reasons why European vines have so generally failed with us. *American Naturalist*: 6, 622–31.

Sahut, F. (1888) *De l'Adaptation des Vignes Américaines au Sol et au Climat*. Coulet, Montpellier.

Saintpierre, C. (1860) *De la Fermentation et de la Putréfaction*. Typographie de Boehm & fils, Montpellier.

Signoret, V. (1870) *Phylloxéra Vastatrix Cause Prétendue de la Maladie Actuelle de la Vigne*. Félix Malteste, Paris.

Société Linnéenne de Bordeaux (1869a) *Rapport sur la maladie nouvelle de la vigne*. Coderc, Degréteau et Poujol, Bordeaux.

Société Linnéenne de Bordeaux (1869b) *Deuxième Rapport sur la Maladie Nouvelle de la Vigne*. Coderc, Degréteau et Poujol, Bordeaux.

Sullivan, V. (1996) New rootstocks stop vineyard pest for now. *California Agriculture*: 50, 7–8.

Trimoulet, A.-H. (1875) *5ème Mémoire sur la maladie de la vigne: remèdes preconisés par les phylloxéristes et les antiphylloxéristes*. Imprimerie F. Degréteau, Bordeaux.

Vialla, L. (1869) Rapport sur le nouvelle maladie du vigne. *Journal d'Agriculture*: 4, 341–60.

Whetzel, H. H. (1918) *An Outline of the History of Phytopathology*. W. B. Saunders, Philadelphia.

Zacharewicz, C. M. (1932) Cinquantenaire de la Reconstitution en Vaucluse. *Revue de Viticulture*: 77, 270–85.

# 5 Wine and heart disease: a statistical approach

*M. Bobak and M. Marmot*

## Introduction

The notion that wine is protective against heart disease has become part of conventional wisdom. Many lay people see wine as different from other alcoholic beverages, rather than as one of them. By implication, the protective effect against heart disease is seen as a property of wine, rather than of alcohol in general. In this chapter, we will review the evidence linking the consumption of wine, and of alcohol in general, with a reduced risk of coronary heart disease. We will start in the reverse order, with alcohol in general, and will then explore whether wine might have any protective effect above that of alcohol alone.

Throughout this chapter, we will be confronted with the issue of the pattern of drinking. In general, the vast majority of studies to date have relied on some indicator of 'average' consumption (mean intake or number of drinks per day or week, etc.). This is, however, a crude measure of drinking. Two people consuming the same amount, say 140 g of alcohol per week, may drink very differently. One can drink 20 g of alcohol each day (regular or sustained drinking), while the second may consume the whole amount in one evening (episodic or binge-drinking pattern). While this state of affairs has been recognized for a long time (Edwards *et al.* 1994), researchers have only recently addressed more systematically the question of whether regular and episodic drinking may have different health effects (Grant and Litvak 1998; Rehm *et al.* 2001). Wherever possible, we too will comment on this aspect. We will also briefly address the question of whether the risk of death related to drinking differs by other factors, such as diet or additional risk factors (i.e. whether alcohol interacts with other variables).

## Alcohol and all-cause mortality

Perhaps the most reliable information on the effects of alcohol came from studies on mortality. For all-cause mortality, virtually all prospective studies have found a U-shaped or J-shaped association with alcohol consumption (Marmot 1984; Marmot and Brunner 1991; Royal College of Physicians

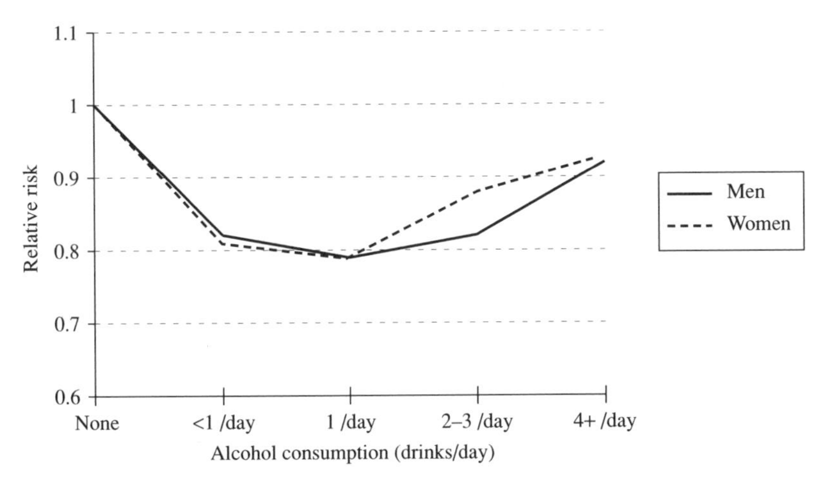


*Figure 5.1* Alcohol consumption and all-cause mortality in US men and women (based on Thun *et al.* 1997).

1995; Rimm *et al.* 1996; Corrao *et al.* 2000). In general, moderate drinkers have lower mortality than non-drinkers; there seems to be little association between consumption and mortality within moderate intake levels. At higher intake levels, the mortality risk curve starts to rise again above that of non-drinkers. This is well illustrated by the American Cancer Prevention Study, the largest study on alcohol and mortality in individuals to date. It examined the drinking habits and other life-style factors in nearly half a million US adults, and recorded deaths occurring in the subsequent nine years (Figure 5.1) (Thun *et al.* 1997). The risk of death was found to be lowest among men and women who drank about one or two drinks a day. Above this level of consumption, the risk of death increased but, even in the highest consumption categories, remained below that of non-drinkers. In many other studies, however, the risk of death from all causes among heavier drinkers (above, say, 50–60 g of alcohol per day) was some 20% to 50% higher than among non-drinkers (Royal College of Physicians 1995).

Most studies of alcohol mortality have focused on middle-aged and elderly subjects; consequently little is known about younger people. The few studies including younger subjects suggest that the protective effect of moderate intake is not present among younger people (Andreasson *et al.* 1988; Gronbaek 2001). This is because the association between alcohol and all-cause mortality is an aggregate of the effects of alcohol on different causes of death.

Death from injury and accidents is more common among younger people, while cancers and cardiovascular disease are more common in older age groups. Alcohol was found to be associated with increased mortality from several types of cancer, injuries and accidents, liver cirrhosis and other health outcomes. The association with injury deaths among younger people is particularly strong. In addition, heavy drinking increases the risk of haemorrhagic stroke, and may also increase that of ischaemic stroke and coronary heart disease; on the other hand, moderate regular drinking seems protective against circulatory diseases, such as heart disease and ischaemic stroke. The sum of these effects is the familiar U-shaped or J-shaped curve.

## Alcohol and cardiovascular disease

The vast majority of cardiovascular deaths in Europe are from coronary heart disease and from cerebrovascular disease (stroke), with the proportion of each varying between countries. Stroke can be caused either by a blood clot in the cerebral circulation (ischaemic strokes) or by brain haemorrhage (haemorrhagic strokes). Cardiovascular diseases deserve more attention for two reasons: first, cardiovascular diseases are the most common cause of death in middle-aged and older groups in most countries; second, the association between alcohol consumption and cardiovascular diseases is complex and not yet fully understood.

### *Regular moderate drinking*

Results of the numerous studies of alcohol and coronary heart disease are remarkably consistent (for reviews, see Marmot 1984; Marmot and Brunner 1991; Royal College of Physicians 1995; Rimm *et al.* 1996; Doll 1997; Fagrell *et al.* 1999; Corrao *et al.* 2000). The evidence suggests that the relation between alcohol and both coronary heart disease and stroke follows a U-shaped or L-shaped curve. In nearly all studies, the lowest risk of cardiovascular and coronary heart disease was found among moderate drinkers. The risk is higher in non-drinkers and in some, but not all, studies also in heavier drinkers.

Figures 5.2 and 5.3 show mortality from coronary heart disease according to levels of alcohol intake in the above mentioned American Cancer Prevention Study (Thun *et al.* 1997). There was an L-shaped association with mortality from coronary heart disease in subjects with pre-existing cardiovascular disease, and an approximately U-shaped association in subjects without pre-existing disease, but the risk in the highest category did not exceed that of non-drinkers. The risk of all types of cardiovascular diseases combined was lower in drinkers than in non-drinkers, and there was no apparent excess risk in the highest drinking category.

It is difficult to estimate the levels of alcohol intake associated with the maximum protection against coronary heart disease because different studies used different measures and definitions, and also because exact measurement of alcohol intake is difficult. A recent meta-analysis of 51 studies found that the risk of coronary heart disease was lowest (25% lower than among non-drinkers) at a consumption of 25 g of alcohol per day (Corrao *et al.* 2000). In a subset of 28 cohort studies judged as good quality, the maximum protection (20% lower risk) was seen at a daily consumption of 20 g of alcohol. It thus

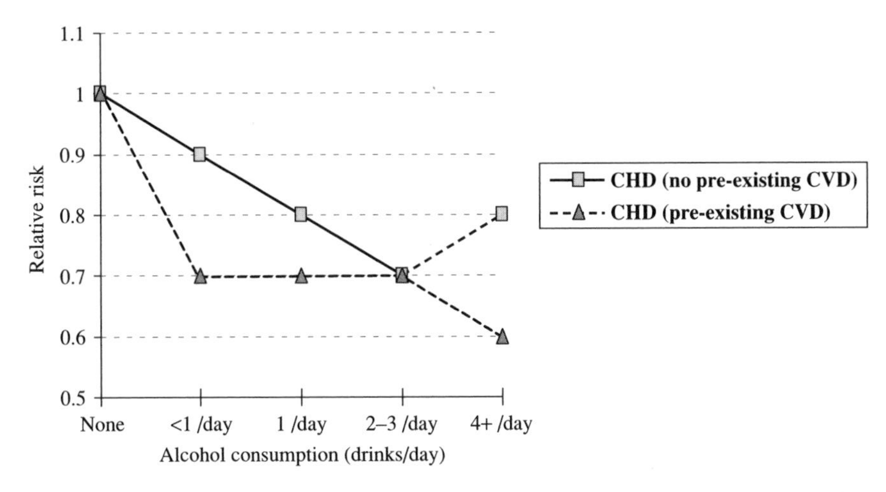


*Figure 5.2* Alcohol consumption and mortality from coronary heart disease in US men (based on Thun *et al.* 1997). CHD, coronary heart disease; CVD, cardiovascular disease.

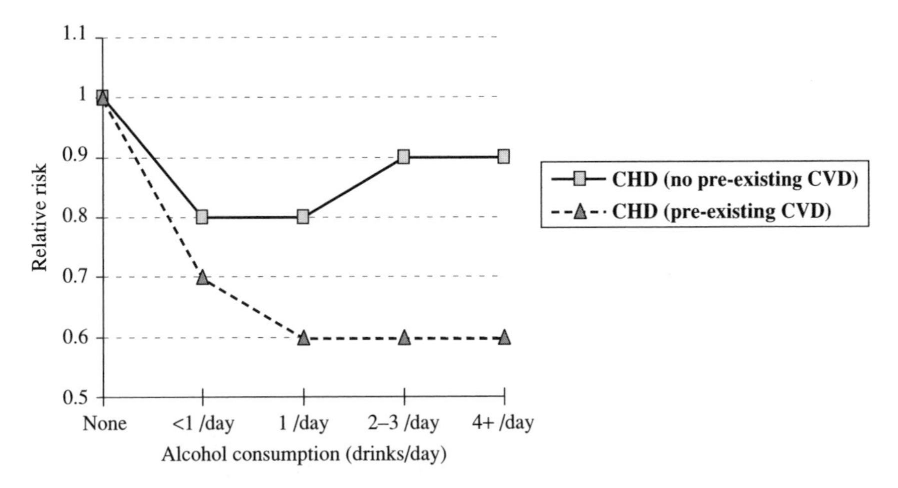


*Figure 5.3* Alcohol consumption and mortality from coronary heart disease in US women (based on Thun *et al.* 1997). CHD, coronary heart disease; CVD, cardiovascular disease.

appears that consumption of 20–30 g of alcohol daily, corresponding to two to three drinks, is associated with maximum protection.

A U-shaped curve also seems to exist for moderate alcohol intake and risk of sudden cardiac death. Sudden cardiac deaths are a sub-group of deaths coded as coronary heart diseases; they account for about half of all cardiovascular deaths (Huikuri *et al.* 2001). The majority of sudden cardiac deaths are caused by arrhythmias triggered by acute coronary events, such as myocardial infarction. The US Physicians Health Study found that compared with non-drinkers, men drinking two to four drinks per week and five to six drinks per week had a relative risk of sudden cardiac death of 0.4 and 0.2, respectively; the risk was approaching unity at two drinks per day (Albert *et al.* 1999).

While the majority of studies produce consistent results, several issues remain unresolved. Some authors have suggested that all or part of the J-shaped curve may be due to a health selection bias. In addition, the biological mechanisms of the protective effect of alcohol, the possible interaction of alcohol with conventional risk factors, and the effects of heavy drinking and of binge-drinking remain controversial. These issues are briefly reviewed below.

### *Health selection*

It has been argued that the protective effect is an artefact due to the fact that people with health problems caused by alcohol may have stopped drinking; this would lead to an artificially higher mortality among non-drinkers. This explanation is not specific to cardiovascular diseases; it would also apply to other causes of death and to all-cause mortality. However, available data do not support this hypothesis (Edwards *et al.* 1994; Klatsky 2001). Several groups have separated ex-drinkers and 'never drinkers', and have not confirmed the health selection hypothesis. Moreover, exclusion of deaths occurring in the first years of follow-up (which are more likely to result from pre-existing alcohol-related disease) do not change the results of the major studies (Gronbaek 2001). There is now a consensus that most of the protective effects of moderate drinking are genuine, not least because good evidence now exists of the biological mechanisms involved in the protective effect.

### *Biological mechanisms in the protective effect*

One of the criteria for accepting an association between an exposure and a disease as causal requires that biological mechanisms exist for the association. This is the case in the protective effects of moderate alcohol intake. There have been numerous experimental studies of the effect of alcohol on different physiological mechanisms. The pooled analysis of these studies (Rimm *et al.* 1999) is summarized in Table 5.1. Intake of alcohol is significantly associated with several biological risk factors of coronary heart disease. Using published associations between these risk factors and coronary heart disease, the authors estimated that the intake of 30 g of pure alcohol per day would reduce the risk of coronary heart disease by 25% (Rimm *et al.* 1999). This predicted reduction is consistent with the protective effect observed in prospective studies.

*Table 5.1* Change of biological risk factors associated with an intake of 30 g of alcohol per day (based on Rimm *et al.* 1999)

| *Biological factor* | *Change (95% confidence interval)* |
|---|---|
| High density lipoprotein (HDL) cholesterol (mg/dl) | + 3.99 (3.25; 4.76) |
| Apolipoprotein A I (mg/dl) | + 8.82 (7.79; 9.86) |
| Triglycerides (mg/dl) | + 5.69 (2.49; 8.89) |
| Plasminogen (% of standard) | + 1.47% (−1.18; 4.42) |
| Fibrinogen (mg/dl) | − 7.5 (−17.7; 32.7) |
| Lp(a) lipoprotein (mg/dl) | − 0.70 (−3.38; 1.99) |
| Tissue-type plasminogen activator antigen (ng/ml) | +1.25 (−0.31; 2.81) |

## *Effect modification by coronary risk*

Available evidence suggests that the protective effect of alcohol is more pronounced among subjects at high risk of coronary heart disease. For example, the J-shaped curve persists into older ages, although for virtually all other coronary risk factors the relative risk reduces with increasing age. Fuchs *et al.* (1995) found that the effect of moderate alcohol intake was more pronounced among women with a higher prevalence of coronary risk factors than among women with more favourable risk factor profile. The American Cancer Prevention cohort found similar results (Thun *et al.* 1997). The authors classified subjects into three categories by cardiovascular risk: 'low risk' (age less than 60, no history of heart disease or hypertension and no medication for coronary heart disease, hypertension, stroke or diabetes); 'medium risk' (younger than 60 but with presence of risk factors); and 'high risk' (aged 60 or more and with pre-existing risk factors). There was a J-shaped association between alcohol and mortality in the low risk group, a U-shaped association in the intermediate group, and an L-shaped association in the high risk group. Similar results were obtained when subjects were stratified by the presence of pre-existing coronary heart disease (Figures 5.2 and 5.3). However, the benefits of alcohol did not compensate for the large increase in mortality caused by smoking.

## *Public health implications*

Accepting the protective effect of alcohol against cardiovascular diseases as genuine is an important issue. Because cardiovascular diseases are the major cause of death worldwide, even a small protective effect has major implications for estimating the total burden of disease attributable to alcohol. The harmful effect on accidents or cancers could be counterbalanced by the positive effects on cardiovascular mortality. Calculations conducted for the World Bank/ WHO Global Burden of Disease Project suggest that the impact of alcohol on mortality differs by age group. In younger age groups (below 44 years), the impact is clearly negative, mainly through deaths from injuries and accidents. Above the age of 70, alcohol seems to prevent more deaths (mainly from cardiovascular disease) than it causes (Murray and Lopez 1996). Despite such calculations, however, alcohol poses a number of health hazards, and it is not to be recommended as a prevention strategy (Goldberg *et al.* 2001).

### *Heavy drinking and binge-drinking*

While there is little doubt that regular moderate drinking has a protective effect against cardiovascular disease, the issue of heavy drinking, either sustained or episodic, has not been resolved. This issue should not be underestimated. Despite the large number of studies of the effect of alcohol on cardiovascular diseases, it is likely that heavy drinkers are underrepresented in epidemiological studies. Even in populations with high alcohol consumption, problem drinkers are particularly difficult to reach. It is therefore difficult to extrapolate the findings on 'heavy drinkers' included in epidemiological studies to heavy drinkers not so included.

The effects of binge-drinking on cardiovascular diseases have also not been systematically studied. It has been proposed that such drinkers have an elevated risk of sudden cardiac death (possibly resulting from ventricular arrhythmias) and cardiomyopathies (McKee and Britton 1998), and there is some evidence that binge-drinking may increase the risk of myocardial infarction (Kauhanen *et al.* 1997). This question may be important to explain the high and fluctuating mortality rates in Russia and other parts of the former Soviet Union (Leon *et al.* 1997). It has been proposed that the (suspected) predominant drinking pattern – binge drinking – may underlie the recent dramatic increase in cardiovascular mortality in the former Soviet Union (Britton and McKee 2000).

Britton and colleagues conducted a systematic review of the literature on heavy alcohol consumption and cardiovascular diseases (Britton and McKee 2000). This review confirmed the protective effect of moderate regular drinking described above, but highlighted the scarcity of studies on heavy or episodic drinking. For example, from the 42 prospective studies that fulfilled the inclusion criteria, only seven (in fact only six, as one study was represented by two papers) examined drinking pattern or heavy drinking.

The results of these seven papers differed markedly from the remaining reports. Six reported a positive association between heavy alcohol consumption and cardiovascular diseases, and one study found no association. The indices of heavy alcohol intake included: binge-drinking (at least six bottles of beer in one session), frequency of hangovers or intoxication, being registered as a problem drinker (in Sweden), being admitted for alcohol treatment, and being unable to perform at work properly because of problems with alcohol. The relative risk of cardiac death among these heavy drinkers, compared with non-drinkers, ranged from 2 to 6, and seemed to be higher for sudden cardiac death than for myocardial infarction.

An additional cohort study published recently also supports this view. A US-based investigation found that light drinkers who had episodes of heavy drinking (binges) had a mortality almost twice as high as light drinkers without heavy drinking episodes (Rehm *et al.* 2001). This further illustrates the importance of drinking pattern in understanding the effects of drinking.

The findings of the prospective studies are largely (but not entirely) supported by case-control studies and case series (Britton *et al.* 1998). In general, the risk of sudden cardiac death, myocardial infarction, atrial fibrillation, and supraventricular arrhythmia seemed to be increased among heavy drinkers. There was some suggestion that people who died from sudden cardiac death were more likely to have consumed alcohol during the last hours before their death than people who died from other cardiovascular causes.

In the meta-analysis by Corrao *et al.* (2000) mentioned above, it was estimated that daily consumption of 113 g alcohol and more was associated with a risk of coronary heart disease significantly higher than that among abstainers. In the subset of cohort studies, the risk of coronary mortality was significantly increased among those consuming 89 g or more daily. Although most studies included only a small number of subjects consuming high amounts of alcohol regularly, it appears that the protective effect is absent among such drinkers. Heavy drinkers may, in fact, have a higher risk of coronary death than abstainers.

The British Regional Heart Study has specifically addressed the question of whether the risk of sudden cardiac death is related to heavy drinking. The study found that the incidence of sudden cardiac death was about double among heavy drinkers (more than six drinks daily) compared with the rest (Wannamethee and Shaper 1992). The increased incidence of sudden death in heavy drinkers was similar in all social groups, was more pronounced in older men, and was strongest among men with no evidence of pre-existing coronary heart disease. Interestingly, however, there was an inverse relationship between non-sudden cardiac death and alcohol intake. As a result, the combined sudden and non-sudden deaths from heart disease showed an approximately U-shaped association with alcohol.

The elevated risk of coronary death among heavy drinkers may not be universal. Recently published results of a cohort of heavy drinkers in Italy did not find an increased risk of cardiovascular deaths compared with the general population (Cipriani *et al.* 2002). However, there was also no protective effect on cardiovascular deaths, and there was a significantly increased risk of cardiac death from arrhythmias. In the meta-analysis by Corrao *et al.* (2000), the risk of coronary heart disease in heavy drinkers was higher than that of abstainers in non-Mediterranean countries but it was slightly lower than in abstainers in Mediterranean populations. It is possible that the adverse cardiovascular effects of heavy drinking are contingent on the type of alcohol, other factors (such as nutrition), or on the rates of heart disease in the population.

Interestingly, the binge-drinking pattern may also be associated with stroke (Hillbom 1998). A Swedish prospective study of more than 15 000 middle-aged men and women found that the risk of ischaemic (but not haemorrhagic) stroke was increased among men who drank infrequently but substantially. No such association, however, was found in women (Hansagi *et al.* 1995).

### *Biological mechanisms for the adverse effects*

It has been proposed that sustained or episodic heavy drinking may have a different effect on biological risk factors compared with moderate regular drinking. The data, however, provide only weak support for this view. The review by McKee and Britton (1998) mentioned above found that in animal models, heavy alcohol intake did reduce the cardioprotective HDL (high density lipoprotein) cholesterol and increased LDL (low density lipoprotein) cholesterol concentrations, but this was not confirmed by research on humans. There are few data regarding the effect of heavy drinking on blood clotting, but withdrawal may lead to an increased risk of thrombosis (Hillbom *et al.* 1985; Renaud and Ruf 1996). In animal models, high doses of alcohol were found to disrupt the cardiac conducting system and to damage heart muscle. Again, it is not known whether these effects are present at blood concentrations of alcohol that occur in humans. Blood pressure was found to be elevated in the intoxication phase during a binge and in sustained heavy drinkers (Seppa *et al.* 1994; Seppa and Sillanaukee 1999).

There is no general consensus on the cardiovascular effects of sustained heavy or episodic binge-drinking. There is, however, sufficient evidence in the literature to raise this question and to issue a warning that heavy drinking may, in fact, involve an increased risk of cardiovascular death, and to justify more research (Poikolainen 1998; Puddey *et al.* 1999).

## Wine and the heart: is wine more cardioprotective than ethanol?

The evidence summarized above demonstrates that light-to-moderate drinking is associated with a lower risk of coronary heart disease compared with non-drinkers, and that the protective effect appears to be genuine, independent of known biases and confounding factors. The evidence also suggests that heavy drinking does not provide protection against heart disease, and that it may, in fact, be associated with a higher risk compared with non-drinkers. The question remains whether, among moderate drinkers, wine is associated with a higher degree of protection against heart disease compared with other alcoholic beverages. This problem is addressed in this section.

## *The French paradox*

The hypothesis that wine protects against heart disease was originally proposed to explain the comparatively low mortality from coronary heart disease in France despite relatively high levels of known coronary risk factors, such as smoking, high blood pressure, cholesterol, fat intake or obesity (the 'French paradox'). One proposed explanation was that the low rates of coronary heart disease are due to a high intake of wine (particularly red wine, which contains various substances with possible cardioprotective effects). Several studies based on international data on mortality rates and alcohol (wine) intake have supported this interpretation. Mortality from coronary heart disease was, in general, lower in countries with higher per capita intakes of alcohol, and the link with alcohol appears stronger for wine intake than for alcohol (ethanol) in general or for other beverages (LaPorte *et al.* 1980; Criqui and Ringel 1994; Leger *et al.* 2002).

The fact that mortality from heart disease is low and wine intake is high in France does not, of course, prove that wine reduces heart disease mortality. This is a purely ecological (geographical) association based on comparing countries rather than individuals. All sorts of biases and confounders cannot be controlled in such analyses. For example, identification and coding of different causes of death may vary by country; socioeconomic characteristics must be taken into account; health behaviour patterns are different in different regions; dietary habits vary markedly; medical care also differs between countries, etc. There is, however, a more fundamental problem with this type of 'evidence': the so-called 'ecological fallacy' (Piantadosi *et al.* 1988). This bias is specific to ecological studies, and in principle it means that we cannot be sure that *individuals* who have a low risk of heart disease in countries with high wine intake are those who consume a lot of wine. In other words, these studies imply an extrapolation from populations (countries) to individuals but since there are no data on individuals this extrapolation is necessarily speculative. It needs to be confirmed by studies in individuals.

## *Studies in individuals*

There have been very few studies directly comparing the effect of different types of alcoholic beverages on the risk of coronary heart disease. In the mid-1990s, Rimm and colleagues reviewed the literature and addressed this question (Rimm *et al.* 1996). They found clear evidence of a strong inverse relationship between moderate total alcohol intake and coronary heart disease, but there was no consistent pattern of differential effects of different beverages. Out of 10 prospective studies, four found a significant protective effect of wine, four found this association with beer, and four found it for spirits. It appeared that the maximal protective effect was related to the beverage most frequently consumed in a given population, suggesting that it is the usual (not-excessive and regular) drinking that is protective, rather than a specific beverage. The authors concluded that if there were any extra cardioprotective effect of any specific beverage, it would be 'modest at best or possibly restricted to a certain sub-population'.

Two studies in particular provided some support for the wine hypothesis. In a prospective study of almost 130 000 US adults, Klatsky and Armstrong (1993) investigated whether different alcoholic beverages confer a varying risk of heart disease. The results suggested a tendency towards a lower risk of heart disease among those who preferred wine than among those who preferred beer or spirits. However, in later analyses of hospitalization for heart disease in the same cohort, the lowest risk was found among men preferring beer and (but not statistically significantly) women preferring wine (Klatsky *et al.* 1997). Red wine was no more protective than other types of wine.

The Copenhagen City Heart Study also examined directly the hypothesis that wine may be associated with a protective effect against heart disease that is greater than that provided by other alcoholic beverages (Gronbaek *et al.* 2000). In this study, more than 13 000 men and 11 500 women were followed for an average of 10.5 years. The study found an L-shaped association between total alcohol intake and mortality from heart disease. However, when subjects were divided into wine-drinkers and non-wine-drinkers, light drinkers who avoided wine had a 24% lower risk of death from heart disease than non-drinkers while light drinkers who drank wine had 42% lower mortality. Compared with subjects who never drank wine, those who drank three to five glasses of wine per day had half of the risk of death from all causes (Gronbaek *et al.* 2000).

There is a potential measurement error here. Few people consume strictly one type of alcohol only. Most people drink several types of alcohol during a typical week (and often during one day). Classifying them into wine- or non-wine-drinkers is clearly a simplification. It is even more difficult to exclude confounding by other risk factors. This is discussed in the next subsection.

Theoretically, it is also possible to compare studies in different populations. If the protective effect were specific to wine only, studies in non-wine-drinking populations should not find any protective effect of alcohol. This is clearly not the case. For example, studies in beer-drinking populations found a clear U-shaped association between alcohol (beer) intake and heart disease (Keil *et al.* 1997; Bobak *et al.* 2000; Brenner *et al.* 2001). The magnitude of the effect was entirely consistent with studies in other populations, including wine-drinking populations. A more detailed quantitative comparison of the size of the protective effects between studies in different populations (wine- and non-wine-drinking) would be problematic, for reasons described above for ecological studies, in addition to other important issues related to poor comparability between studies (study design, selection of subjects, alcohol measurement, measurement of other risk factors and other covariates, ascertainment of coronary heart disease, etc.).

### *Problems of studies in individuals*

Studies in individuals – such as the Kaiser Permanente or the Copenhagen City Heart Study – are more reliable than ecological studies. Nevertheless, they are also prone to methodological problems that are illustrated here.

One can start with a question about an ideal study to examine the differential effects of different type of alcoholic beverages on heart disease. Such a study would be a randomized trial, where subjects would be randomly assigned to consumption of a standard dose of different types of alcohol. Unfortunately, such a trial is impossible, not only for ethical reasons, but also because people have different preferences for alcoholic beverages and would probably not comply with the allocated 'intervention'. The second best design would be an observational study in a population where preferences for different types of alcohol (e.g. wine, beer, spirits) would be distributed randomly. And this illustrates the problem: the preferences for alcoholic beverages are not distributed randomly in a given population, at least not in most Western populations.

In the Whitehall II Study of British civil servants, for example, wine-drinkers had higher social status, higher income, higher education, more favourable diet, and a more favourable risk profile than those who consumed predominantly beer or spirits (unpublished data). Analysis of the Copenhagen City Heart Study showed that wine-drinkers were less likely to smoke, had higher education and lower body mass index than beer and spirits drinkers (Gronbaek *et al.* 2000). Another large Danish population study found a similar pattern: wine-drinkers had a more healthy diet and risk-factor profile (Tjonneland *et al.* 1999). In the cohort analysed by Klatsky and Armstrong (1993), wine-preferrers were more often non-smokers, college graduates, and had a more favourable coronary-risk profile. All these factors are also related to heart disease, and can confound the comparison between wine and other beverages (Klatsky 1999). For example, if wine-drinkers are less likely to smoke than beer-drinkers, one would expect a lower risk of heart disease among wine-drinkers just because they smoke less. It is possible to take such confounding factors into account and control, or adjust for them, usually in a statistical analysis. However, it is very difficult to remove all confounding, because it is virtually impossible to measure all potential confounding factors sufficiently precisely (Klatsky 1999). Thus it is plausible that the apparently greater protective effect of wine compared with other beverages is due to residual confounding.

## Conclusions

There is clear evidence that mortality from coronary heart disease is some 20–30% lower among moderate regular drinkers compared with non-drinkers. This reduction of coronary risk appears to be due to known biological effects of ethanol, largely on blood lipids and blood coagulation.

There is little evidence that wine provides a greater protection against heart disease than other alcoholic beverages. It is likely that the apparently beneficial effect of wine compared with spirits or beer found in a few studies is due to residual confounding by other factors. The much publicized beneficial biological effects of substances in red wine (polyphenols, antioxidants, etc.) found in cross-sectional studies and in animal or *in vitro* experiments do not provide definite evidence that wine is more beneficial than other types of alcohol. Existence of biological mechanisms (plausibility) is a weak criterion of causality. It has been proposed that beer also contains additional substances that can confer extra protection against heart disease (Ubbink *et al.* 1998; van der Gaag *et al.* 2000) but the researchers remain cautious and the media have not exaggerated these reports.

Finally, there is growing evidence that heavy drinking is not associated with lower rates of coronary heart disease; in fact, recent studies suggest that heavy drinking may increase the risk of cardiac death. While it is not yet clear whether heavy wine-drinking has the same adverse effects – and the meta-analysis of Corrao *et al.* (2000) suggests that this may not be the case – it is unlikely that it does confer the same protection as moderate drinking.

## References


Albert, C. M., Manson, J. E., Cook, N. R., Ajani, U. A., Gaziano, J. M., Hennekens, C. H. (1999) Moderate alcohol consumption and the risk of sudden cardiac death among US male physicians. *Circulation*: **100**, 944–50.

Andreasson, S., Allebeck, P., Romelsjo, A. (1988) Alcohol and mortality among young men: longitudinal study of Swedish conscripts. *British Medical Journal*: **296**, 1021–5.

Bobak, M., Skodova, Z., Marmot, M. (2000) Effect of beer drinking on risk of myocardial infarction: a population based case-control study. *British Medical Journal*: **320**, 1378–9.

Brenner, H., Rothenbacher, D., Bode, G., Marz, W., Hoffmeister, A., Koenig, W. (2001) Coronary heart disease risk reduction in a predominantly beer-drinking population. *Epidemiology*: 12, 390–5.

Britton, A. and McKee, M. (2000) The relation between alcohol and cardiovascular disease in Eastern Europe: explaining the paradox. *Journal of Epidemiology and Community Health*: **54**, 328–32.

Britton, A., McKee, M., Leon, D. A. (1998) *Cardiovascular disease and heavy drinking: a systematic review. A report to the Health and Population Division of the United Kingdom Department of International Development*. London School of Hygiene and Tropical Medicine, London.

Cipriani, F., Cucinelli, M. L., Dimauro, P. E., Angioli, D., Conte, M., Voller, F., Buiatti, E. (2002) Mortality in a cohort of alcoholics from Arezzo in 1979–1997. *Epidemiologia e Prevenzione*: **25**, 63–70.

Corrao, G., Rubiatti, L., Bagnardi, V., Zambon, A., Poikolainen, K. (2000) Alcohol and coronary heart disease: a meta-analysis. *Addiction*: **95**, 1505–23.

Criqui, M. H. and Ringel, B. L. (1994) Does diet or alcohol explain the French paradox? *Lancet*: **344**, 1719–23.

Doll, R. (1997) One for the heart. *British Medical Journal*: **315**, 1664–8.

Edwards, G., Anderson, P., Babor, T. F., Casswell, S. Ferrence, R., Giesbrecht, C., Godfrey, C., Holder, H. D., Lemmens, P., Makela, K., Midanik, L. T., Novstrom, T., Osterberg, A., Room, R., Simpura, J., Skog, O. J. (1994) *Alcohol Policy and the Public Good* Oxford University Press, Oxford.

Fagrell, B., de Faire, U., Bondy, S., Criqui, M., Gaziano, M., Gronbaek, M., Jackson, R., Klatsky, A., Salonen, J., Shaper, A. G. (1999) The effects of light to moderate drinking on cardiovascular diseases. *Journal of Internal Medicine*: **246**, 331–40.

Fuchs, C. S., Stampfer, M. J., Colditz, G. A., Giovannucci, E. L., Manson, J. E., Kawachi, I., Hunter, D. J., Hankinson, S. E., Hennekens, C. H., Rosner, B. (1995) Alcohol consumption and mortality among women. *New England Journal of Medicine*: **332**, 1245–50.

Goldberg, I. J., Mosca, L., Piano, M. R., Fisher, E. A. (2001) Wine and your heart: A Science Advisory for Healthcare Professionals from the Nutrition Committee, Council on Epidemiology and Prevention, and Council on Cardiovascular Nursing of the American Heart Association. *Circulation*: **103**, 472–5.

Grant, M. and Litvak, J. (1998) *Drinking patterns and their consequences*. Taylor & Francis, Washington, DC.

Gronbaek, M. (2001) Factors influencing the relation between alcohol and mortality – with focus on wine. *Journal of Internal Medicine*: **250**, 291–308.

Gronbaek, M., Becker, U., Johansen, D., Gottschau, A., Schnohr, P., Hein, H. O., Jensen, G., Sorensen, T. I. A. (2000) Types of alcohol consumed and mortality from all causes, coronary heart disease and cancer. *Annals of Internal Medicine*: **133**, 411–19.

Hansagi, H., Romelsjo, A., Gerhardsson de Verdier, M., Andreasson, S., Leifman, A. (1995) Alcohol consumption and stroke mortality. 20-year follow-up of 15 077 men and women. *Stroke*: **26**, 1768–73.

Hillbom, M. (1998) Alcohol consumption and stroke: benefits and risks. *Alcoholism: Clinical and Experimental Research*: **22 (suppl.)**, 352S–358S.

Hillbom, M., Kangasaho, M., Lowbeer, C., Kaste, M., Muuronen, A., Numminen, H. (1985) Effect of ethanol on platelet function. *Alcohol*: **2**, 429–32.

Huikuri, H. V., Castellanos, A., Myerburg, R. J. (2001) Sudden death due to cardiac arrhythmias. *New England Journal of Medicine*: **345**, 1473–82.

Kauhanen, J., Kaplan, G. A., Goldberg, D. E., Salonen, J. T. (1997) Beer binging and mortality: results from the Kuopio ischaemic heart disease risk factors study, a prospective population based study. *British Medical Journal*: **315**, 846–51.

Keil, U., Chambless, L. E., Doering, A., Filipiak, B., Stieber, J. (1997) The relation of alcohol intake to coronary heart disease and all-cause mortality in a beer-drinking population. *Epidemiology*: **8**, 150–6.

Klatsky, A. L. (1999) Is it the drink or the drinker? Circumstantial evidence only raises a probability. *American Journal of Clinical Nutrition*: **69**, 2–3.

Klatsky, A. L. (2001) Diet, alcohol, and health: a story of connections, confounders, and cofactors (editorial). *American Journal of Clinical Nutrition*: **74**, 279–80.

Klatsky, A. L. and Armstrong, M. A. (1993) Alcoholic beverage choice and risk of coronary artery disease mortality: do red wine drinkers fare best? *American Journal of Cardiology*: **71**, 467–9.

Klatsky, A. L., Armstrong, M. A., Friedman, G. D. (1997) Red wine, white wine, liquor, beer, and risk of coronary artery disease hospitalization. *American Journal of Cardiology*: **80**, 416–20.

LaPorte, R. E., Cresanta, J. L., Kuller, L. H. (1980) The relationship of alcohol consumption to atherosclerotic heart disease. *Preventive Medicine*: **9**, 22–40.

Leger, A. S., Cochrane, A. L., Moore, F. (2002) Factors associated with cardiac mortality in developed contries, with particular reference to the consumption of wine. *Lancet*: **i**, 1017–20.

Leon, D. A., Chenet, L., Shkolnikov, V., Zakharov, S., Shapiro, J., Rakhmanova, G., Vassin, S., McKee, M. (1997) Huge variation in Russian mortality rates 1984–94: artefact, alcohol, or what? *Lancet*: **350**, 383–8.

Marmot, M. G. (1984) Alcohol and coronary heart disease. *International Journal of Epidemiology*: **13**, 160–7.

Marmot, M. and Brunner, E. (1991) Alcohol and cardiovascular disease: the status of the U shaped curve. *British Medical Journal*: **303**, 565–8.

McKee, M. and Britton, A. (1998) The positive relationship between alcohol and heart disease in eastern Europe: potential physiological mechanisms. *Journal of the Royal Society of Medicine*: **91**, 402–7.

Murray, C. J. L. and Lopez, A. D. (1996) Quantifying the burden of disease and injury attributable to ten major risk factors, in *The global burden of disease* (eds C. J. L. Murray and A. D. Lopez). WHO, Geneva, pp. 295–324.

Piantadosi, S., Byar, D. P., Green, S. B. (1988) The ecological fallacy. *American Journal of Epidemiology*: **127**, 893–904.

Poikolainen, K. (1998), It can be bad for the heart, too – drinking patterns and coronary heart disease (editorial). *Addiction*: **93**, 1757–9.

Puddey, I. B., Rakic, V., Dimmitt, S. B., Beilin, L. J. (1999) Influence of pattern of drinking on cardiovascular disease and cardiovascular risk factors: a review. *Addiction*: **94**, 649–63.

Rehm, J., Greenfield, T. K., Rogers, J. D. (2001) Average volume of alcohol consumption, patterns of drinking, and all-cause mortality: results from the US National Alcohol Study. *American Journal of Epidemiology*: **153**, 64–71.

Renaud, S. C. and Ruf, J. C. (1996) Effects of alcohol on platelet functions. *Clinica Chimica Acta*: **246**, 77–89.

Rimm, E. B., Klatsky, A., Grobbee, D., Stampfer, M. J. (1996) Review of moderate alcohol consumption and reduced risk of coronary heart disease: is the effect due to beer, wine, or spirits? *British Medical Journal*: **312**, 731–6.

Rimm, E. B., Williams, P., Fosher, K., Criqui, M., Stampfer, M. J. (1999) Moderate alcohol intake and lower risk of coronary heart disease: meta-analysis of effects on lipids and haemostatic factors. *British Medical Journal*: **319**, 1523–8.

Royal College of Physicians, Royal College of Psychiatrists, and Royal College of General Practitioners (1995) *Alcohol and the heart in perspective. Sensible limits reaffirmed. Report of a working group*. Royal College of Physicians, London.

Seppa, K. and Sillanaukee, P. (1999) Binge drinking and ambulatory blood pressure. *Hypertension*: **33**, 79–82.

Seppa, K., Laippala, P., Sillanaukee, P. (1994) Drinking pattern and blood pressure. *American Journal of Hypertension*: **7**, 249–54.

Thun, M. J., Peto, R., Lopez, A. D., Monaco, J. H., Henley, J., Heath, C. W., Doll, R. (1997) Alcohol consumption and mortality among middle-aged and elderly US adults. *New England Journal of Medicine*: **337**, 1705–14.

Tjonneland, A., Gronbaek, M., Stripp, C., Overvad, K. (1999) Wine intake and diet in a random sample of 48 763 Danish men and women. *American Journal of Clinical Nutrition*: **69**, 49–54.

Ubbink, J. B., Fehily, A. M., Pickering, J., Elwood, P. C., Vermaak, W. J. (1998) Homocysteine and ischaemic heart disease in the Caerphilly cohort. *Atherosclerosis*: **140**, 349–56.

van der Gaag, M. S., Ubbink, J. B., Sillanaukee, P., Nikkari, S., Hendriks, H. F. (2000) Effect of consumption of red wine, spirits, and beer on serum homocysteine. *Lancet*: **356**, 512.

Wannamethee, G. and Shaper, A. G. (1992) Alcohol and sudden cardiac death. *British Heart Journal*: **68**, 443–8.

# 6 Wine, alcohol and cardiovascular diseases

*A. L. Klatsky*

## Introduction

Since scientific evidence for the 'cardioprotective' effects of light-to-moderate alcohol drinking started to appear several decades ago, great interest in the concept has developed. Fascination by the scientific community and the general public is easy to understand. First, there is personal involvement, as most people in developed countries are light-to-moderate alcohol-drinkers who drink for sensory pleasure. Second, is a puritanical cultural context in many countries, which causes many to have slight twinges of guilt; this leads to a special delight at hearing that there are possible health benefits to their behaviour. Third, is the novelty of hearing about benefit from a pleasurable personal habit, since there are many with adverse effects. Fourth, is the converse of the second reason offered: there are many who enjoy rebutting those who equate pleasure with sin, and medical benefit from light drinking provides an effective basis for such refutation. Finally, of course, is scientific interest.

Possible differences between the beverage types have become a major secondary focus of interest. Spurred on by the catchiness of the 'French paradox' term, and by intense media attention, the idea that wine, particularly red wine, is *the* beneficial alcoholic beverage has caught on with the public. It is possible that the beverage preferences of scientific investigators help to popularize this idea. Non-alcohol potentially beneficial substances in wine have stimulated much research. This area is in constant need of objective evaluation to assess if the evidence that preferential beneficial health effects of specific beverages is still largely hypothetical.

The term 'cardioprotective' and reference to the benefits of light–moderate drinking for 'cardiovascular disease' mask the diversity of alcohol relations to cardiovascular diseases. There are disparities that pertain both to the various cardiovascular conditions and to differences between light-moderate and heavy drinking. The beneficial effects apply primarily to the atherothrombotic conditions, coronary heart disease (CHD), ischaemic stroke and peripheral vascular disease. People with CHD probably represent the majority of all those with cardiovascular disease (CVD). Thus, CHD dominates the statistics for all CVD, i.e. the relationship of alcohol (or any other trait) to CHD and CVD are automatically somewhat similar. But it is a given that scientists should avoid sloppy generalizations and should be as accurate and specific as possible.

This chapter will first review adverse and beneficial relations of both light–moderate and heavy alcohol drinking to several non-atherothrombotic cardiovascular conditions. Then an overview of relations to CHD, stroke and peripheral arterial disease will be presented.

## Operational definition of moderate and heavy drinking

While all definitions are arbitrary, there is some consensus about a boundary at daily consumption of more than two standard-sized drinks. Above this drinking level net harm is usually seen in survey-based epidemiological studies. Sex, age, and individual factors lower and raise the boundary for individuals. There is some tendency to lower the boundary of light–moderate intake in women to one drink per day, but it is not clear that data about harm support this.

Because heavier drinkers often lie about their alcohol intake, survey-based data usually lower the apparent threshold for harmful alcohol effects. This is best understood by examining as an example alcoholic liver cirrhosis, a condition considered to be related to heavy drinking but not to light drinking. Consider a study of alcoholic liver cirrhosis risk in a population in relation to the amount of drinking reported on a survey. If some proportion of heavy drinkers report light drinking, light drinking will erroneously appear to be related to alcoholic cirrhosis.

Another related categorization problem is that queries on surveys usually require averaging. A person who takes one drink each day has a weekly total of seven drinks. Another person taking seven drinks on Saturday night and none from Sunday to Friday also takes the same weekly total of seven drinks. However, these patterns may have quite different health implications. The amount of alcohol is approximately the same in a standard-sized drink of each major beverage type. Thus, a 4 oz glass of table wine (120 ml @ 12.5% alcohol) = 15 ml alcohol; 1¼ oz of distilled spirits (37.5 ml @ 40% alcohol) = 15 ml alcohol; and 12 oz of US beer (360 ml @ 4% alcohol) = 14.4 ml alcohol. A 'standard drink' thus comprises 15 ml of ethyl alcohol or approximately 12.5 g. Since most people think in terms of 'drinks' rather than millilitres or grams of alcohol, it may best enhance communication to describe alcohol relations by daily or weekly 'standard' drinks. Health professionals need to remember the importance of defining the size of drinks.

## Disparities in alcohol–cardiovascular disease relations

The major disparities in relations between drinking alcoholic beverages and various cardiovascular (CV) conditions (Davidson 1989; Klatsky 1995) make it desirable to consider several disorders separately. It is relevant to include a historical review because of past diversions in understanding alcohol–CV relationships. For several reasons, especially to emphasize the inappropriateness of generalizing the effects of heavy drinking to all 'alcohol use', it is crucial to evaluate separately the roles of light and heavy drinking in each of several conditions. The following areas will be considered.

- *Alcoholic cardiomyopathy*. Although recognized 150 years ago, understanding of alcoholic cardiomyopathy was clouded by confusion with beriberi and attribution of toxicity from alcohol to arsenic or cobalt.
- *Systemic hypertension (HTN)*. A report of a link between heavy drinking and HTN in French soldiers in World War I was apparently ignored for more than 50 years. Epidemiological and intervention studies have now firmly established this association, but a mechanism remains elusive.
- *Heart rhythm disturbances*. An increased risk of supraventricular rhythm disturbances in binge drinkers, the 'holiday heart syndrome', has been recognized by clinicians for 25 years. Data remain sparse about the total role of heavier drinking in cardiac rhythm disturbances.
- *Cerebrovascular disease (stroke)*. Failure of earlier studies to distinguish types of stroke impeded understanding; it now seems probable that alcohol-drinking increases risk of haemorrhagic stroke and lowers risk of ischaemic stroke.
- *Atherosclerotic coronary heart disease (CHD)*. In 1786 William Heberden reported angina pectoris relief by alcohol, and pathologists observed an inverse alcohol–atherosclerosis association in the early 1900s. Recent population studies and plausible mechanisms support a protective effect of alcohol drinking against CHD. International comparisons dating back to 1819 suggest beverage choice as a factor, but this issue remains incompletely resolved.

## Alcoholic cardiomyopathy

### *Definition*

Definitions vary, but many, including this author, use the term 'cardiomyopathy' (CM) to include heart muscle diseases independent of the valves, coronary arteries, pericardium and congenital malformations. There are multiple known causes of *cardiomyopathies*, but a large proportion of cases are idiopathic, with genetic and viral factors suspected. Sustained heavy alcohol drinking is believed (Kasper *et al.* 1994) to be one of the causes of *dilated cardiomyopathy*, a type of cardiomyopathy characterized by an enlarged, weakened, poorly contracting heart. The spectrum of dilated cardiomyopathy ranges from abnormalities detectable only by testing ('subclinical'), to severe illness with heart failure and high mortality rate. CHD has no direct relation to alcoholic cardiomyopathy.

### *History*

An apparent relation between chronic intake of large amounts of alcohol and heart disease was noted by several clinicians and pathologists in the nineteenth century (Friedreich 1861; Walsche 1873; Strumpel 1890; Steell 1893; Osler 1899). A German pathologist (Böllinger 1884) described the 'Münchener Bierherz' among Bavarian beer drinkers, who reportedly averaged 432 litres per year. In 1900, an epidemic of heart disease due to arsenic-contaminated beer occurred in Manchester, UK. This led to the confusion of arsenic toxicity with damage from alcohol (Steell 1906). The great cardiologist William MacKenzie (MacKenzie 1902) described cases of heart failure attributed to alcohol and first used the term 'alcoholic heart disease'. Early in the twentieth century, there was general doubt that alcohol had a direct role in producing heart muscle disease, although some (Vaquez 1921) took a strong view in favour of such a relationship. After the appearance of detailed descriptions of cardiovascular beriberi (Aalsmeer and Wenckebach 1929; Keefer 1930), the concept of 'beriberi heart disease' dominated thinking about the effects of alcohol on the heart for several decades (see below).

### *Evidence*

The concept of the direct toxicity of alcohol on the myocardial cells and the existence of alcoholic CM has now become solidly established (Evans 1961; Sanders 1963; Alexander 1966a,b; Ferrans 1966; Gould *et al.* 1969; Urbano-Marquez *et al.* 1989; Moushmoush and Abi-Mansour 1991; Richardson *et al.* 1998). Varying proportions of chronic heavy alcohol users have been reported in clinical series, probably dependent mostly on the drinking habits of the study population. The absence of diagnostic tests has been a major impediment to epidemiological study, since the entity has been indistinguishable from other forms of dilated cardiomyopathy. The proportion of heavy drinkers who develop cardiomyopathy is not known, but is smaller than the 18–20% proportion that develop liver cirrhosis. Also unknown is the proportion that demonstrates regression with abstinence, but data supporting such regression exist (Demakis *et al.* 1974; Ballester *et al.* 1997). The most convincing evidence for alcoholic cardiomyopathy consists of non-specific functional and structural abnormalities related to alcohol (Alexander 1966a,b; Regan *et al.* 1966; Lochner *et al.* 1969; Regan *et al.* 1969; Bulloch *et al.* 1972; Mathews *et al.* 1981; Regan 1984; Bertolet *et al.* 1991; Gillet *et al.* 1992; Thomas *et al.* 1994). Subclinical abnormalities of function and structure may precede evident illness for years.

An important study (Urbano-Marquez *et al.* 1989) showed a clear relation in alcoholics of lifetime alcohol consumption to structural and functional myocardial and skeletal muscle abnormalities. The large amounts of alcohol needed – equivalent to 120 g alcohol/day for 20 years – make the term 'cirrhosis of the heart' (Walsche 1873) appropriate. Another study (Ballester *et al.* 1997) using indium-111 labelled monoclonal antimyosin antibodies confirmed prior clinical observations that suggested improvement in myocardial damage with abstinence and variability in myocardial susceptibility to alcohol.

A possible non-oxidative metabolic pathway for alcohol has been reported (Lapasota and Lange 1986) in the heart, muscle, pancreas and brain, related to fatty acid metabolism. Fatty acid ethyl ester accumulation was related to blood alcohol levels and mitochondrial metabolism. Other increased enzymatic activity in myocardial cells has also been reported, including alpha-hydroxybutyric dehydrogenase, creatine kinase, lactic dehydrogenase and malic dehydrogenase (Richardson *et al.* 1998). It is not clear whether the reported enzymatic activity reflects causative processes or an adaptive reaction. The histological findings include evidence of inflammation, lipid deposits, focal or diffuse fibrosis, and mitochondrial damage (Alexander 1966b; Burch *et al.* 1971; Bulloch *et al.* 1972; Urbano-Marquez *et al.* 1989). The histology has not been generally considered sufficiently characteristic for specific diagnosis, although some (Richardson *et al.* 1998) believe that hypertrophy, fibrosis and cell nuclear disruption are greater in alcoholic cardiomyopathy than in dilated cardiomyopathy from other causes. The diagnosis depends upon the combination of heart muscle disease without other evident cause and a compatible alcohol-drinking history. The key aspect of the latter is the large amount of lifetime intake involved; there is no evidence that moderate or intermediate amounts of drinking can cause cardiomyopathy. Only amount of alcohol, not choice of wine, spirits or beer seem to be crucial. Onset may be insidious, but sometimes seems subacute. Early manifestations are non-specific electrocardiographic findings and rhythm disturbances. The late picture includes heart failure, chronic rhythm disturbances, conduction abnormalities, systemic emboli and death (Burch *et al.* 1966; Regan 1984).

### *Alcohol and heart failure*

It is clear that chronic very substantial alcohol intake can lead to heart failure via cardiomyopathy. Since a diagnosis of alcoholic cardiomyopathy requires exclusion of other types of heart disease, the possible role of alcohol as a contributing factor remains less known but of concern. It seems plausible that amounts of drinking that are substantially less than needed to produce cardiomyopathy might act in concert with other conditions or cofactors to cause heart muscle dysfunction. Actually, the limited evidence available (Cooper *et al.* 2000; Abramson *et al.* 2001; Klatsky 2001) suggests that light–moderate drinking not only does not increase heart failure risk, but that risk is decreased, possibly independent of CHD.

### *Possible cofactors with alcohol in cardiomyopathy*

In connection with susceptibility to cardiomyopathy, it seems appropriate to consider further the arsenic and cobalt beer-drinker episodes and beriberi heart disease due to thiamine (cocarboxylase or Vitamin B1) deficiency.

- *Arsenic-beer-drinkers' disease* refers to a localized epidemic in Manchester, UK, in 1900, which proved to be due to contamination of beer by arsenic with prominent cardiovascular manifestations, especially heart failure (Reynolds 1901; Gowers 1901; Royal Commission Appointed to Inquire into Arsenical Poisoning from the Consumption of Beer and other Articles of Feed or Drink 1903). It was determined that the affected beer had 2–4 parts per million of arsenic, in itself not an amount likely to cause serious toxicity (Gowers 1901), and that some people seemed to have a 'peculiar idiosyncrasy' (Reynolds 1901).
- *Cobalt-beer-drinkers' disease*, recognized 65 years after the arsenic-beer episode, was similar in some respects (Alexander 1969). In the mid-1960s, reports appeared of heart failure epidemics among beer drinkers in Omaha and Minneapolis in the USA, Quebec, in Canada, and Leuven, in Belgium. The explanation proved to be the addition of small amounts of cobalt chloride by certain breweries to improve the foaming qualities of beer (Morin and Daniel, 1967). Because of the location of the investigators who solved the puzzle, the condition became known as Quebec beer-drinkers' cardiomyopathy. Removal of the cobalt additive ended the epidemic in all locations. It was established that both cobalt and substantial amounts of alcohol seemed to be needed to produce this condition. Biochemical mechanisms were not established. The arsenic and cobalt episodes raise the possibility of other cofactors in alcoholic cardiomyopathy, but none has been established.
- *Cardiovascular beriberi* dominated thinking about alcohol and cardiovascular disease for many years. The classical description (Aalsmeer and Wenckebach 1929) defined high-output heart failure in Javanese polished rice-eaters. It became assumed that heart failure in heavy alcohol drinkers in the West was due to associated nutritional deficiency states. But most heart failure cases in North American and European alcoholics did not fit this pattern. They had low output heart failure, were well-nourished, and responded poorly to thiamine (Blankenhorn 1945; Blankenhorn *et al.* 1946). Blacket and Palmer (1960) clarified the situation: 'It [beriberi] responds completely to thiamine, but merges imperceptibly into another disease, called alcoholic cardiomyopathy, which doesn't respond to thiamine.' It is evident that many cases previously called 'cardiovascular beriberi' would now be called 'alcoholic heart disease'.

## Hypertension (HTN)

### *Background*

First reported in 1915 in middle-aged French servicemen (Lian, 1915), an association between heavy drinking and HTN was then ignored for more than 50 years before receiving further attention (Clark *et al.* 1967; Dyer *et al.* 1977; Klatsky *et al.* 1977). Many cross-sectional and prospective epidemiological studies reported since the mid-1970s have solidly established an empiric alcohol–HTN link (MacMahon 1987; Keil *et al.* 1993; Klatsky 2000). Clinical experiments have been confirmatory. So far, a mechanism has not been demonstrated. The evidence is sufficient that clinicians should consider heavy alcohol-drinking to be a probable HTN risk factor. The importance of this relation is due to the substantial prevalence of HTN and to the fact that a relation of increased blood pressure to moderate alcohol drinking is considered unresolved by some.

### *Epidemiological studies*

Dozens of cross-sectional studies almost unanimously show higher mean blood pressures and/or higher HTN prevalence with increasing alcohol-drinking. Reviews (MacMahon 1987; Keil *et al.* 1993; Klatsky 1995, 2000) detail this observation in North American, European, Australian and Japanese populations and show its independence from adiposity, salt intake, education, cigarette smoking, and several other potential confounders. Most studies show no increase in blood pressure at light-moderate alcohol drinking, but there are some that raise the question of a continuous relation at all drinking levels to higher blood pressure. This may be an example of a spuriously low threshold for harm resulting from systematic underreporting of drinking. Several studies show an unexplained J-shaped curve in women (Klatsky *et al.* 1977; Criqui *et al.* 1981; Witteman *et al.* 1990), with lowest pressures in lighter drinkers. An early Kaiser Permanente study (Figure 6.1) (Klatsky *et al.* 1977; Criqui *et al.* 1981; Witteman *et al.* 1990) shows these relationships in the two sexes in each of three racial groups. A later Kaiser Permanente study (Klatsky *et al.* 1986a) again showed a J-curve in women, but a continuous relationship in men starting at one to two drinks per day. In this later study, the data showed that ex-drinkers had similar blood pressures to those of non-drinkers, and that elevated blood pressures regressed within a week upon abstinence from alcohol. In both studies, HTN prevalence was approximately doubled among the heaviest drinkers (≥ 6 drinks daily), compared with abstainers or light drinkers. Data from prospective studies (Dyer *et al.* 1977; Witteman *et al.* 1990; Gordon and Kannel 1983; Friedman *et al.* 1988; Ascherio *et al.* 1992) show higher risk of HTN development among heavier alcohol drinkers. At least two of them (Witteman *et al.* 1990; Ascherio *et al.* 1992) were well-controlled for multiple nutritional factors.

### *Clinical experiments*

A landmark study (Potter and Beevers 1984) showed in hospitalized hypertensive men that three to four days of an intake of four pints of beer raised blood pressure and that three to four days of abstinence resulted in lower pressures. A 12-week crossover design trial (Puddey *et al.* 1985) showed similar results in ambulatory normotensives, and the observation was later

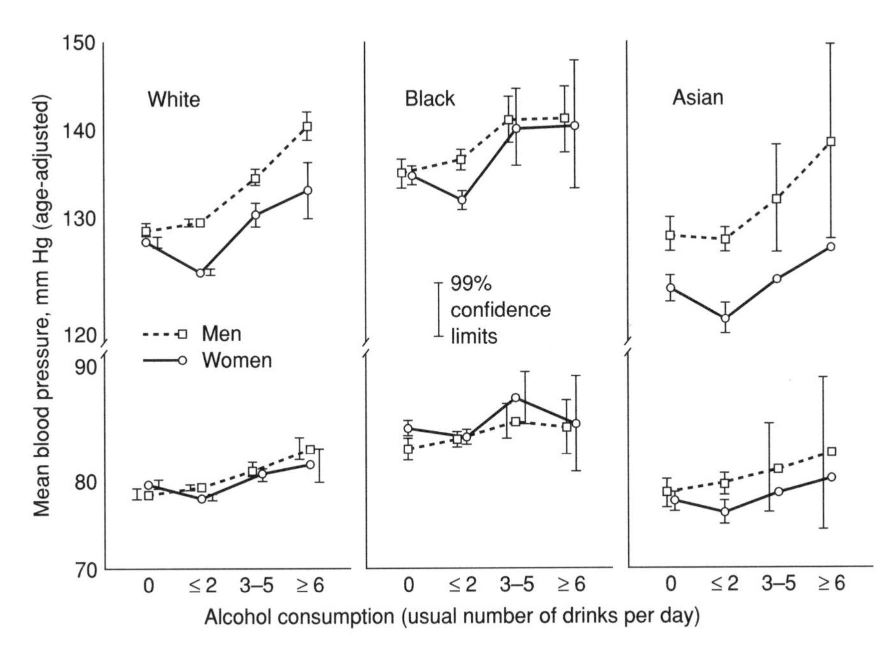


*Figure 6.1* Mean systolic blood pressures (upper half) and mean diastolic blood pressures (lower half) for white, black or Asian men and women with known drinking habits. Small circles represent data based on fewer than 30 people. (From Klatsky *et al.* 1977. Used by permission.)

confirmed in hypertensives (Puddey *et al.* 1987). Other studies show that heavier alcohol intake interferes with drug treatment of HTN (Beevers 1990) and that moderation or avoidance of alcohol supplements or improves on other non-pharmacological interventions for blood pressure-lowering, such as weight reduction (Puddey *et al.* 1990), exercise (Cox *et al.* 1990) or sodium restriction (Parker *et al.* 1990). The clinical experiments suggest neither that light–moderate drinking raises blood pressure nor that alcohol withdrawal – at the drinking levels studied – is responsible for the alcohol-associated increased blood pressures.

### *Possible mechanisms*

The alcohol–HTN relation is a subacute one, developing in days to weeks (Potter and Beevers 1984; Puddey *et al.* 1985, 1987, 1990; Beevers 1990; Cox *et al.* 1990; Parker *et al.* 1990). Acute human and animal experiments show no consistent immediate increase in blood pressure from alcohol administration (Grollman 1930; MacMahon 1987; Keil *et al.* 1993; Klatsky 2000). Ambulatory monitoring has shown a depressor effect of a substantial dinner-time alcohol dose, lasting up to eight hours, with a pressor effect the next morning (Kawano *et al.* 1992). Much work has failed to uncover a biological mechanism. There seem to be no consistent relationships to plasma renin, aldosterone, cortisol, catecholamines or insulin (Arkwright *et al.* 1983; MacMahon 1987; Potter and Beevers 1991; Keil *et al.* 1993; Kojima *et al.* 1993; Klatsky 2000). Experiments suggest independence from acetaldehyde-induced flushing (Kawano *et al.* 1992). Speculations raise the possibility of changes in intracellular sodium metabolism (Coca *et al.* 1992; Kojima *et al.* 1993), a direct smooth muscle effect via calcium transport mechanisms, impaired insulin sensitivity, impaired baroreflex activity, magnesium depletion, and a heightened responsiveness of the sympathetic nervous system (Coca *et al.* 1992; Randin *et al.* 1995). An overactive sympathetic nervous system exists during the alcohol withdrawal state, but, because the clinical experiments do not show a blood pressure rise after alcohol cessation, this is not the likely explanation for the alcohol–HTN relation.

### *Sequelae of alcohol-associated HTN*

Complex interactions of alcohol, various cardiovascular conditions, and risk factors make the study of this important subject difficult. Since CHD and strokes are major sequelae of HTN, the lower risk of CHD and ischaemic stroke among drinkers confounds study of this aspect. A counterbalancing role of HTN has been observed in two alcohol–CHD studies (Criqui 1987; Langer *et al.* 1992). An attempt to study whether alcohol-associated HTN had the same prognosis as HTN not so associated led to the conclusion that alcohol's harmful and beneficial effects so dominated the outcome that the basic question could not be answered (Klatsky 2000).

### *Interpretive problems and conclusions*

A satisfactory long-term clinical trial of alcohol, HTN, and HTN sequelae has not been performed and is, perhaps, unlikely to be. Thus, the closest practical alternatives are prospective observational studies and short-to-intermediate-term clinical trials. The intrinsic problems in studies of alcohol and health effects are complex and well known (Klatsky 2000). Under-reporting of heavier alcohol intake is one of these, but is incorrectly cited as a factor in the alcohol–HTN relation. As already stated, the major effect of such under-reporting would be an apparent, but spurious, relationship of HTN to lighter drinking. In fact, the true threshold for the relation could be higher than is suggested by the epidemiological data.

Because many traits are related to alcohol-drinking or to HTN, it is difficult to rule out all indirect explanations. Psychological or social stress is especially difficult to exclude, but some data show independence from several measures of such stress (Arkwright *et al.* 1983). The intervention studies provide good evidence against most indirect explanations. Except for the failure, so far, to demonstrate a biological mechanism, other criteria for causality are satisfactorily fulfilled. It is the author's opinion that the relationship between heavier drinking and higher risk of HTN is causal and that alcohol-related HTN is the commonest reversible form of HTN. Alcoholic beverage type (wine, beer or spirits) seems to be a minor factor. Estimates of the proportion of HTN due to heavy drinking vary with the population involved; the contribution of alcohol depends substantially upon the drinking habits of the group under study. Among the lowest estimates are 5% (Friedman *et al.* 1983) or 7% (MacMahon 1987) of hypertension, considering both sexes together. This translates into 1–2 million people with alcohol-associated HTN in the USA, using 20–40 million hypertensives as the denominator. It is probable that alcohol restriction plays a major role in HTN management and prevention (Joint National Committee on Detection Evaluation and Treatment of High Blood Pressure 1993).

## Cardiac arrhythmias

An association of heavier alcohol consumption with atrial arrhythmias has been observed for decades, typically occurring after a large meal accompanied by much alcohol. This has been called the 'holiday heart phenomenon' on the basis of the observation (Ettinger *et al.* 1978) that supraventricular arrhythmias in alcoholics without overt cardiomyopathy were most likely to occur on Mondays or between Christmas and New Year's Day. Various atrial rhythm disturbances have been reported to be associated with binge-drinking; atrial fibrillation is the commonest manifestation. The problem typically resolves with abstinence, with or without other specific treatment. A Kaiser Permanente study (Cohen *et al.* 1988) compared atrial arrhythmias in 1322 people reporting six or more drinks per day with arrhythmias in 2644 light drinkers. The relative risk in the heavier drinkers was at least doubled for each of the following: atrial fibrillation, atrial flutter, supraventricular tachycardia, and atrial premature complexes.

Increased ventricular ectopic activity has been documented after ingestion of substantial amounts of alcohol, although epidemiological studies have not shown a higher risk of sudden death in drinkers (Siscovick *et al.* 1986). Speculation about the mechanisms for the relation between heavier drinking and arrhythmias include myocardial damage, electrolyte disturbances or other metabolic effects, vagal reflexes, effects upon nerve or muscle cell conduction/refractory times, catecholamines, and acetaldehyde. A Finnish report (Maki *et al.* 1998) involved study of men with recurrent alcohol-associated atrial fibrillation. In controlled analyses of a number of tests and measurements, there was some evidence for exaggerated sympathetic nervous system reaction in these people.

## Cerebrovascular disease

### *Background, definitions and problems*

Older studies of the relationship of stroke to alcohol-drinking were hampered by imprecise diagnosis of stroke type before modern imaging techniques improved diagnostic accuracy. Since risk factors differ somewhat for the two major stroke types, proper diagnosis is quite important. *Haemorrhagic strokes* are due to ruptured blood vessels on the brain surface (subarachnoid haemorrhage) and in the brain substance (intracerebral haemorrhage). *Ischaemic strokes* are due to blockage of blood vessels by thrombosis in the brain blood vessels, emboli to the brain from the heart or elsewhere, or blockage of blood vessels outside the brain (most notably the carotid arteries). All studies of alcohol and stroke are complicated by the complex and disparate relations of both stroke and alcohol to other CV conditions. Age, cigarette smoking and HTN are important risk factors for both major stroke types. Several CV conditions predispose to cardio-embolic ischaemic stroke; these include mitral valvular disease, CHD, cardiomyopathy and atrial fibrillation. Anticoagulant therapy, often employed in CV problems, increases the risk of haemorrhagic stroke. These considerations, plus the disparities in alcohol–CV relationships and the lighter/heavier/binge-drinking differences, create almost Byzantine complexity when considering overall alcohol–stroke relationships.

Since disparities between haemorrhagic stroke and ischaemic stroke mandate separate analysis, it is fortunate that diagnostic accuracy is good in the current brain imaging era. But, in evaluating alcohol–stroke studies, another important factor is study type. Without randomized controlled trials, observational prospective studies represent the best available data. Stroke, generally a sudden catastrophic event, is especially susceptible to biased retrospective recall of recent drinking. Thus, studies of recent drinking are less valuable than prospective analyses of risk. Hypertension in heavy drinkers may be in the causal chain for increased risk, and analyses should not control for this (Klatsky 2000). Smoking, presumably not caused by drinking, should be controlled, or increased risk due to smoking might spuriously be attributed to alcohol.

### *Epidemiological data*

Heavy drinkers showed increased risk of haemorrhagic stroke in most reported prospective and retrospective studies (Van Gign *et al.* 1993; Klatsky *et al.* 2002); thus, this relation seems clear. But the relation of light–moderate drinking to haemorrhagic stroke is unclear; some prospective studies show increased risk and others do not (Van Gign *et al.* 1993; Camargo 1996; Thrift *et al.* 1999; Klatsky *et al.* 2002).

The relation of heavy drinking and ischaemic stroke risk is also unclear (Van Gign *et al.* 1993; Camargo 1996; Klatsky *et al.* 2001). Prospective studies show conflicting relations; most retrospective studies show increased risk. For lighter drinking and ischaemic stroke, most prospective studies and retrospective studies show decreased risk; since none show increased risk, the case for an inverse relation is strong (Van Gign *et al.* 1993; Camargo 1996; Sacco *et al.* 1999; Klatsky *et al.* 2001).

Several studies show these disparities in a single cohort. For example, the Nurse's Health Study (Stampfer *et al.* 1988) showed drinkers to be at higher risk of subarachnoid haemorrhage, but lower risk of occlusive stroke. Large Kaiser Permanente studies looked at the relations between reported alcohol use and the incidence of hospitalization for several types of cerebrovascular disease (Klatsky *et al.* 1989; Klatsky *et al.* 2001, 2002). Heavy, but not lighter, drinking was related to higher hospitalization rates for haemorrhagic stroke; higher blood pressure appeared to be a partial mediator of this relation. Alcohol use was associated with lower hospitalization rates for ischaemic stroke, an inverse relation present in both sexes, whites, blacks, and Asians (Klatsky *et al.* 2001).

### *Mechanisms*

In stroke risk, the antithrombotic (or anticlotting) actions of alcohol may be important (Van Gign *et al.* 1993; Zakhari and Gordis 1999; Klatsky *et al.* 2000), increasing the risk of haemorrhagic strokes and decreasing the risk of ischaemic strokes; these include lower fibrinogen levels, decreased platelet activity, and lessened thromboxane A activity. Other potential mechanisms for protection by alcohol against risk of ischaemic stroke are similar to those for CHD (see below) and include increased high density lipoprotein (HDL) cholesterol, increased prostacyclin and endogenous thromboplastin plasminogen activator (t-PA), and diminished insulin resistance.

### *Beverage choice*

There are limited data about any role for beverage choice (wine, beer or spirits) and the risk of ischaemic stroke. A Danish study (Truelsen *et al.* 1998) showed that people who reported 'any wine' v. 'none' had a lower risk of stroke (any type) after complete adjustment for confounding variables. In that study there was no association between intake of beer or spirits on stroke risk. The investigators concluded that compounds in wine additional to ethanol are responsible for the protective effect on the risk of stroke. As discussed below with respect to CHD, decreased low density lipoprotein cholesterol (LDL) oxidation caused by antioxidants in wine are a hypothetical non-alcohol mechanism for protection against atherothrombotic conditions.

Other studies with relevant data suggest no independent relation to beverage choice. This was the case for ischaemic stroke in the large Northern Manhattan Stroke Study (Sacco *et al.* 1999). In the latest Kaiser Permanente study of haemorrhagic stroke (Klatsky *et al.* 2002), there was no independent relationship of choice of wine, beer or spirits in analyses controlled or not controlled for total alcohol intake; separate analyses of drinkers of each wine type (red only, white only, etc.) also showed no independent relationships. In the Kaiser Permanente prospective study of ischaemic stroke (Klatsky

*Table 6.1* Relationship of light and heavy drinking to major stroke types

| *Type of stroke* | *Approximate proportion of total* | *Light/moderate drinking** | *Heavy drinking** |
|---|---|---|---|
| Haemorrhagic | 15% | No relation | ↑ Risk |
| Ischaemic | 85% | ↓ Risk | Unclear |

* $\geq$ 3 standard drinks/day is defined as boundary between light–moderate and heavy drinking.

*et al.* 2001), the drinking of spirits was associated with slightly higher risk, but a slightly lower risk for wine-drinking was not statistically significant. Again in this study, separate analyses of drinkers of each wine type (red only, white only, etc.) also showed no independent relations.

### *Summary*

At this time there is no consensus about the relations of alcohol drinking to the various types of stroke. Table 6.1 summarizes the author's opinion about the current evidence.

## Coronary heart disease

### *Background and history*

This condition is responsible for the majority of cardiovascular deaths. Although incidence is decreasing, CHD remains the leading cause of death in men and women in developed countries. It thus dominates statistics for cardiovascular mortality and has an impact on total mortality. A number of CHD risk factors have been uncovered by epidemiological studies. Among the established risk factors with probable causal effect are cigarette smoking, HTN, diabetes mellitus, increased LDL cholesterol, and decreased HDL cholesterol. Atherosclerotic narrowing of major epicardial coronary arteries is the usual basis of CHD. Thrombosis in narrowed vessels plays a critical role in major events, such as acute myocardial infarction or sudden death. The biological state of the atherosclerotic plaque and the endothelial lining of the coronary vessels plays an important role in the risk of these major CHD events.

Another major clinical expression of CHD is the symptom angina pectoris, which represents the perception of inadequate oxygen supply to the heart muscle. Since the classical description of angina relief by alcohol (Heberden 1786), there has been widespread, but erroneous, assumption that alcohol is a coronary vasodilator. Actually, exercise electrocardiographic data (Orlando *et al.* 1976) suggest that alcohol's effect on angina is purely subjective. Thus, it is probably dangerous for CHD patients to drink before exercise, as they might thereby lose the benefit of a warning symptom.

In the early 1900s, reports appeared of an inverse relation between heavy alcohol drinking and atherosclerotic disease (Cabot 1904; Hultgen 1910), but others (Wilens 1947; Ruebner *et al.* 1961) explained this as an artefact due to the premature deaths of many heavy drinkers. Early studies of alcoholics and problem drinkers suggested a high CHD rate (Wilens 1947; Ruebner *et al.* 1961), but some of these studies did not allow for the role of traits associated with alcoholism, such as cigarette smoking. Such studies can tell nothing about the role of light–moderate drinking.

### *Epidemiological studies*

These have consistently shown a reduced risk of acute myocardial infarction and death from CHD in light–moderate drinkers compared with alcohol-abstainers (Moore and Pearson 1986; Marmot and Brunner 1991; Maclure 1993; Renaud *et al.* 1993; Klatsky 1994; Rehm *et al.* 1997). Because the symptom angina pectoris is subjective, it is difficult to quantify. Thus, angina is less used as a study endpoint and has been relatively little studied epidemiologically in relation to alcohol. Studies showing increased risk of CHD in alcohol-abstainers than in drinkers include international comparisons (St Leger *et al.* 1979; Renaud and de Lorgeril 1992; Criqui and Ringel 1994), time-trend analyses of CHD over many years (LaPorte *et al.* 1980), case-control studies (Klatsky *et al.* 1974; Stason *et al.* 1976; Hennekens *et al.* 1978; Kozararevic *et al.* 1980; Siscovick *et al.* 1986; Kono *et al.* 1986, 1991; Jackson *et al.* 1991) prospective population studies (Dyer *et al.* 1980; Klatsky *et al.* 1981a,b, 1986b, 1990, 1992, 1997; Yano *et al.* 1977, 1984; Stampfer *et al.* 1988; Boffetta and Garfinkel 1990; Rimm *et al.* 1991; Gaziano *et al.* 1993; Doll *et al.* 1994; Goldberg *et al.* 1995; Fuchs *et al.* 1995; Hein *et al.* 1996; Camargo *et al.* 1997; Keil *et al.* 1997; Rehm *et al.* 1997; Thun *et al.* 1997; Kitamura *et al.* 1998; Renaud *et al.* 1998), and studies of coronary arteriograms (Barboriak *et al.* 1979; Handa *et al.* 1990). The most convincing data are those from prospective population studies in which alcohol use is determined before the CHD events. The various studies show reduced CHD risk among light–moderate drinkers in both men and women, whites, blacks, and Asians, various age groups, and other subsets. The apparent benefit is seen in people with and without known CHD, HTN or diabetes mellitus. Not surprisingly, the impact on total mortality risk is strongest among persons of $\geq$ 50 years of age (Klatsky *et al.* 1992; Fuchs *et al.* 1995).

There is much less consistency in the studies with respect to heavy drinking and CHD risk. Many population studies of non-fatal infarction show that both lighter and heavier drinkers are at lower risk than abstainers, but studies of CHD mortality mostly show a U-curve or J-curve relation to the amount of alcohol intake, with abstainers and heavier drinkers at higher risk than light–moderate drinkers (Moore and Pearson 1986; Marmot and Brunner 1991; Maclure 1993; Renaud *et al.* 1993; Klatsky 1994). Reasons for the

*Table 6.2* Relative risk of CHD hospitalization or death according to alcohol

| *Alcohol use* | *Hospitalization** *RR (95% CI)*† | *Death*‡ *RR (95% CI)*† |
|---|---|---|
| Abstainer | 1.0 (ref.) | 1.0 (ref.) |
| Ex-drinker | 1.0 (0.9–1.2) | 1.1 (1.0–1.3) |
| Drinkers | | |
| < 1/month | 1.0 (0.9–1.1) | 1.0 (0.9–1.0) |
| < 1/day, > 1/month | 0.9 (0.8–1.0) | 0.8 (0.7–0.9) |
| 1–2/day | 0.7 (0.6–0.8) | 0.7 (0.6–0.8) |
| 3–5/day | 0.7 (0.8–0.8) | 0.8 (0.7–0.9) |
| ≥ 6/day | 0.6 (0.5–0.8) | 1.0 (0.8–1.4) |

* First for any CHD diagnosis ($n$ = 3869 through 1994).
† Computed from coefficients estimated by Cox proportional hazards model; covariates include sex, age, race, smoking, education and coffee.
‡ Through 1998 ($n$ = 3001).
RR, relative risk; CI, confidence interval.

increased risk of heavier drinkers (the upper limb of the U- or J-curve) may include the effects of spree-drinking, alcohol-associated HTN or arrhythmias, misdiagnosis of other conditions (e.g. dilated cardiomyopathy) as CHD, or a truly different effect of heavier drinking on CHD.

Deficiencies in some of the population studies included failure to separate lifelong abstainers from ex-drinkers or inadequate control for baseline CHD risk. Thus, a counterhypothesis to benefit by alcohol, sometimes called the 'sick quitter' hypothesis developed. This stated that the non-drinking referent groups in these studies might be at higher risk for reasons other than abstinence (Shaper *et al.* 1988). Total independence from indirect, or confounded, explanation is possible only with controlled, randomized, blinded experiments. However, the counterhypothesis is effectively refuted by prospective studies that separate ex-drinkers from lifelong abstainers and that also control well for baseline CHD (Moore and Pearson 1986; Marmot and Brunner 1991; Maclure 1993; Renaud *et al.* 1993; Klatsky 1994). A few examples follow.

- Updated statistics from Kaiser Permanente studies about both hospitalizations and mortality are summarized in Table 6.2. Analysis of alcohol habits in relation to CHD hospitalizations (Klatsky *et al.* 1990) showed that ex-drinkers and infrequent (< 1/month) drinkers were at a risk similar to that of lifelong abstainers, with a lower CHD risk present among all other drinkers, independent of a number of potential confounders including baseline CHD risk at examination and beverage choice. Data for CHD mortality also show no relations of past and very light drinking. A U-shaped curve, with a nadir at 1–2 and 3–5 drinks/day and independent of multiple potential confounders and baseline risk, related alcohol amount to CHD deaths.

- A large prospective study among female nurses free of evident CHD and with detailed control of nutrient intake (Stampfer *et al.* 1988) showed a progressive inverse relation of alcohol use to CHD events, independent of prior reduction in alcohol intake. Further analysis of these data (Fuchs *et al.* 1995) demonstrated that the adverse effects of drinking limited the net beneficial effects of moderate alcohol use in women to people at above-average CHD risk (essentially, those older than 50 years).
- The 12-year prospective American Cancer Society Study (Boffetta and Garfinkel 1990) of 276 802 men showed a U-curve for CHD mortality, with an RR of 0.8 (v. abstainers) at 1–2 drinks/day.
- In a study of 51 529 non-physician health professionals, well controlled for dietary habits, newly diagnosed CHD was inversely related to increasing alcohol intake (Rimm *et al.* 1991).
- The Auckland Heart Study was partially designed to examine the hypothesis that people at high CHD risk are likely to become non-drinkers. In both sexes the data showed that moderate drinkers had lower CHD risk than both lifelong abstainers and ex-drinkers (Jackson *et al.* 1991). Many other examples could be cited, but perhaps more convincing are the data from a recent meta-analysis (Corrao *et al.* 2000). The selection criteria for inclusion were rigorous; of 196 articles screened, 51 (43 cohort studies and eight case-control studies) met them. Most of the studies included used CHD mortality or combined mortality/morbidity as endpoints. There was decreased CHD risk at light–moderate drinking levels, maximal at 25 g/day of alcohol. At more than 90 g/day (approximately seven standard drinks), risk rose above that of the abstainers. Seven drinks/day is well above all usual definitions of moderate drinking.

## *Possible mechanisms for CHD protection by alcohol*

### *Effects on blood lipid factors*

The most studied mechanism for protection by alcohol against CHD and other atherosclerotic disease and, perhaps, the most plausible, is a link via blood lipid factors. These factors play a central role in the development of CHD (Criqui and Golomb 1998).

- There is a progressive positive relation between higher levels of LDL cholesterol, the so-called 'bad cholesterol' and CHD. Alcohol may be associated with lower LDL levels (Hein *et al.* 1996), but it is unclear that this is independent of other dietary factors.
- HDL cholesterol (the 'good cholesterol') levels are inversely related to CHD risk (Rifkind 1990; Renaud *et al.* 1993; Criqui and Golomb 1998), possibly acting by abetting removal of lipid deposits in large blood vessels. HDL binds with cholesterol in the tissues and may aid in preventing tissue oxidation of LDL cholesterol; it then carries it back to

the liver for elimination or reprocessing. The net effect is reduction of cholesterol build-up in the walls of large blood vessels, such as the coronary arteries. The evidence is solid that, in the absence of severe liver impairment, alcohol ingestion raises HDL levels (Castelli *et al.* 1977; Hulley and Gordon 1981; Rimm *et al.* 1991; Renaud *et al.* 1993; Gaziano *et al.* 1993; Criqui and Golomb 1998). Among HDL subspecies, some data suggest that $HDL_2$ may be more protective (Haskell *et al.* 1984), but other studies suggest that both $HDL_2$ and $HDL_3$ are protective (Salonen *et al.* 1988; Diehl *et al.* 1988; Sweetham *et al.* 1989; Rimm *et al.* 1991; Gaziano *et al.* 1993). The biochemical pathways for alcohol's effect on the HDL cholesterol are poorly understood. It may be pertinent that some data show elevation by alcohol of apolipoproteins $A_1$ and $A_2$, associated with HDL particle formation (Camargo *et al.* 1985; Masarei *et al.* 1986; Moore *et al.* 1988).

- A subset of heavier drinkers has a substantial increase in triglyceride levels, but this is infrequently seen with lighter–moderate drinking. There is increasing feeling that triglycerides may play an independent role in CHD risk. Some feel that the ratio between total cholesterol and HDL cholesterol, which indirectly incorporates data about LDL, HDL, and triglycerides, may be the best single CHD risk indicator (Criqui and Golomb 1998).
- The case for a lipid link for alcohol's protection against CHD rests primarily on HDL effects. Three separate studies (Criqui *et al.* 1987; Suh *et al.* 1992; Gaziano *et al.* 1993) have quantitatively examined the hypothesis that the apparent protective effect of alcohol against CHD is mediated by higher HDL cholesterol levels in drinkers. The findings were similar in all three analyses in suggesting that higher HDL levels in drinkers mediated about half of the lower CHD risk. Data from one of these studies (Gaziano *et al.* 1993) suggested that both $HDL_2$ and $HDL_3$ were involved.

*Via antithrombotic mechanisms*

Various aspects of clotting may be inhibited by alcohol (Klatsky and Armstrong 1993; Renaud *et al.* 1993; Hendriks and van der Gang 1998; Booyse and Parks 2001). An antithrombotic action of alcohol could partially account for the lower CHD risk at very light drinking levels (e.g., several drinks per week) seen in several of the epidemiological studies. Some data support the following mechanisms:

- decreased platelet stickiness (Mikhailidis *et al.* 1983; Jakubowski *et al.* 1988; Renaud *et al.* 1992)
- increased thromboxane/prostacyclin ratio (Landolfi and Steiner 1984)
- lowered fibrinogen levels (Meade *et al.* 1979; Hendriks *et al.* 1994; Hendriks and van der Gang 1998)

- interaction with aspirin in prolonging bleeding time (Deykin *et al.* 1982)
- increased release of plasminogen activator (Gaziano *et al.* 1993; Laug 1983).

Perhaps the evidence about the anticlotting effects of alcohol is best for the fibrinogen lowering effects of alcohol (Hendriks and van der Gang 1998).

*Via glucose metabolism*

Although alcohol-drinking, especially heavier intake, has been associated with higher blood glucose levels (Gerard *et al.* 1977), lighter drinking has been associated with possible beneficial changes in insulin and glucose metabolism. Lower blood insulin levels and enhanced insulin sensitivity in lighter drinkers have been reported (Mayer *et al.* 1993; Facchini *et al.* 1994; Kiechl *et al.* 1996; Lazarus *et al.* 1997). Studies suggest that light–moderate drinking is associated with reduced risk of type 2 diabetes mellitus (Ajani *et al.* 2000a,b; Wei and Bulkley 1982; Hu *et al.* 2001). Increased insulin resistance, glucose intolerance and diabetes mellitus appear to be stages of a continuous progressive process. All are powerful CHD predictors. The effects of lighter drinking on these phenomena are likely to play a role in protection by alcohol against CHD. It should be noted that the risk of CHD is reduced in lighter drinkers whether they have diabetes or not (Criqui and Golomb 1999; Solomon *et al.* 2000; Ajani *et al.* 2000a).

*Via stress reduction*

Hypothetical considerations about a possible benefit from the anti-anxiety or stress-reducing effects of alcohol have no good supporting data.

***Role of beverage choice (wine, spirits or beer)***

In 1819 Dr Samuel Black, a perceptive Irish physician with a great interest in angina pectoris, wrote, to this author's knowledge, the first commentary pertinent to the 'French paradox'. Noting apparent angina disparity between Ireland and France, he attributed the low prevalence in the latter to 'the French habits and modes of living, coinciding with the benignity of their climate and the peculiar character of their moral affections' (Black 1819). It was to be 160 years before data were presented from the first international comparison study suggesting less CHD in wine-drinking countries than in beer- or spirits-drinking countries (St Leger *et al.* 1979). Several confirmatory international comparison studies have followed (Renaud and de Lorgeril 1992; Criqui and Ringel 1994; Renaud *et al.* 1998). The 'French paradox' concept has arisen from these data. The 'paradox' is the fact that France tends to be an outlier on graphs of mean dietary fat intake v. CHD mortality unless adjusted for wine alcohol intake (Criqui and Ringel 1994; Renaud *et al.* 1998). Reports of non-alcohol antioxidant phenolic compounds (Siemann and Creasy 1992; Frankel *et al.* 1993; Maxwell *et al.* 1994; Booyse and Parks 2001), endothelial relaxants (Fitzpatrick *et al.* 1993) or antithrombotic substances (Kluft *et al.* 1990; Demrow *et al.* 1995; Booyse and Parks 2001) in wine, especially red wine, have appeared. Inhibition of oxidative modification of LDL cholesterol is probably anti-atherogenic, although prospective clinical trials of antioxidant supplements are not yet conclusive (Diaz *et al.* 1997; Virtamo *et al.* 1998).

A Kaiser Permanente cohort study of 221 people who died of CHD (Klatsky and Armstrong, 1993) and who took 80–90% of their beverage alcohol as one preferred beverage type, showed that, compared with non-drinkers, CHD risk was significantly lower among preferrers of each beverage type. When the CHD risks of the beverage preference groups were compared with each other, there was a gradient of apparently increasing protection from spirits to beer to wine. However, there were substantial differences in traits between the preference groups (Klatsky *et al.* 1990). The wine-drinkers had the most favourable CHD risk profiles, leading to the hypothesis that favourable uncontrolled traits (e.g. dietary habits, physical exercise, use of antioxidant supplements) of wine-preferrers might explain the findings.

An analysis of the role of beverage choice among 3931 people hospitalized for coronary disease used a proxy variable for reported frequency of drinking each beverage type, enabling use of all available beverage choice data (Klatsky *et al.* 1997). Adjusted analyses, not controlled for total alcohol intake, showed inverse relationships to CHD risk for each beverage type, weakest for spirits-use. In sex specific data this inverse relation was significant for beer-use in men and for wine-use in women. When controlled for total alcohol intake, only beer-use in men remained significantly related. There were no significant differences in risk between drinkers of red, white, both red and white, and other types of wine. It was concluded that all beverage types protect against CHD, with additional protection by specific beverages likely to be minor.

Although antioxidant and other substances in wine are an attractive hypothetical explanation for CHD protection, the prospective population studies provide no consensus that wine has additional benefits, and various studies show benefit for wine, beer, spirits, or all three major beverage types (Klatsky *et al.* 1992; Renaud *et al.* 1993; Klatsky 1994; Rimm *et al.* 1996; Klatsky *et al.* 1997). People drinking the beverage types differ in user traits, with wine-drinkers having the most favourable CHD risk profile (Klatsky *et al.* 1990). Drinking pattern differences among the beverage types are probably another factor. The wine/spirits/beer issue is unresolved at this time, but it seems likely that ethyl alcohol is the major factor with respect to lower CHD risk. As yet, there seem to be no compelling health-related data that preclude personal preference as the best guide to the choice of beverage.

### *Is the alcohol–CHD relation a causal protective effect?*

It remains theoretically possible that lifelong abstainers could differ from drinkers in psychological traits, dietary habits, physical exercise habits, or in some other way that could be related to CHD risk, but there is no good evidence for such a trait. The various studies indicate that such a correlate would need to be present in people of both sexes, various countries, and multiple racial groups. While it remains possible that other factors play a role, a causal, protective effect of alcohol is a simpler and more plausible explanation.

## Peripheral vascular disease

This term is usually used to refer to atherosclerotic disease in the aorta or large extremity arteries, and is far more common in the vessels to the legs than in those to the arms. The risk factors are similar to those for CHD, except that smoking and diabetes are especially strong predictors (Hirsch *et al.* 2001). The prognosis is related to the very high prevalence of concomitant CHD and cerebrovascular disease. There are sparse data about the role of alcohol, but data from the Framingham Heart Study (Djousse *et al.* 2000) show a U-shaped relation of alcohol intake to risk of leg claudication, a common symptom of peripheral vascular disease. As for CHD, these data suggest a protective effect of moderate alcohol intake.

## Conclusion and public health implication

This chapter has reviewed the evidence for disparity in the relations of alcohol and CV disorders. These are summarized in Table 6.3, with specific emphasis on the disparity between the overall favourable relations of lighter drinking and the overall unfavourable relations of heavier drinking. Counselling people who are concerned about the health effects of alcohol-drinking needs to be individualized according to each person's specific medical history and risks (Friedman and Klatsky 1993; Pearson and Terry 1994; Klatsky 2001). Advice to the general public is more problematic because of the substantially unpredictable risk of progression to problem drinking. A few rules seem sensible:

1. The overall health risks of a heavy drinker will be reduced by reduction of drinking or abstinence.
2. Because of the unknown risk of progression to heavier drinking, abstainers should not be indiscriminately advised to drink for CV health benefit.
3. Light to moderate drinkers, who constitute the majority of people, need no change in drinking habits for health reasons, except in special circumstances.

*Table 6.3* Relationship of alcohol drinking to cardiovascular conditions

| *Condition* | *Relation by amount of alcohol drinking* | | *Comment* |
|---|---|---|---|
| | *Small* | *Large* | |
| Dilated cardiomyopathy | No relation | One (of many) causes | Unknown cofactors |
| Beriberi | No direct relation | No direct relation | Thiamine deficiency |
| A/c beer disease* | No relation | Synergistic | Probable examples of cardiomyopathy cofactors |
| Hypertension | Little or none | Probably a causal factor | Mechanism unknown |
| Coronary disease | Protective | ? Protective | Via HDL, antithrombotic effects, not beverage-specific; possible additional benefit from wine |
| Arrhythmia | ? None | Probably a causal factor | ? Susceptibility factors |
| Haemorrhagic stroke | No relation or ? increased risk | Increased risk | Via higher blood pressure, antithrombotic actions |
| Ischaemic stroke | Protective | ? Protective | Complex interactions with other conditions |

* Arsenic-/cobalt-beer-drinkers' disease.

## References

Aalsmeer, W. C. and Wenckebach, K. F. (1929) Herz und Kreislauf bei der Beri-Beri Krankheit. *Weiner Archiv für Innere Medizin*: **16**, 193–272.

Abramson, J. L., Williams, S. A., Krumholz, H. M., Vaccarino, V. (2001) Moderate alcohol consumption and risk of heart failure among older persons. *Journal of the American Medical Association*: **285**, 1971–7.

Ajani, U. A., Gaziano, J. M., Lotufo, P. A., Liu, S., Hennekens, C. H., Buring, J. E., Manson, J. E. (2000a) Alcohol consumption and risk of coronary heart disease by diabetes status. *Circulation*: **102**, 500–5.

Ajani, U. A., Hennekens, C. H., Spelsberg, A., Manson, J. E. (2000b) Alcohol consumption and risk of type 2 diabetes mellitus among US male physicians. *Archives of Internal Medicine*: **160**, 1025–30.

Alexander, C. S. (1966a) Idiopathic heart disease. I. Analysis of 100 cases, with special reference to chronic alcoholism. *American Journal of Medicine*: **41**, 213–28.

Alexander, C. S. (1966b) Idiopathic heart disease. II. Electron microscopic examination of myocardial biopsy specimens in alcoholic heart disease. *American Journal of Medicine*: **41**, 229–34.

Alexander, C. S. (1969) Cobalt and the heart. *Annals of Internal Medicine*: **70**, 411–3.

Arkwright, P. D., Beilin, L. J., Rouse, I. L., Vandongen, R. (1983) Alcohol, personality, and predisposition to essential hypertension. *Journal of Hypertension*: 1, 365–71.

Ascherio, A., Rimm, E. B., Giovannucci, E. L., Colditz, G. A., Rosner, B., Willett, W. C., Sacks, F., Stampfer, M. J. (1992) A prospective study of nutritional factors and hypertension among US men. *Circulation*: **86**, 1475–84.

Ballester, M., Marti, V., Carrio, I., Obrador, D., Moya, C., Pons-Llado, G., Berna, L., Lamich, R., Aymat, M. R., Barbanoj, M., Guardia, J., Carreras, F., Udina, C., Auge, J. M., Marrugat, J., Permanyer, G., Caralps-Riera, J. M. (1997) Spectrum of alcohol-induced myocardial damage detected by indium-111-labeled monoclonal antimyosin antibodies. *Journal of the American College of Cardiology*: **29**, 160–7.

Barboriak, J. J., Anderson, A. J., Rimm, A. A. and Tristani, F. E. (1979) Alcohol and coronary arteries. *Alcoholism Clinical and Experimental Research*: **3**, 29–32.

Beevers, D. G. (1990) Alcohol, blood pressure and antihypertensive drugs (Editorial). *Journal of Clinical Pharmacology and Therapeutics*: **15**, 395–7.

Bertolet, B. D., Freund, G., Martin, C. A., Perchalski, D. L., Williams, C. M. and Pepine, C. J. (1991) Unrecognized left ventricular dysfunction in an apparently healthy alcohol abuse population. *Drug and Alcohol Dependence*: **28**, 113–19.

Black, S. (1819) Clinical and Pathological Reports (ed. A. Wilkinson). Newry, pp. 1–47.

Blacket, R. B. and Palmer, A. J. (1960) Haemodynamic studies in high output beriberi. *British Heart Journal*: 22, 483–501.

Blankenhorn, M. A. (1945) The diagnosis of beriberi heart disease. *Annals of Internal Medicine*: **23**, 398–404.

Blankenhorn, M. A., Viter, C. F., Scheinker, I. M., Austin, R. S. (1946) Occidental beriberi heart disease. *Journal of the American Medical Association*: **131**, 717–27.

Boffetta, P. and Garfinkel, L. (1990) Alcohol drinking and mortality among men enrolled in an American Cancer Society prospective study. *Epidemiology*: 1, 342–8.

Böllinger, O. (1884) Über die Haussigkeit und Ursachen der idiopathischen Herzhypertrophy in Menschen. *Deutsche Medizinische Wochenschrift*: **10**, 180.

Booyse, F. M. and Parks, D. A. (2001) Moderate wine and alcohol consumption: beneficial effects on cardiovascular disease. *Thrombosis and Haemostasis*: **86**, 517–28.

Bulloch, R. T., Pearce, M. B., Murphy, M. L., Jenkins, B. J., Davis, J. L. (1972) Myocardial lesions in idiopathic and alcoholic cardiomyopathy. Study by ventricular septal biopsy. *American Journal of Cardiology*: 29, 15–25.

Burch, G. E., Phillips, Jr, J. H., and Ferrans, V. J. (1966) Alcoholic cardiomyopathy. *American Journal of Medical Science*: **252**, 89–123.

Burch, G. E., Colcolough, H. L., Harb, J. M., Tsui, C. Y. (1971) The effect of ingestion of ethyl alcohol, wine and beer on the myocardium of mice. *American Journal of Cardiology*: 27, 522–8.

Cabot, R. C. (1904) Relation of alcohol to arteriosclerosis. *Journal of the American Medical Association*: 44, 774–5.

Camargo, Jr, C. A. (1996) Case-control and cohort studies of moderate alcohol consumption and stroke. *Clinica Chimica Acta*: **246**, 107–19.

Camargo, Jr, C. A., Williams, P. T., Vranizan, K. M., Albers, J. J., Wood, P. D. (1985) The effect of moderate alcohol intake on serum apolipoproteins A-I and A-II: A controlled study. *Journal of the American Medical Association*: **253**, 2854–5.

Camargo, Jr, C. A., Stampfer, M. J., Glynn, R. J., Grodstein, F., Gaziano, J. M., Manson, J. E., Buring, J. E., Hennekens, C. H. (1997) Moderate alcohol consumption and risk for angina pectoris or myocardial infarction in U.S. male physicians. *Annals of Internal Medicine*: **126**, 372–5.

Castelli, W. P., Doyle, J. T., Gordon, T., Hames, C. G., Hjortland, M. C., Hulley, S. B., Kagan, A., Zukel, W. J. (1977) Alcohol and blood lipids: The Cooperative Lipoprotein Phenotyping Study. *Lancet*: 2, 153–5.

Clark, V. A., Chapman, J. M. and Coulson, A. H. (1967) Effects of various factors on systolic and diastolic blood pressure in the Los Angeles heart study. *Journal of Chronic Diseases*: **20**, 571–81.

Coca, A., De la Sierra, A., Sanchez, M., Picado, M. J., Lluch, M. M., Urbano-Marquez, A. (1992) Chronic alcohol intake induces reversible disturbances on cellular Na+ metabolism in humans: Its relationship with changes in blood pressure. *Alcoholism Clinical and Experimental Research*: **16**, 714–20.

Cohen, E. J., Klatsky, A. L. and Armstrong, M. A. (1988) Alcohol use and supraventricular arrhythmia. *American Journal of Cardiology*: **62**, 971–3.

Cooper, H. A., Exner, D. V. and Domanski, M. J. (2000) Light-to-moderate alcohol consumption and prognosis in patients with left ventricular systolic dysfunction. *Journal of the American College of Cardiology*: **35**, 1753–9.

Corrao, G., Rubbiati, L., Bagnardi, V., Zambon, A., Poikolainen, K. (2000) Alcohol and coronary heart disease: a meta-analysis. *Addiction*: **95**, 1505–23.

Cox, K. L., Puddey, I. B., Morton, A. R., Masarei, J. R., Vandongen, R., Beilin, L. J. (1990) A controlled comparison of the effects of exercise and alcohol on blood pressure and serum high density lipoprotein cholesterol in sedentary males. *Clinical Experiments in Pharmacologic Physiology*: **17**, 251–6.

Criqui, M. H. (1987) Alcohol and hypertension: new insights from population studies. *European Heart Journal*: **8** (suppl B), 19–26.

Criqui, M. H. and Golomb, B. A. (1998) Epidemiologic aspects of lipid abnormalities. *American Journal of Medicine*: **105**, 48S–57S.

Criqui, M. H. and Golomb, B. A. (1999) Should patients with diabetes drink to their health? *Journal of the American Medical Association*: **282**, 279–80.

Criqui, M. H. and Ringel, B. L. (1994) Does diet or alcohol explain the French paradox? (See comments.) *Lancet*: **344**, 1719–23.

Criqui, M. H., Wallace, R. B., Mishkel, M., Barrett-Connor, E., Heiss, G. (1981) Alcohol consumption and blood pressure: The Lipid Research Clinics Prevalence Study. *Hypertension*: **3**, 557–65.

Criqui, M. H., Cowan, L. D., Tyroler, H. A., Bangdiwala, S., Heiss, G., Wallace, R. B., Cohn, R. (1987) Lipoproteins as mediators for the effects of alcohol consumption and cigarette smoking on cardiovascular mortality: results form the Lipid Research Clinics Follow-up Study. *American Journal of Epidemiology*: **126**, 629–37.

Davidson, D. M. (1989) Cardiovascular effects of alcohol. *Western Journal of Medicine*: **151**, 430–9.

Demakis, J. G., Proskey, A., Rahimtoola, S. H., Jamil, M., Sutton, G. C., Rosen, K. M., Gunnar, R. M., Tobin, J. R. (1974) The natural course of alcoholic cardiomyopathy. *Annals of Internal Medicine*: **80**, 293–7.

Demrow, H. S., Slane, P. R. and Folts, J. D. (1995) Administration of wine and grape juice inhibits *in vivo* platelet activity and thrombosis in stenosed canine coronary arteries. *Circulation*: **91**, 1182–8.

Deykin, D., Janson, P. and McMahon, L. (1982) Ethanol potentiation of aspirin-induced prolongation of the bleeding time. *New England Journal of Medicine*: **306**, 852–4.

Diaz, M. N., Frei, B., Vita, J. A., Keaney, Jr, J. F. (1997) Antioxidants and atherosclerotic heart disease. *New England Journal of Medicine*: **337**, 408–16.

Diehl, A. K., Fuller, J. H., Mattock, M. B., Salter, A. M., el-Gohari, R., Keen, H. (1988) The relationship of high density lipoprotein subfractions to alcohol consumption, other lifestyle factors, and coronary heart disease. *Atherosclerosis*: **69**, 145–53.

Djousse, L., Levy, D., Murabito, J. M., Cupples, L. A., Ellison, R. C. (2000) Alcohol consumption and risk of intermittent claudication in the Framingham Heart Study. *Circulation*: **102**, 3092–7.

Doll, R., Peto, R., Hall, E., Wheatley, K., Gray, R. (1994) Mortality in relation to consumption of alcohol: 13 years' observation on male British doctors. *British Medical Journal*: **309**, 911–18.

Dyer, A. R., Stamler, J., Paul, O., Berkson, D. M., Lepper, M. H., McKean, H., Shekelle, R. B., Lindberg, H. A., Garside, D. (1977) Alcohol consumption, cardiovascular risk factors, and mortality in two Chicago epidemiologic studies. *Circulation*: **56**, 1067–74.

Dyer, A. R., Stamler, J., Paul, O., Lepper, M., Shekelle, R. B., McKean, H., Garside, D. (1980) Alcohol consumption and 17-year mortality in the Chicago Western Electric Company Study. *Preventive Medicine*: **9**, 78–90.

Ettinger, P. O., Wu, C. F., De La Cruz, Jr, C., Weisse, A. B., Ahmed, S. S., Regan, T. J. (1978) Arrhythmias and the 'Holiday Heart': alcohol-associated cardiac rhythm disorders. *American Heart Journal*: **95**, 555–62.

Evans, W. (1961) Alcoholic cardiomyopathy. *American Heart Journal*: **61**, 556–7.

Facchini, F., Chen, Y. D. and Reaven, G. M. (1994) Light-to-moderate alcohol intake is associated with enhanced insulin sensitivity. *Diabetes Care*: **17**, 115–19.

Ferrans, V. J. (1966) Alcoholic cardiomyopathy. *American Journal of Medical Science*: **252**, 89–104.

Fitzpatrick, D. F., Hirschfield, D. L. and Coffey, R. G. (1993) Endothelium-dependent vasorelaxing activity of wine and other grape products. *American Journal of Physiology*: **265**, H774–8.

Frankel, E. N., Kanner, J., German, J. B., Parks, E., Kinsella, J. E. I. (1993) Inhibition *in vitro* of oxidation of human low-density lipoprotein by phenolic substances in red wine. *Lancet*: **342**, 454–7.

Friedman, G. D. and Klatsky, A. L. (1993) Is alcohol good for your health? (Editorial.) *New England Journal of Medicine*: **329**, 1882–3.

Friedman, G. D., Fireman, B. H., Petitti, D. B., Siegelaub, A. B., Ury, H. K., Klatsky, A. L. (1983) Psychological question score, cigarette smoking, and myocardial infarction score: a continuing enigma. *Preventive Medicine*: **12**, 533–40.

Friedman, G. D., Selby, J. V., Quesenberry, C. P. J., Armstrong, M. A., Klatsky, A. L. (1988) Precursors of essential hypertension: The role of body weight, reported alcohol and salt use, and parental history of hypertension. *Preventive Medicine*: **17**, 387–402.

Friedreich, N. (1861) *Handbuch der speziellen Pathologie und Therapie. Sektion 5, Krankheiten des Herzens*. Ferdinand Enke, Erlangen.

Fuchs, C. S., Stampfer, M. J., Colditz, G. A., Giovannucci, E. L., Manson, J. E., Kawachi, I., Hunter, D. J., Hankinson, S. E., Hennekens, C. H., Rosner, B.

(1995) Alcohol consumption and mortality among women. *New England Journal of Medicine*: **332**, 1245–50.

Gaziano, J. M., Buring, J. E., Breslow, J. L., Goldhaber, S. Z., Rosner, B., VanDenburgh, M., Willett, W., Hennekens, C. H. (1993) Moderate alcohol intake, increased levels of high density lipoprotein and its subfractions, and decreased risk of myocardial infarction. *New England Journal of Medicine*: **329**, 1829–34.

Gerard, M. J., Klatsky, A. L., Friedman, G. D., Siegelaub, A. B., Feldman, R. (1977) Serum glucose levels and alcohol consumption habits in a large population. *Diabetes*: **26**, 780–5.

Gillet, C., Juilliere, Y., Pirollet, P., Aubin, H. J., Thouvenin, A., Danchin, N., Cherrier, F., Paille, F. (1992) Alcohol consumption and biological markers for alcoholism in idiopathic dilated cardiomyopathy: a case-controlled study. *Alcohol*: **27**, 353–8.

Goldberg, D. M., Hahn, S. E. and Parkes, J. G. (1995) Beyond alcohol: beverage consumption and cardiovascular mortality. *Clinica Chimica Acta*: **237**, 155–87.

Gordon, T. and Kannel, W. B. (1983) Drinking and its relation to smoking, BP, blood lipids, and uric acid. *Archives of Internal Medicine*: **143**, 1366–74.

Gould, L., Zahir, M., Shariff, M., DiLieto, M. (1969) Cardiac hemodynamics in alcoholic heart disease. *Annals of Internal Medicine*: **71**, 543–54.

Gowers, W. R. (1901) In Royal Medical and Chirurgical Society. Epidemic of arsenical poisoning in beer-drinkers in the north of England during the year 1900. *Lancet*: **1**, 98–100.

Grollman, A. (1930) The action of alcohol, caffeine, and tobacco on the cardiac output (and its related functions) of normal man. *Journal of Pharmacological Experiments and Therapies*: **39**, 313–27.

Handa, K., Sasaki, J., Saku, K., Kono, S., Arakawa, K. (1990) Alcohol consumption, serum lipids and severity of angiographically determined coronary artery disease. *American Journal of Cardiology*: **65**, 287–9.

Haskell, W. L., Camargo, C. J. and Williams, P. T. (1984) The effect of cessation of and resumption of moderate alcohol intake on serum high-density-lipoprotein subfractions: A controlled study. *New England Journal of Medicine*: **310**, 805–10.

Heberden, W. (1786) Some account of a disorder of the breast. *Medical Transactions of the Royal College of Physicians, London*: **2**, 59–67.

Hein, H. O., Suadicani, P. and Gyntelberg, F. (1996) Alcohol consumption, serum low density lipoprotein cholesterol concentration, and risk of ischaemic heart disease: six year follow up in the Copenhagen male study. *British Medical Journal*: **312**, 736–741.

Hendriks, F. J. and van der Gang, M. S. (1998) Alcohol, anticoagulation and fibrinolysis, in *Alcohol and Cardiovascular Diseases* (Novartis Foundation Symposium 216), (eds, Chadwick, D. J. and Goode, J. A.). Wiley, Chichester, pp. 111–24.

Hendriks, H. F., Veenstra, J., Velthuis-te Wierik, E. J., Schaafsma, G., Kluft, C. (1994) Effect of moderate dose of alcohol with evening meal on fibrinolytic factors. *British Medical Journal*: **308**, 1003–6.

Hennekens, C. H., Rosner, B. and Cole, D. S. (1978) Daily alcohol consumption and fatal coronary heart disease. *American Journal of Epidemiology*: **107**, 196–200.

Hirsch, A. T., Criqui, M. H., Treat-Jacobson, D., Regensteiner, J. G., Creager, M. A., Olin, J. W., Krook, S. H., Hunninghake, D. B., Comerota, A. J., Walsh, M. E., McDermott, M. M., Hiatt, W. R. (2001) Peripheral arterial disease detection, awareness, and treatment in primary care. *Journal of the American Medical Association*: **286**, 1317–24.

Hu, F. B., Manson, J. E., Stampfer, M. J., Colditz, G., Liu, S., Solomon, C. G., Willett, W. C. (2001) Diet, lifestyle, and the risk of type 2 diabetes mellitus in women. *New England Journal of Medicine*: **345**, 790–7.

Hulley, S. B. and Gordon, S. (1981) Alcohol and high-density lipoprotein cholesterol: Causal inference from diverse study designs. *Circulation*: **64** (suppl. III), III-57–III-63.

Hultgen, J. F. (1910) Alcohol and nephritis: Clinical study of 460 cases of chronic alcoholism. *Journal of the American Medical Association*: **55**, 279–81.

Jackson, R., Scragg, R. and Beaglehole, R. (1991) Alcohol consumption and risk of coronary heart disease. *British Medical Journal*: **303**, 211–16.

Jakubowski, J. A., Vaillancourt, R. and Deykin, D. (1988) Interaction of ethanol, prostacyclin, and aspirin in determining human platelet reactivity *in vitro*. *Arteriosclerosis*: **8**, 436–41.

Joint National Committee on Detection Evaluation and Treatment of High Blood Pressure (1993) The Fifth Report to the Joint National Committee. *Archives of Internal Medicine*: **153**, 158–83.

Kasper, E. K., Willem, W. R. P., Hutchins, G. M., Deckers, J. W., Hare, J. M., Baughman, K. L. (1994) The causes of dilated cardiomyopathy: a clinicopathologic review of 673 consecutive patients. *Journal of the American College of Cardiology*: **23**, 586–90.

Kawano, Y., Abe, H., Kojima, S., Ashida, T., Yoshida, K., Imanishi, M., Yoshima, H., Kimura, G., Kuramochi, M., Omae, T. (1992) Acute depressor effect of alcohol in patients with essential hypertension. *Hypertension*: **20**, 219–26.

Keefer, C. S. (1930) The beri-beri heart. *Archives of Internal Medicine*: **45**, 1–22.

Keil, U., Swales, J. D. and Grobbee, D. E. (1993) Alcohol intake and its relation to hypertension, in *Health Issues Related to Alcohol Consumption* (ed. P. M. Verschuren). ILSI Press, Washington, DC, pp. 17–42.

Keil, U., Chambless, L. E., Doring, A., Filipiak, B., Stieber, J. (1997) The relation of alcohol intake to coronary heart disease and all-cause mortality in a beer-drinking population. (See comments.) *Epidemiology*: **8**, 150–6.

Kiechl, S., Willeit, J., Poewe, W., Egger, G., Oberhollenzer, F., Muggeo, M., Bonora, E. (1996) Insulin sensitivity and regular alcohol consumption: large, prospective, cross sectional population study (Bruneck study). (See comments.) *British Medical Journal*: **313**, 1040–4.

Kitamura, A., Iso, H., Sankai, T., Naito, Y., Sato, S., Kiyama, M., Okamura, T., Nakagawa, Y., Iida, M., Shimamoto, T., Komachi, Y. (1998) Alcohol intake and premature coronary heart disease in urban Japanese men. *American Journal of Epidemiology*: **147**, 59–65.

Klatsky, A. L. (1994) Epidemiology of coronary heart disease – influence of alcohol. *Alcoholism, Clinical and Experimental Research*: **18**, 88–96.

Klatsky, A. L. (1995) Blood pressure and alcohol intake, in *Hypertension: Pathophysiology, Diagnosis, and Management* (eds J. H. Laragh and B. M. Brenner), 2nd edn. Raven Press, New York, pp. 2649–67.

Klatsky, A. (2000) Alcohol and hypertension, in *Hypertension* (eds S. Operil and M. Weber). W. B. Saunders, Philadelphia, pp. 211–20.

Klatsky, A. L. (2001) Should patients with heart disease drink alcohol? *Journal of the American Medical Association*: **285**, 2004–6.

Klatsky, A. L. and Armstrong, M. A. (1993) Alcoholic beverage choice and risk of coronary artery disease mortality: Do red wine drinkers fare best? *American Journal of Cardiology*: 71, 467–9.

Klatsky, A. L., Friedman, G. D. and Siegelaub, A. B. (1974) Alcohol consumption before myocardial infarction: Results from the Kaiser-Permanente epidemiologic study of myocardial infarction. *Annals of Internal Medicine*: **81**, 294–301.

Klatsky, A. L., Friedman, G. D., Siegelaub, A. B., Gerard, M. J. (1977) Alcohol consumption and blood pressure: Kaiser-Permanente multiphasic health examination data.' *New England Journal of Medicine*: **296**, 1194–2000.

Klatsky, A. L., Friedman, G. D. and Siegelaub, A. B. (1981a) Alcohol and mortality. A ten-year Kaiser Permanente experience. *Annals of Internal Medicine*: **95**, 139–45.

Klatsky, A. L., Friedman, G. D. and Siegelaub, A. B. (1981b) Alcohol use and cardiovascular disease: the Kaiser Permanente experience. *Circulation*: **64** (suppl. III), 32–41.

Klatsky, A. L., Armstrong, M. A. and Friedman, G. D. (1986a) Relations of alcoholic beverage use to subsequent coronary artery disease hospitalizations. *American Journal of Cardiology*: **58**, 710–14.

Klatsky, A. L., Friedman, G. D. and Armstrong, M. A. (1986b) The relationship between alcoholic beverage use and other traits to blood pressure: A new Kaiser Permanente study. *Circulation*: **73**, 628–36.

Klatsky, A. L., Armstrong, M. A. and Friedman, G. D. (1989) Alcohol use and subsequent cerebrovascular disease hospitalizations. *Stroke*: **20**, 741–6.

Klatsky, A. L., Armstrong, M. A. and Friedman, G. D. (1990) Risk of cardiovascular mortality in alcohol drinkers, ex-drinkers and nondrinkers. *American Journal of Cardiology*: **66**, 1237–42.

Klatsky, A. L., Armstrong, M. A. and Friedman, G. D. (1992) Alcohol and mortality. *Annals of Internal Medicine*: **117**, 646–54.

Klatsky, A. L., Armstrong, M. A. and Friedman, G. D. (1997) Red wine, white wine, liquor, beer, and risk for coronary artery disease hospitalization. *American Journal of Cardiology*: **80**, 416–20.

Klatsky, A. L., Armstrong, M. A. and Poggi, J. (2000) Risk of pulmonary embolism and/or deep venous thrombosis in Asian-Americans. *American Journal of Cardiology*: **85**, 1334–7.

Klatsky, A. L., Armstrong, M. A., Friedman, G. D., Sidney, S. (2001) Alcohol drinking and risk of hospitalization for ischemic stroke. *American Journal of Cardiology*: **88**, 703–6.

Klatsky, A. L., Armstrong, M. A., Sidney, S., Friedman, G. D. (2002) Alcohol drinking and risk of hemorrhagic stroke. *Neuroepidemiology*: **21**, 115–22.

Kluft, C., Veenstra, J., Schaafsma, G., Pikaar, N. A. (1990) Regular moderate wine consumption for five weeks increases plasma activity of the plasminogen activator inhibitor-1 (PAI-1) in healthy young volunteers. *Fibrinolysis*: 4 (suppl. 2), 69–70.

Kojima, S., Kawano, Y., Abe, H., Sanai, T., Yoshida, K., Imanishi, M., Ashida, T., Kimura, G., Yoshimi, H., Matsuoka, H., Omae, T. (1993) Acute effects of alcohol on blood pressure and erythrocyte sodium concentration. *Journal of Hypertension*: **11**, 185–90.

Kono, S., Ikeda, M., Tokudome, S., Nishizumi, M., Kuratsune, M. (1986) Alcohol and mortality: a cohort study of male Japanese physicians. *International Journal of Epidemiology*: **15**, 527–32.

Kono, S., Handa, K., Kawano, T., Hiroki, T., Ishihara, Y., Arakawa, K. (1991) Alcohol intake and nonfatal acute myocardial infarction in Japan. *American Journal of Cardiology*: **68**, 1011–14.

Kozararevic, D., McGee, D., Vojvodic, N., Racic, Z., Dawber, T., Gordon, T., Zukel, W. (1980) Frequency of alcohol consumption and morbidity and mortality: The Yugoslavia Cardiovascular Disease Study. *Lancet*: 1, 613–16.

Landolfi, R. and Steiner, M. (1984) Ethanol raises prostacyclin *in vivo* and *in vitro*. *Blood*: **64**, 679–82.

Langer, R. D., Criqui, M. H. and Reed, D. M. (1992) Lipoproteins and blood pressure as biologic pathways for the effect of moderate alcohol consumption on coronary heart disease. *Circulation*: **85**, 910–15.

Lapasota, E. A. and Lange, L. G. (1986) Presence of nonoxidative ethanol metabolism in human organs commonly damaged by ethanol abuse.' *Science*: **231**, 497–9.

LaPorte, R. E., Cresanta, J. L. and Kuller, L. H. (1980) The relationship of alcohol consumption to atherosclerotic heart disease. *Preventive Medicine*: **9**, 22–40.

Laug, W. E. (1983) Ethyl alcohol enhances plasminogen activator secretion by endothelial cells. *Journal of the American Medical Association*: **250**, 772–6.

Lazarus, R., Sparrow, D. and Weiss, S. T. (1997) Alcohol intake and insulin levels. The Normative Aging Study. *American Journal of Epidemiology*: **145**, 909–16.

Lian, C. (1915) L'alcoolisme causé d'hypertension arterielle. *Bulletin de l' Academie de Médecine (Paris)*: **74**, 525–8.

Lochner, A., Cowley, R. and Brink, A. J. (1969) Effect of ethanol on metabolism and function of perfused rat heart. *American Heart Journal*: **78**, 770–80.

MacKenzie, J. (1902) *The Study of the Pulse*. Y. J. Pentland, Edinburgh.

Maclure, M. (1993) Demonstration of deductive meta-analysis: Ethanol intake and risk of myocardial infarction. *Epidemiology Review*: **15**, 328–51.

MacMahon, S. (1987) Alcohol consumption and hypertension. *Hypertension*: **9**, 111–21.

Maki, T., Toivonen, L., Koskinen, P., Naveri, H., Harkonen, M., Leinonen, H. (1998) Effect of ethanol drinking, hangover, and exercise on adrenergic activity and heart rate variability in patients with a history of alcohol-induced atrial fibrillation. *American Journal of Cardiology*: **82**, 317–22.

Marmot, M. and Brunner, E. (1991) Alcohol and cardiovascular disease: The status of the U-shaped curve. *British Medical Journal*: **303**, 565–8.

Masarei, J. R. L., Puddey, I. B., Rouse, I. L., Lynch, W. J., Vandongen, R., Beilin, L. (1986) Effects of alcohol consumption on serum lipoprotein-lipid and apolipoprotein concentrations. *Atherosclerosis*: **60**, 79–87.

Mathews, Jr, E. C., Gardin, J. M., Henry, W. L., Del Negro, A. A., Fletcher, R. D., Snow, J. A., Epstein, S. E. (1981) Echocardiographic abnormalities in chronic alcoholics with and without overt congestive heart failure. *American Journal of Cardiology*: **47**, 570–8.

Maxwell, S., Cruickshank, A. and Thorpe, G. (1994) Red wine and antioxidant activity in serum. (Letter.) *Lancet*: **344**, 193–4.

Mayer, E. J., Newman, B., Quesenberry, C. P. J., Friedman, G. D., Selby, J. V. (1993) Alcohol consumption and insulin concentrations. Role of insulin in associations of alcohol intake with high-density lipoprotein cholesterol and triglycerides. *Circulation*: **88**, 2190–7.

Meade, T. W., Chakrabarti, R., Haines, A. P., North, W. R., Stirling, Y. (1979) Characteristics affecting fibrinolytic activity and plasma fibrinogen concentrations. *British Medical Journal*: 1, 153–6.

Mikhailidis, D. P., Jeremy, J. Y., Barradas, M. A., Green, N., Dandona, P. (1983) Effects of ethanol on vascular prostacyclin (prostaglandin) I2 synthesis, platelet aggregation, and platelet thromboxane release. *British Medical Journal*: **287**, 1495–8.

Moore, R. D. and Pearson, T. A. (1986) Moderate alcohol consumption and coronary artery disease. A review. *Medicine*: **65**, 242–67.

Moore, R. D., Smith, C. R., Kwiterovich, P. O., Pearson, T. A. (1988) Effect of low-dose alcohol use versus abstention on apolipoproteins A-I and B. *American Journal of Medicine*: **84**, 884–96.

Morin, Y. and Daniel, P. (1967) Quebec beer-drinkers' cardiomyopathy: etiological considerations. *Canadian Medical Association Journal*: **97**, 926–8.

Moushmoush, B. and Abi-Mansour, P. (1991) Alcohol and the heart. The long-term effects of alcohol on the cardiovascular system. *Archives of Internal Medicine*: **151**, 36–42.

Orlando, J. F., Aronow, W. S. and Cassidy, J. (1976) Effect of ethanol on angina pectoris. *Annals of Internal Medicine*: **842**, 652–5.

Osler, W. (1899) *The Principles and Practice of Medicine*. Appleton, New York.

Parker, M., Puddey, I. B., Beilin, L. J., Vandongen, R. (1990) A 2-way factorial study of alcohol and salt restriction in treated hypertensive men. *Hypertension*: **16**, 398–406.

Pearson, T. A. and Terry, P. (1994) What to advise patients about drinking alcohol. *Journal of the American Medical Association*: **272**, 957–8.

Potter, J. F. and Beevers, D. G. (1984) Pressor effect of alcohol in hypertension. *Lancet*: **1**, 119–22.

Potter, J. F. and Beevers, D. G. (1991) Factors determining the acute pressor response to alcohol. *Clinical and Experimeutal Hypertension A*: **13**, 13–34.

Puddey, I. B., Beilin, I. J., Vandongen, R., Rouse, I. L., Rogers, P. (1985) Evidence of a direct effect of alcohol consumption on blood pressure in normotensive men: A randomized controlled trial. *Hypertension*: 7, 707–13.

Puddey, I. B., Beilin, L. J. and Vandongen, R. (1987) Regular alcohol use raises blood pressure in treated hypertensive subjects. A randomized controlled trial. *Lancet*: **i**, 647–51.

Puddey, I. B., Parker, M. and Beilin, L. J. (1990) Alcohol restriction and weight reduction have independent and additive effects in lowering blood pressure – a randomized controlled intervention study in overweight men. (Abstract.) *Journal of Hypertension*: **8** (suppl. 3), S 33.

Randin, D., Vollenweider, P., Tappy, L., Jequier, E., Nicod, P., Scherrer, U. (1995) Suppression of alcohol-induced hypertension by dexamethasone. *New England Journal of Medicine*: **332**, 1733–7.

Regan, T. J. (1984) Alcoholic cardiomyopathy. *Progress in Cardiovascular Diseases*: **27**, 141–52.

Regan, T. J., Koroxenidis, G., Moschos, C. B., Oldewurtel, H. A., Lehan, P. H., Hellems, H. K. (1966) The acute metabolic and hemodynamic responses of the left ventricle to ethanol. *Journal of Clinical Investigation*: **45**, 270–80.

Regan, T. J., Levinson, G. E., Oldewurtel, H. A., Frank, M. J., Weisse, A. B. and Moschos, C. B. (1969) Ventricular function in noncardiacs with alcoholic fatty liver: role of ethanol in the production of cardiomyopathy. *Joural of Clinical Investigation*: **48**, 397–407.

Rehm, J. T., Bondy, S. J., Sempos, C. T., Vuong, C. V. (1997) Alcohol consumption and coronary heart disease morbidity and mortality. *American Journal of Epidemiology*: **146**, 495–501.

Renaud, S. and de Lorgeril, M. (1992) Wine, alcohol, platelets, and the French paradox for coronary heart disease. *Lancet*: **339**, 1523–6.

Renaud, S., Beswick, A. D., Fehily, A. M., Sharp, D. S., Elwood, P. C. (1992) Alcohol and platelet aggregation: The Caerphilly Prospective Heart Disease Study. *American Journal of Clinical Nutrition*: **55**, 1012–17.

Renaud, S., Criqui, M. H., Farchi, G., Veenstra, J. (1993) Alcohol drinking and coronary heart disease, in *Health Issues Related to Alcohol Consumption* (ed. P. M.). Verschuren ILSI Press, Washington, DC, pp. 81–124.

Renaud, S. C., Gueguen, R., Schenker, J., d'Houtaud, A. (1998) Alcohol and mortality in middle-aged men from eastern France. *Epidemiology*: **9**, 184–8.

Reynolds, E. S. (1901) An account of the epidemic outbreak of arsenical poisoning occurring in beer drinkers in the North of England and the Midland Counties in 1900. *Lancet*: 1, 166–70.

Richardson, P. J., Patel, V. B. and Preedy, V. R. (1998) Alcohol and the myocardium, in *Alcohol and Cardiovascular Diseases (Novartis Foundation Symposium 216)* (eds D. J. Chadwick and J. A. Goode), Wiley, Chichester, pp. 2–18.

Rifkind, B. M. (1990) High-density lipoprotein cholesterol and coronary artery disease: Survey of the evidence. *American Journal of Cardiology*: **66**, 3A–6A.

Rimm, E., Giovannucci, E. L., Willett, W. C., Colditz, G. A., Ascherio, A., Rosner, B., Stampfer, M. J. (1991) Prospective study of alcohol consumption and risk of coronary heart disease in men. *Lancet*: **388**, 464–8.

Rimm, E., Klatsky, A. L., Grobbee, D., Stampfer, M. J. (1996) Review of moderate alcohol consumption and reduced risk of coronary heart disease: Is the effect due to beer, wine, or spirits? *British Medical Journal*: **312**, 731–6.

Royal Commission Appointed to Inquire into Arsenical Poisoning from the Consumption of Beer and other Articles of Feed or Drink (1903) *Final Report, Part I*. Wyman and Sons, London.

Ruebner, B. H., Miyai, K. and Abbey, H. (1961) The low incidence of myocardial infarction in hepatic cirrhosis – A statistical artefact? *Lancet*: 2, 1435–6.

Sacco, R. L., Elkind, M., Boden-Albala, B., Lin, I. F., Kargman, D. E., Hauser, W. A., Shea, S., Paik, M. C. (1999) The protective effect of moderate alcohol consumption on ischemic stroke. *Journal of the American Medical Association*: **281**, 53–60.

Salonen, J. T., Seppanen, K. and Rauramma, R. (1988) Serum high density lipoprotein cholesterol subfractions and the risk of acute myocardial infarction: A population study in Eastern Finland. *Circulation*: **78** (suppl. II, abstract), II–281.

Sanders, V. (1963) Idiopathic disease of myocardium. *Archives of Internal Medicine*: **112**, 661–76.

Shaper, A. G., Wannamethee, G. and Walker, M. (1988) Alcohol and mortality in British men: Explaining the U-shaped curve. *Lancet*: 2, 1267–73.

Siemann, E. H. and Creasy, L. L. (1992) Concentration of the phytoalexin resveratrol in wine. *American Journal of Enology and Viticulture*: **43**, 1–4.

Siscovick, D. S., Weiss, N. S. and Fox, N. (1986) Moderate alcohol consumption and primary cardiac arrest. *American Journal of Epidemiology*: **123**, 499–503.

Solomon, C. G., Hu, F. B., Stampfer, M. J., Colditz, G. A., Speizer, F. E., Rimm, E. B., Willett, W. C. and Manson, J. E. (2000) Moderate alcohol consumption and risk of coronary heart disease among women with type 2 diabetes mellitus. *Circulation*: **102**, 494–9.

Stampfer, M. J., Colditz, G. A., Willett, W. C., Speizer, F. E., Hennekens, C. H. (1988) Prospective study of moderate alcohol consumption and the risk of coronary disease and stroke in women. *New England Journal of Medicine*: **319**, 267–73.

Stason, W. B., Neff, R. K., Miettinen, O. S., Jick, H. (1976) Alcohol consumption and nonfatal myocardial infarction. *American Journal of Epidemiology*: **104**, 603–8.

Steell, G. (1893) Heart failure as a result of chronic alcoholism. *Medical Chronicles of Manchester*: **18**, 1–22.

Steell, G. (1906) *Textbook on Diseases of the Heart*. Blakiston, Philadelphia.

St Leger, A. S., Cochrane, A. L. and Moore, F. (1979) Factors associated with cardiac mortality in developed countries with particular reference to the consumption of wine. *Lancet*: **1**, 1017–20.

Strumpel, A. (1890) *A Textbook of Medicine*. Appleton, New York.

Suh, I., Shaten, J., Cutler, J. A., Kuller, L. (1992) Alcohol use and mortality from coronary heart disease: The role of high-density-lipoprotein cholesterol. *Annals of Internal Medicine*: **116**, 881–7.

Sweetham, P., Bainton, D. and Baker, I. (1989) High density lipoproteins (HDL) subclass and coronary heart disease (CHD) risk factors in British men: The Speedwell Study. *Circulation*: **80** (suppl. II, abstract), II–207.

Thomas, A. P., Rozanski, D. J., Renard, D. C., Rubin, E. (1994) Effects of ethanol on the contractile function of the heart: a review. *Alcoholism, Clinical and Experimental Research*: **18**, 121–31.

Thrift, A. G., Donnan, G. A. and McNeil, J. J. (1999) Heavy drinking, but not moderate or intermediate drinking, increases the risk of intracerebral hemorrhage. *Epidemiology*: **10**, 307–12.

Thun, M. J., Peto, R., Lopez, A. D., Monaco, J. H., Henley, S. J., Heath, Jr, C. W., Doll, R. (1997) Alcohol consumption and mortality among middle-aged and elderly U.S. adults. *New England Journal of Medicine*: **337**, 1705–14.

Truelsen, T., Gronbaek, M., Schnohr, P., Boysen, G. (1998) Intake of beer, wine, and spirits and risk of stroke: the Copenhagen city heart study. *Stroke*: **29**, 2467–72.

Urbano-Marquez, A., Estrich, R., Navarro-Lopez, F., Grau, J. M., Rubin, E. (1989) The effects of alcoholism on skeletal and cardiac muscle. *New England Journal of Medicine*: **320**, 409–15.

Van Gign, J., Stampfer, M. J., Wolfe, C., Algra, A. (1993) The association between alcohol consumption and stroke, in *Health Issues Related to Alcohol Consumption* (ed. P. M. Verschuren), ILSI Press, Washington, DC, pp. 43–80.

Vaquez, H. (1921) In *Maladies du Coeur*, Baillière et fils, Paris, pp. 308.

Virtamo, J., Rapola, J. M., Ripatti, S., Heinonen, O. P., Taylor, P. R., Albanes, D., Huttunen, J. K. (1998) Effect of vitamin E and beta carotene on the incidence of primary nonfatal myocardial infarction and fatal coronary heart disease. *Archives of Internal Medicine*: **158**, 668–75.

Walsche, W. H. (1873) *Diseases of the heart and great vessels*. Smith, Elder, London.

Wei, J. Y. and Bulkley, B. H. (1982) Myocardial infarction before age 36 years in women: Predominance of apparent nonatherosclerotic events. *American Heart Journal*: **104**, 561–6.

Wilens, S. L. (1947) Relationship of chronic alcoholism to atherosclerosis. *Journal of the American Medical Association*: **135**, 1136–9.

Witteman, J. C., Willett, W. C., Stampfer, M. J., Colditz, G. A., Kok, F. J., Sacks, F. M., Speizer, F. E., Rosner, B., Hennekens, C. H. (1990) Relation of moderate alcohol consumption and risk of systemic hypertension in women. *American Journal of Cardiology*: **65**, 633–7.

Yano, K., Reed, D. M. and McGee, D. L. (1984) Ten-year incidence of coronary heart disease in the Honolulu Heart Program: Relationship to biologic and lifestyle characteristics. *American Journal of Epidemiology*: **119**, 653–66.

Yano, K., Rhoads, G. G. and Kagan, A. (1977) Coffee, alcohol, and risk of coronary heart disease among Japanese men living in Hawaii. *New England Journal of Medicine*: **297**, 405–9.

Zakhari, S. and Gordis, E. (1999) Moderate drinking and cardiovascular health. *Proceedings of the Association of American Physicians*: **111**, 148–58.

# 7 Wine flavonoids, LDL cholesterol oxidation and atherosclerosis

*M. Aviram and B. Fuhrman*

## LDL cholesterol oxidation and atherosclerosis

Atherosclerosis is the leading cause of morbidity and mortality in the Western world. The early atherosclerotic lesion is characterized by arterial foam cells, which are derived from cholesterol-loaded macrophages (Schaffner *et al.* 1980; Gerrity 1981). Most of the accumulated cholesterol in foam cells originates from plasma low-density lipoprotein (LDL), which is internalized into the cells via the LDL receptor. However, native LDL does not induce cellular cholesterol accumulation, since LDL receptor activity is down-regulated by the cellular cholesterol content (Brown and Goldstein 1986; Goldstein and Brown 1990). LDL has to undergo some modifications, such as oxidation, in order to be taken up by macrophages at an enhanced rate via the macrophage scavenger receptors, which are not subjected to down-regulation by increased cellular cholesterol content (Steinberg *et al.* 1989; Aviram 1993 a,b).

The oxidative modification hypothesis of atherosclerosis proposes that LDL oxidation plays a pivotal role in early atherogenesis (Witztum and Steinberg 1991; Parthasarathy and Rankin 1992; Aviram 1995, 1996; Berliner and Heinecke 1996; Jialal and Devaraj 1996; Steinberg 1997; Parthasarathy *et al.* 1998; Kaplan and Aviram 1999). This hypothesis is supported by evidence that LDL oxidation occurs *in vivo* (Aviram 1996; Herttuala 1998) and contributes to the clinical manifestation of atherosclerosis. Oxidized LDL (Ox-LDL) is more atherogenic than native LDL, since it contributes to cellular accumulation of cholesterol and oxidized lipids, and to foam cell formation (Steinberg *et al.* 1989; Aviram 1991, 1993 a, b; Herttuala 1998;). In addition, Ox-LDL atherogenicity is related to recruitment of monocytes into the intima (Kim *et al.* 1994), to stimulation of monocyte adhesion to the endothelium (Khan *et al.* 1995), and to cytotoxicity toward arterial cells (Penn and Chisolm 1994; Rangaswamy *et al.* 1997).

The process of LDL oxidation is unlikely to occur in plasma because the plasma contains high concentrations of antioxidants and metal ion chelators. Therefore, LDL oxidation is more likely to occur within the artery wall, which is an environment depleted of antioxidants. All major cells of the arterial wall, including endothelial cells, smooth muscle cells and monocyte- derived macrophages can oxidize LDL (Heinecke *et al.* 1984; Witztum and Steinberg 1984; Parthasarathy *et al.* 1986; Aviram and Rosenblat 1994; Aviram *et al.* 1996). Macrophage-mediated oxidation of LDL is a key event during early atherosclerosis and requires the binding of LDL to the macrophage LDL receptor under oxidative stress (Aviram and Rosenblat 1994). Macrophage-mediated oxidation of LDL is considerably affected by the oxidative state in the cells, which depends on the balance between cellular oxygenases and macrophage-associated antioxidants (Aviram and Fuhrman 1998a). Macrophage-mediated oxidation of LDL can also result from an initial cellular lipid peroxidation. When cultured macrophages were exposed to oxidative stress, such as ferrous ions, cellular lipid peroxidation was shown to occur (Fuhrman *et al.* 1994, 1997). These 'oxidized macrophages' could easily oxidize the LDL lipids, even in the absence of any added transition metal ions. LDL oxidation by oxidized macrophages can also result from the transfer of peroxidized lipids from the cell membranes to the LDL particle.

The oxidative state of LDL is also affected by paraoxonase. Human serum paraoxonase (PON1) is an enzyme with esterase activity, which is physically associated with high-density lipoprotein (HDL), and is also distributed in tissues such as liver, kidney and intestine (La Du *et al.* 1993; Mackness *et al.* 1996). HDL-associated PON1 has recently been shown to protect LDL, as well as HDL particles, against oxidation induced by either copper ions or by free radical generators (Aviram *et al.* 1998 a,b). This effect of PON1 may be relevant to its beneficial properties against cardiovascular disease (Aviram 1999; La Du *et al.* 1999), since human serum paraoxonase activity has been shown to be inversely related to the risk of cardiovascular disease (Aviram 1999; La Du *et al.* 1999). On the other hand, PON1 was found to be inactivated by lipid peroxides, and antioxidants were shown to preserve PON1's activity because they decrease the formation of lipid peroxides (Aviram *et al.* 1999).

## Flavonoids and coronary artery disease

Consumption of flavonoids in the diet was previously shown to be inversely associated with morbidity and mortality from coronary heart disease (Hertog *et al.* 1995). The average daily human intake of flavonoids varies between as low as 25 mg to as high as 1g (Hertog *et al.* 1993a; Leibovitz and Mueller 1993; Hollman 1997; Bravo 1998; de Vries *et al.* 1998). Following oral intake, some of the ingested flavonoids are absorbed from the gastrointestinal tract, and some of the absorbed flavonoids are metabolized by the gastrointestinal microflora. The bioavailability and metabolic modifications of flavonoids determine the antioxidative capacity of these potent antioxidants *in vivo*. Different classes of flavonoids are present in different fruits and vegetables, and also in beverages such as tea and wine. Flavonoids may prevent coronary artery disease by inhibiting LDL oxidation, macrophage foam cell formation and atherosclerosis (Rice-Evans *et al.* 1995; Catapano 1997; Aviram and

Fuhrman 1998b; Fuhrman and Aviram 2001a, b, c). Flavonoids can reduce LDL lipid peroxidation by acting as free radical scavengers, metal ion chelators, or by sparing LDL-associated antioxidants. Flavonoids can also reduce macrophage oxidative stress by inhibition of cellular oxygenases such as the NADPH oxidase and also by activating cellular antioxidants such as the glutathione system (Elliott *et al.* 1992; Rosenblat *et al.* 1999). The effect of flavonoids on cell-mediated oxidation of LDL is determined by their accumulation in lipoprotein on the one hand, and in arterial cells, such as macrophages, on the other (Hsiech *et al.* 1988; Luiz da Silva *et al.* 1998; Aviram and Fuhrman 1998b; Rosenblat *et al.* 1999). The antioxidant activity of the flavonoids is related to their chemical structure (Rice-Evans *et al.* 1996; van Acker *et al.* 1996).

## Wine polyphenols inhibit LDL oxidation and cardiovascular diseases

The 'French Paradox', i.e. a low incidence of cardiovascular events despite a diet high in saturated fat, was attributed to the regular drinking of red wine in southern France (Renaud and de Lorgeril 1992). Wine has been part of the human culture for over 6000 years, serving dietary and socioreligious functions. The beneficial effect of red wine consumption against the development of atherosclerosis was attributed in part to its alcohol, but mostly to the antioxidant activity of its polyphenols. In addition to ethanol, red wine contains a wide range of polyphenols with important biological activities (Hertog *et al.* 1993b; Soleas *et al.* 1997), including the flavonols, quercetin and myricetin (10–20 mg/l), the flavanols, catechin and epi(gallo)catechin (up to 270 mg/l), gallic acid (95 mg/l), condensed tannins (catechin and epicatechin polymers, 2500 mg/l) and also polymeric anthocyanidins (Figure 7.1).

### In vitro *studies*

Phenolic substances in red wine were shown to inhibit LDL oxidation *in vitro* (Frankel *et al.* 1993b). In previous studies, red wine-derived phenolic acids (Nardini *et al.* 1995; Abu-Amsha *et al.* 1996), resveratrol (Frankel *et al.* 1993a), flavonols (quercetin, myricetin) (Manach *et al.* 1995; Vinson *et al.* 1995; Luiz da Silva *et al.* 1998), catechins (Hsiech *et al.* 1988; Salah *et al.* 1995) and the grape extract itself (Lanninghamfoster *et al.* 1995; Rao and Yatcilla 2000) have been shown to possess antioxidant properties. The finding that ethanol and wine stripped of phenols did not affect LDL oxidation, further confirmed that the active antioxidant components in red wine are phenolic compounds (Kerry and Abbey 1997). Red wine fractionation showed monomeric catechins, procyanidins, monomeric anthocyanidins and phenolic acids to possess major antioxidative potency (Kerry and Abbey 1997).

The phenolic compounds in red wine are derived from the grape's skin, as well as from its seeds, stems and pulp, all of which are important sources of

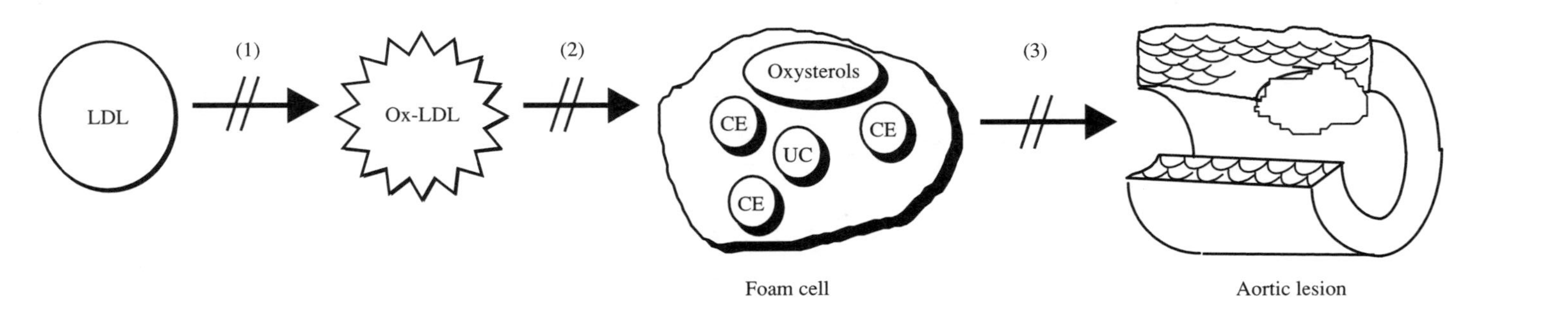


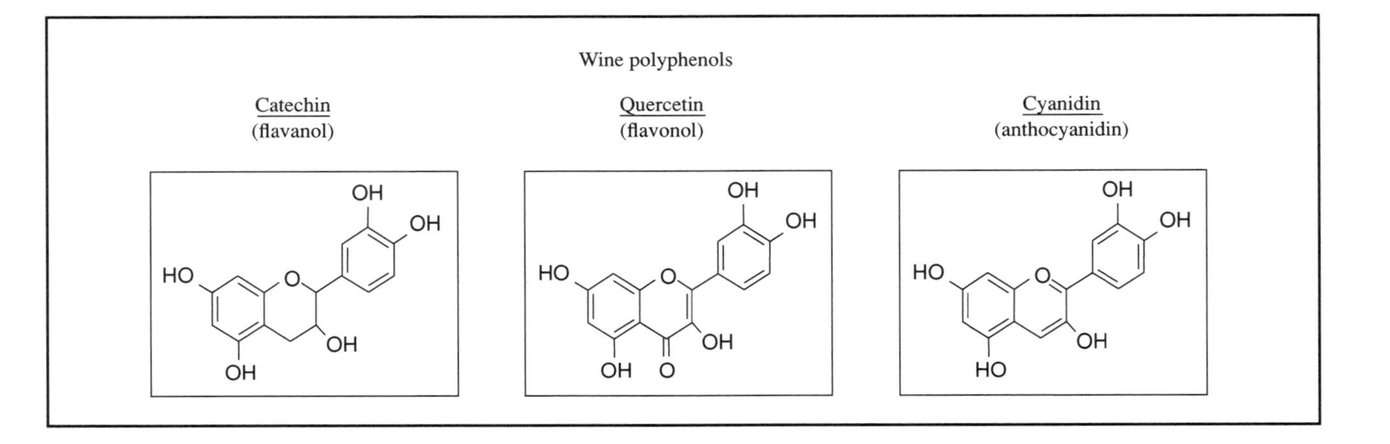


*Figure 7.1* Antiatherogenic effects of wine polyphenols. Wine polyphenol consumption inhibits LDL oxidation (1), macrophage–foam cell formation (2) and the development of aortic atherosclerotic lesions (3). Major polyphenols in red wine are shown, along with their chemical structure. Ox-LDL, oxidized LDL; UC, unesterified cholesterol; CE, cholesteryl ester.

flavanols that are transferred to the wine during initial preparation, together with the grape juice at the first stage of wine fermentation. On the other hand, white wines are usually made from the free-running juice, without the grape mash or contact with the skin. This is thought to be the main reason for the relatively low polyphenol content and for the low antioxidant activity of white wine compared with red (Vinson and Hontz 1995; Fuhrman and Aviram 1996; Serafini *et al.* 1998; Lamuela-Raventos and de la Torre-Boronat 1999; Paganga *et al.* 1999; Rifici *et al.* 1999).

In previous studies, red wine, which contains much higher concentrations of polyphenols than white wine, was shown to be more effective in inhibiting LDL oxidation (Frankel *et al.* 1993b; Frankel *et al.* 1995; Fuhrman *et al.* 1995; Lairon and Amiot 1999). Recently, we produced white wine with red wine-like antioxidant characteristics by increasing the polyphenol content of the white wine (Fuhrman *et al.* 2001). This was achieved by allowing grape skin contact for a short period of time in the presence of added alcohol, in order to facilitate the extraction of grape skin polyphenols into the wine. We have analysed the antioxidant capacity of white wine samples obtained from whole squeezed grapes that were stored for increasing periods of time before removal of the skin, or from whole squeezed grapes to which increasing concentrations of alcohol were added. White wine obtained from whole squeezed grapes that were incubated for 18 h with 18% alcohol, contained 60% higher concentration of polyphenols than untreated white wine, and exhibited a significant antioxidant capacity against LDL oxidation, almost similar to that of red wine. A maximal free radical scavenging capacity (as measured by a 79% reduction in the optical density at 517 nm of a 1,1-diphenyl-2-picryl-hydrazyl (DPPH) solution) was induced by the white wine sample derived from whole squeezed grapes that were pre-incubated with 18% alcohol. This effect was similar to the free radical scavenging capacity exhibited by a similar concentration of red wine (Figure 7.2A). Furthermore, a maximal inhibition by 87% of copper ion-induced LDL oxidation was induced by wine samples derived from whole squeezed grapes that were pre-incubated with 18% alcohol, a very similar inhibition (94%) to that exhibited by red wine (Figure 7.2B). The antioxidant capacity of this white wine was directly proportional to its polyphenol content (Figure 7.2C), and these results are in agreement with previous reports (Lamuela-Raventos and de la Torre-Boronat 1999; Tubaro *et al.* 1999).

Possible mechanisms for the antioxidative effect of wine-polyphenols against LDL oxidation include the binding capacity of the LDL particle and the scavenging of free radicals.

To find out whether the red wine-derived polyphenols, catechin or quercetin provide their antioxidative protection to LDL due to their binding to the lipoprotein, we pre-incubated LDL (1 mg of protein/ml) for 18 h at 37°C with 50 μmol of the pure polyphenols, or with 10% red wine, followed by removal of the unbound flavonoids (by dialysis). Using reverse phase-HPLC (high performance liquid chromatography) with ultraviolet (UV) detection,

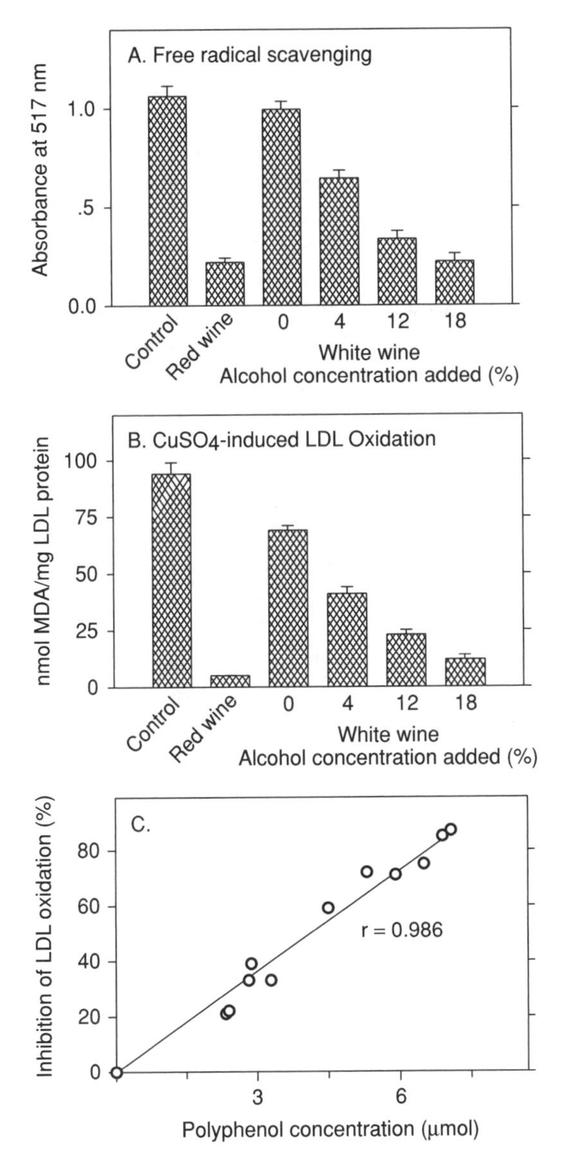


*Figure 7.2* The effect of alcohol added to whole squeezed grapes on the wine polyphenol content and on its antioxidant capacity. Whole squeezed Muscat grapes were incubated for 18 h with increasing concentrations of alcohol up to 18%, after which the juice was separated from grape skins and left to ferment into wine. A. Aliquots of 25 μl/ml from each wine sample were added to DPPH solution (0.1 mmol/l) and the optical density at 517 nm was recorded after 5 min. Results are expressed as mean ± S.D. (n = 3). B. Wine samples at a final concentration of 2 μl/ml were added to LDL (100 mg of protein/l) and incubated with 5 μmol/l $CuSO_4$ for 2 h at 37°C. LDL oxidation was measured by the thiobarbituric acid reactive substances (TBARS) assay. Results are expressed as mean ± S.D. (n = 3). C. Linear regression analysis of total polyphenol concentration in wine and wine-induced inhibition of LDL oxidation.

no significant levels of either catechin or quercetin could be detected in the LDL samples.

As polyphenols interact with surface components of LDL, such as fatty acids (ester bond) or sugar residues (ether bond), we performed an alkaline or acidic hydrolysis of the LDL samples prior to the HPLC analysis of the polyphenols. The alkaline hydrolysis of the LDL samples, did not result in identification of the above flavonoids. However, when acidic hydrolysis was performed on the LDL samples prior to HPLC analysis, both catechin (0.35 nmol/mg LDL protein) and quercetin (1.00 nmol/mg LDL protein) were identified, suggesting a glycosidic bond between the flavonoids and the LDL surface protein/lipid components.

To determine the antioxidant capability of red wine, catechin or quercetin, we performed the DPPH assay. The addition of increasing concentrations of red wine, up to 20 μl/ml, to the DPPH solution, decreased optical absorbance at 517 nm from 1.018 to 0.160 optical density (OD) within 3 min. Addition of 100 μmol of quercetin to the DPPH solution decreased optical absorbance at 517 nm from 1.047 to 0.133 OD within 8 min, whereas during a similar period of time catechin reduced the optical density of the DPPH solution to only 0.676 OD units. These results suggest that red wine possesses free radical scavenging capacity, and that quercetin is more potent in this respect than catechin. These characteristics may contribute to the higher potency of red wine and quercetin, compared with catechin, to inhibit LDL oxidation.

### In vivo *studies in humans*

We have previously demonstrated that LDL derived from human male subjects who consumed red wine (400 ml/day for 2 weeks) was more resistant to lipid peroxidation than LDL obtained from the same subjects before wine consumption (Fuhrman *et al.* 1995). In contrast, the resistance to oxidation of LDL derived from subjects who consumed the same volume of white wine showed no significant change compared with baseline LDL oxidation rates (Fuhrman *et al.* 1995; Fuhrman and Aviram 1996). The administration of 400 ml of red wine to healthy human volunteers for a period of 2 weeks resulted in a substantial prolongation of the lag phase required for the initiation of LDL oxidation (by as much as 130 min) (Figure 7.3A), whereas consumption of a similar volume of white wine had no significant effect on LDL oxidation (Figure 7.3B).

In parallel, the propensity of the volunteers' LDL obtained after wine consumption to copper ion-induced lipid peroxidation was reduced compared with baseline LDL as measured by a 72% decrement in content of lipoprotein-associated lipid peroxides (Figure 7.3C), whereas after white wine consumption no significant effect was observed (Figure 7.3D).

The effects of red wine consumption on LDL oxidation could be related to an elevation in total polyphenol content in the plasma and in the LDL particle. Thus, polyphenolic substances present in red wine but not in white,

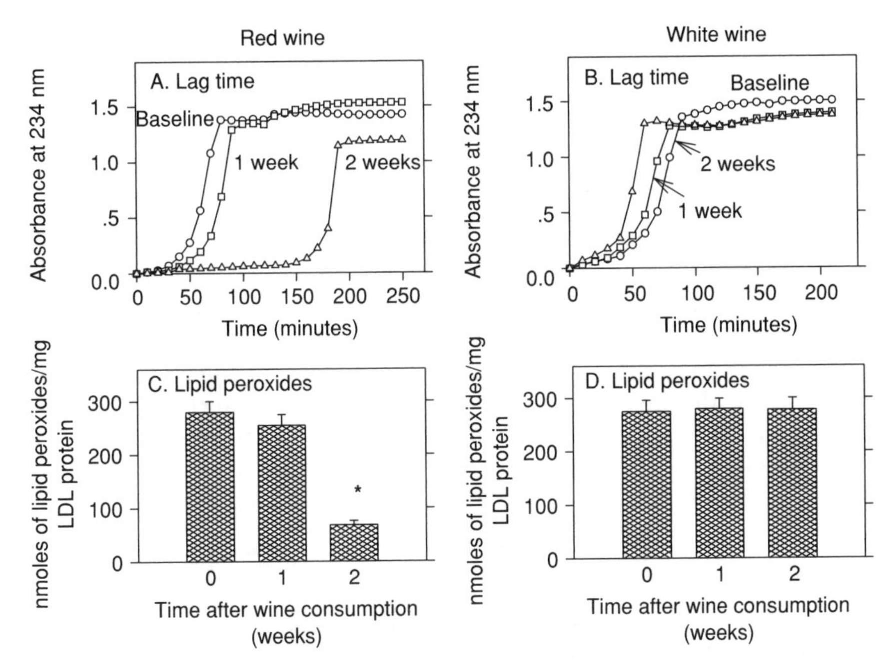


*Figure 7.3* The antioxidative effects of red wine compared with white wine consumption on LDL oxidation. LDL (200 μg of protein/ml) isolated before ('0 time', baseline) or after 1 or 2 weeks of red wine (A, C) or white wine (B, D) consumption, was incubated with $CuSO_4$ (10 μmol/l). A, B: LDL oxidation was kinetically monitored by continuous monitoring of absorbance at 234 nm. C, D: LDL oxidation was determined by measuring the formation of lipid peroxides. Results are expressed as mean ± S.D. (n = 3). *$p < 0.01$ (v. 0 time).

are absorbed, bind to plasma LDL and protect the lipoprotein from oxidation (Fuhrman *et al.* 1995).

Consumption of red wine was shown to inhibit LDL oxidation *ex vivo* in some (Fuhrman *et al.* 1995; Nigdikar *et al.* 1998; Chopra *et al.* 2000), but not in all human intervention studies (de Rijke *et al.* 1996; Caccettea *et al.* 2000); discrepancies in the results may be related to variations in the polyphenol composition of the wines used. In a recent study by Nigdikar *et al.* (1998) consumption of 400 ml of red wine for 2 weeks was shown to increase plasma and LDL-associated polyphenols and to protect LDL against copper ion-induced oxidation, as we have previously shown (Fuhrman *et al.* 1995). In the study of Nigdikar *et al.*, an increased lag time of only 31% was noted, compared with a 290% increase in lag time obtained in our study. We thus hypothesize that the polyphenol composition of wine determines its ability to inhibit LDL oxidation. Comparison of the composition of the red wine used in our studies (Cabernet Sauvignon cultivar grown in Israel) (Fuhrman *et al.* 1995) with that of the red wine used in the study performed in the UK (Cabernet Sauvignon cultivar grown in France) (Nigdikar *et al.* 1998), revealed that although total polyphenol content was similar (1650 and 1800 mg/l, respectively), the major differences were the finding of a sixfold higher content of flavonols (glycosides and aglycones) and of some monomeric anthocyanins in the Israeli red wine compared with the French (Howard *et al.* 2002).

There is a wide variation in the flavonol content of different red wines throughout the world (McDonald *et al.* 1998) and a major determining factor for this phenomenon is related to the amount of sunlight to which the grapes are exposed during cultivation (Price *et al.* 1995). Flavonol synthesis in the skin of the grape is increased in response to sunlight, acting as a yellow filter against the harmful effect of UV light. Thus, the different climatic conditions under which the grapes were grown could explain the increased flavonol content in the Israeli red wines compared with those studied by Nigdikar *et al.* (1998).

### *Studies in atherosclerotic mice in search of mechanisms for wine polyphenol antiatherogenicity*

The direct effect of red wine consumption on the development of atherosclerotic lesions was further studied in atherosclerotic apolipoprotein E-deficient (E°) mice (Hayek *et al.* 1997). Over a period of six weeks, these mice were given in their drinking water a placebo (1.1% alcohol), or 0.5 ml of red wine /day per mouse, or the purified polyphenols catechin or quercetin (50 μg/day/mouse), which are major polyphenols in red wine (Aviram *et al.* 1997; Hayek *et al.* 1997). The atherosclerotic areas in the E° mice consuming red wine, catechin or quercetin were significantly reduced by 40%, 39% and 46%, respectively, compared with the placebo-treated E° mice (Figure 7.4).

The mechanisms responsible for this reduction may be related to red wine-induced inhibition of cellular lipoprotein uptake (Figure 7.5A) or to a reduction in LDL lipid peroxidation, induced either by AAPH (2,2′ azobis, 2-amidinopropane hydrochloride) (Figure 7.5B) or by basal oxidative state (Figure 7.5C). An additional mechanism may be the ability of red wine and quercetin to increase serum paraoxonase activity (Figure 7.6), and thus to stimulate hydrolysis of oxidized lipids in atherosclerotic lesions.

We investigated macrophage uptake of LDL derived from E° mice consuming red wine or red wine-derived polyphenols (Hayek *et al.* 1997). Incubation of J-774 A1 macrophages for 5 h at 37°C with 10 μg of protein/ml of plasma LDL from E° mice obtained after consumption of catechin, quercetin or red wine, resulted in a 31%, 40% or 52% reduction in LDL-induced cellular cholesterol esterification, respectively, compared with LDL derived from the placebo-treated group (Figure 7.5A). These results suggest that reduced atherosclerotic lesion formation in E° mice after polyphenol consumption may be attributed to reduced uptake of their LDL by arterial macrophages, and thus to attenuation of foam cell formation and atherosclerosis.

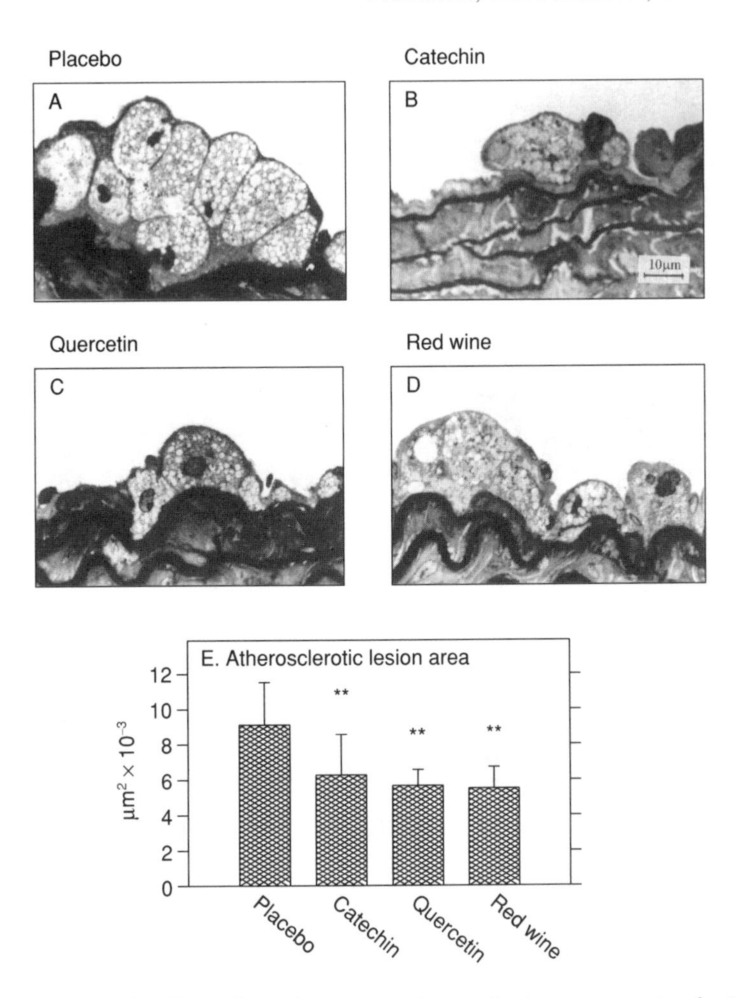


*Figure 7.4* Effects of catechin, quercetin or red wine consumption by E° mice on the size of the atherosclerotic lesion area of their aortic arch. Samples of aortic arch from E° mice that have consumed placebo, catechin, quercetin, or red wine for 6 weeks were analysed. Photomicrographs of typical atherosclerotic lesions of the aortic arch following treatment with placebo (A), catechin (B), quercetin (C), or red wine (D) are shown. The sections were stained with alkaline toluidine blue. All micrographs are at the same magnification. (E): The lesion area is expressed in square micrometers ± S.D. ** $p < 0.05$ v. placebo.

Since enhanced cellular uptake of LDL is also associated with lipoprotein oxidation, we have further investigated the effect of polyphenol consumption on LDL susceptibility to oxidation. LDL (100 μg of protein/ml) derived from E° mice with dietary supplementation for six weeks with placebo, catechin, quercetin or red wine, was incubated with copper ions (10 μmol),

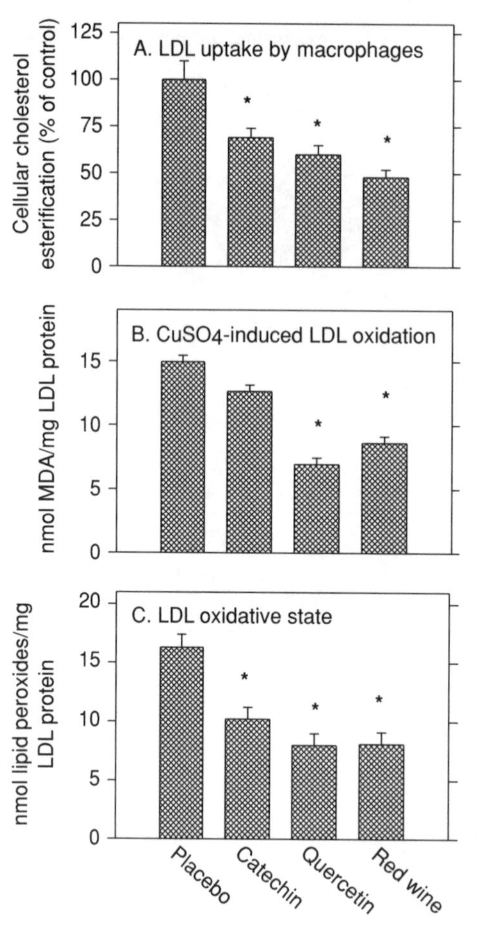


*Figure 7.5* Mechanisms responsible for the antiatherosclerotic effects of red wine consumption in E° mice. A. Macrophage uptake of LDL derived from E° mice consuming catechin, quercetin or red wine. LDL (25 μg of cholesterol/ml) from mice consuming placebo, catechin, quercetin or red wine for 6 weeks, was incubated for 18 h at 37°C with J-774 A1 macrophages. During the last 2 h of incubation, [$^3$H]-oleate complexed with albumin was added to the medium. The rate of [$^3$H]-cholesteryl oleate formation was determined in the lipid extract of the cells after separation by thin layer chromatography. Results are expressed as mean ± S.D. (n = 3). *$p < 0.01$ (v. placebo). B. The effect of catechin, quercetin, or red wine consumption by E° mice on the susceptibility of their LDL to oxidation. LDL (100 μg of protein/ml) derived from E° mice consuming placebo, catechin, quercetin or red wine for 6 weeks, was incubated for 2 h at 37°C with 10 μmol $CuSO_4$. LDL oxidation was determined by measurement of LDL-associated thiobarbituric acid reactive substances (TBARS). Results represent mean ± S.D. (n = 3). *$p < 0.01$ (v. placebo). C. The basal oxidative state of LDL derived from E° mice consuming catechin, quercetin or red wine. Lipid peroxide levels were measured in LDL samples (100 μg of protein/ml) from E° mice consuming placebo, catechin, quercetin, or red wine for 6 weeks. Results are expressed as mean ± S.D. (n = 3). *$p < 0.01$ v. placebo.

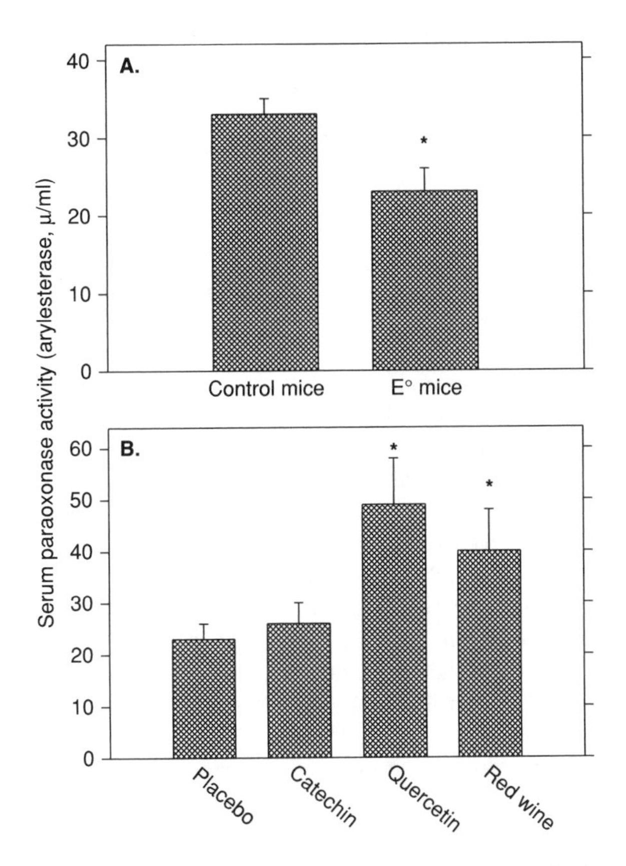


*Figure 7.6* The effect of catechin, quercetin, or red wine consumption by E° mice on their serum paraoxonase activity. Paraoxonase was measured as arylesterase activity. (A) Serum from E° mice was compared with that of control mice; (B) serum from E° mice consuming placebo, catechin, quercetin or red wine for 6 weeks. Results are expressed as mean ± S.D. of three separate determinations. *$p < 0.01$ (v. placebo).

with the free radical initiator AAPH (5 mmol), for 2 at 37°C, or with J-774A1 cultured macrophages under oxidative stress (2 μmol $CuSO_4$). Copper ion-induced oxidation of LDL derived from E° mice after consumption of red wine or quercetin was delayed by 120 min, whereas the onset of lipid peroxidation in LDL derived from E° mice consuming catechin was retarded by only 40 min compared with LDL from the placebo group.

Furthermore, quercetin or red wine consumption resulted in a 54% and 43% reduction in copper ion-induced oxidation, measured as TBARS (Figure 7.5B – see caption), or as lipid peroxide-formation, respectively. Similarly, an 83% and 81% reduction in AAPH-induced oxidation, and a 33% and 30% inhibition in macrophage-mediated oxidation, respectively, were noted. These results suggest that consumption of red wine or its flavonol, quercetin, substantially inhibit LDL oxidative modification by various types of oxidation inducer.

Using E° mice, which are under oxidative stress (their blood contains lipid peroxides even when oxidation has not been chemically induced), we have also analysed the effect of dietary supplementation of polyphenols on the oxidative state of their LDL under basal conditions (not induced by copper ions or by AAPH). LDL derived from E° mice that have consumed catechin, quercetin or red wine for six weeks was less oxidized compared with LDL from mice on placebo. Evidence for this was provided by a 39%, 48% and 49% reduction in levels of LDL-associated lipid peroxides, respectively (Figure 7.5C).

After acid hydrolysis of the LDL samples, we found that LDL obtained after red wine consumption by E° mice contained 3.65 nmol of catechin and 3.00 nmol of quercetin/mg LDL protein. No flavonoids could be identified in LDL samples derived from the placebo-treated mice.

Paraoxonase, the HDL-associated esterase that can hydrolyse specific oxidized lipids, is inactivated under oxidative stress (Aviram *et al.* 1999), and thus potent antioxidants may preserve the enzyme activity. Serum paraoxonase activity in E° mice (the plasma of which is oxidized and highly susceptible to oxidation) was found to be lower by 27% compared with activity in control mice (23 ± 3 units/ml in E° mice v. 33 ± 2 units/ml in control mice) (Figure 7.6A). In serum derived from E° mice after two weeks of catechin, quercetin or red wine consumption, serum paraoxonase activity (measured as arylesterase activity) was higher by 14%, 113% and 75%, respectively, compared with serum obtained from placebo-treated E° mice (Hayek *et al.* 1997) (Figure 7.6B). The increased levels of serum paraoxonase manifested in E° mice that have consumed red wine flavonoids can contribute to the reduction in LDL oxidation, secondary to hydrolysis of LDL-associated lipid peroxides, by paraoxonase.

## Summary

The current view of the mechanisms involved in wine polyphenol protection against macrophage foam cell formation and atherosclerosis is illustrated in Figure 7.7. Following wine ingestion, wine-derived polyphenols bind to the LDL particle and protect the lipoprotein against lipid peroxidation. In the presence of antioxidants, under reduced oxidative stress, the activity of HDL-associated paraoxonase, which can hydrolyse oxidized lipids, is preserved, thus further reducing the oxidative stress on lipoproteins (LDL and HDL). Wine-polyphenols, which accumulate in arterial wall cells, including macrophages, also protect them from lipid peroxidation, thus reducing macrophage atherogenicity (decreased cellular lipoprotein uptake and macrophage-mediated oxidation of LDL).

In conclusion, then, wine polyphenols interact with lipoproteins and with arterial macrophages, reduce foam cell formation and attenuate the development of the atherosclerotic lesion.

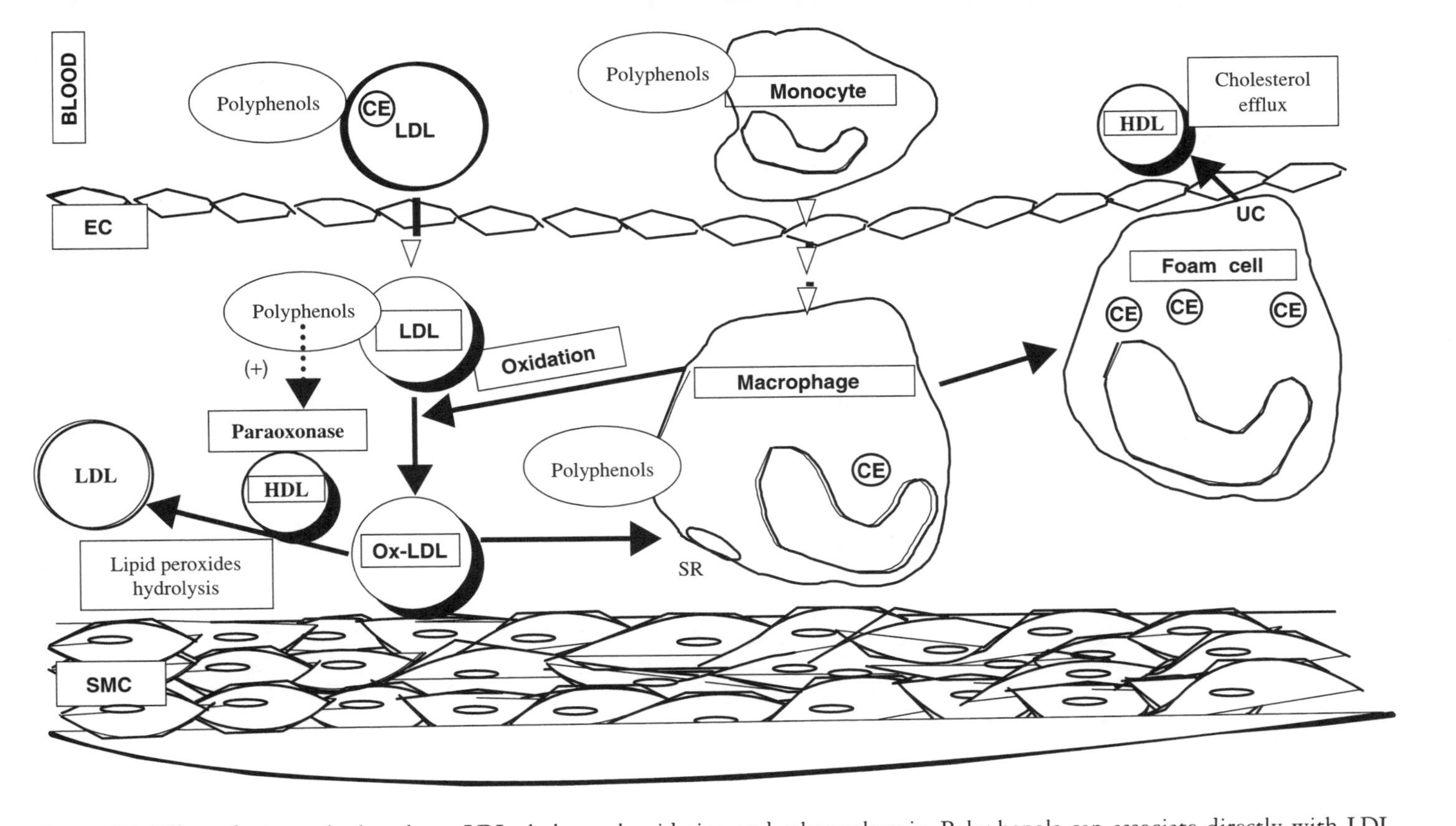


*Figure 7.7* Effect of wine polyphenols on LDL cholesterol oxidation and atherosclerosis. Polyphenols can associate directly with LDL, resulting in the inhibition of LDL oxidation. Polyphenols can also associate with arterial cells such as monocytes/macrophages, resulting in the inhibition of macrophage-mediated oxidation of LDL. Furthermore, polyphenols preserve the activity of the HDL-associated enzyme, paraoxonase (PON1), which further protects lipids in lipoproteins and in cells from oxidation. Altogether, these effects lead to a reduced formation of macrophage-foam cells, and attenuate development of the atherosclerotic lesion.

## References

Abu-Amsha, R., K. D. Croft, I. B. Puddey, J. M. Proudfoot, L. J. Beilin (1996) Phenolic content of various beverages determines the extent of inhibition of human serum and low density lipoprotein oxidation *in vitro*: identification and mechanism of some cinnamic derivatives from red wine. *Clinical Science*: **91**, 449–58.

Aviram, M. (1991) The contribution of the macrophage receptor for oxidized LDL to its cellular uptake. *Biochemical and Biophysical Research Communications*: **179**, 359–65.

Aviram, M. (1993a) Beyond cholesterol: Modifications of lipoproteins and increased atherogenicity, in *Atherosclerosis Inflammation and Thrombosis* (eds Neri Serneri, G. G., Gensini, G. F., Abbate R., and Prisco, D.), pp. 15–36, Scientific Press, Florence, Italy.

Aviram, M. (1993b) Modified forms of low density lipoprotein and atherosclerosis. *Atherosclerosis*: **98**, 1–9.

Aviram, M. (1995) Oxidative modification of low density lipoprotein and atherosclerosis. *Israel Journal of Medical Science*: **31**, 241–9.

Aviram, M. (1996) Interaction of oxidized low density lipoprotein with macrophages in atherosclerosis and the antiatherogenicity of antioxidants. *European Journal of Clinical Chemistry and Clinical Biochemistry*: **34**, 599–608.

Aviram M. (1999) Does paraoxonase play a role in susceptibility to cardiovascular disease? *Molecular Medicine*: **5**, 381–6.

Aviram, M. and B. Fuhrman (1998a) LDL oxidation by arterial wall macrophages depends on the antioxidative status in the lipoprotein and in the cells: role of prooxidants vs. antioxidants. *Molecular and Cellular Biochemistry*: **188**, 149–59.

Aviram, M. and B. Fuhrman (1998b) Polyphenolic flavonoids inhibit macrophage-mediated oxidation of LDL and attenuate atherogenesis. *Atherosclerosis*: **137** (suppl.), S45–S50.

Aviram, M. and M. Rosenblat (1994) Macrophage mediated oxidation of extracellular low density lipoprotein requires an initial binding of the lipoprotein to its receptor. *Journal of Lipid Research*: **35**, 385–98.

Aviram, M., M. Rosenblat, A. Etzioni, A. and R. Levy (1996) Activation of NADPH oxidase is required for macrophage-mediated oxidation of low density lipoprotein. *Metabolism*: **45**, 1069–79.

Aviram, M., T. Hayek, B. Fuhrman (1997) Red wine consumption inhibits LDL oxidation and aggregation in humans and in atherosclerotic mice. *BioFactors*: **6**, 415–19.

Aviram M., M. Rosenblat, C. L. Bisgaier, R. S. Newton, S. L. Primo-Parmo, and B. N. La Du (1998a) Paraoxonase inhibits high density lipoprotein (HDL) oxidation and preserves its functions: A possible peroxidative role for paraoxonase. *Journal of Clinical Investigation*: **101**, 1581–90.

Aviram M., S. Billecke, R. Sorenson, C. Bisgaier, R. Newton, M. Rosenblat, J. Erogul, C. Hsu, C. Dunlop, and B. La Du (1998b) Paraoxonase active site required for protection against LDL oxidation involves its free sulfhydryl group and is different from that required for its arylesterase/paraoxonase activities: selective action of human paraoxonase allozymes Q and R. *Arteriosclerosis, Thrombosis and Vascular Biology*: **18**, 1617–24.

Aviram M., M. Rosenblat, S. Billecke, J. Erogul, R. Sorenson, C. L. Bisgaier, R. S. Newton, and B. La Du (1999) Human serum paraoxonase (PON 1) is inactivated by oxidized low density lipoprotein and preserved by antioxidants. *Free Radical Biology and Medicine*: **26**, 892–904.

Berliner, J. A. and J. W. Heinecke (1996) The role of oxidized lipoproteins in atherosclerosis. *Free Radical Biology and Medicine*: **20**, 707–27.

Bravo, L. (1998) Polyphenols: chemistry, dietary sources, metabolism, and nutritional significance. *Nutrition Reviews*: **56**, 317–33.

Brown, M. S. and J. L. Goldstein (1986) A receptor-mediated pathway for cholesterol homeostasis. *Science*: **232**, 34–47.

Caccettea, R. A., K. D. Croft, L. J. Beilin, I. B. Puddey (2000) Ingestion of red wine significantly increases plasma phenolic acid concentrations but does not acutely affect *ex-vivo* lipoprotein oxidizability. *American Journal of Clinical Nutrition*: **71**, 67–74.

Catapano, A. L. (1997) Antioxidant effect of flavonoids. *Angiology*: **48**, 39–44.

Chopra M., P. E. E. Fitzsimons, J. J. Strain, D. I. Thurnham, A. N. Howard (2000) Non-alcoholic red wine extract and quercetin inhibit LDL oxidation without affecting plasma antioxidant vitamins and carotenoid concentrations. *Clinical Chemistry*: **46**, 1162–70.

de Rijke, Y. B., P. N. Demacker, N. A. Assen, L. M. Sloots, M. B. Katan, A. F. H. Stalenhoef (1996) Red wine consumption does not affect oxidizability of low-density lipoproteins in volunteers. *American Journal of Clinical Nutrition*: **63**, 329–34.

de Vries, J. H., P. C. Hollman, S. Meyboom, M. N. Buysman, P. L. Zock, W. A. van Staveren, M. B. Katan (1998) Plasma concentrations and urinary excretion of the antioxidant flavonols quercetin and kaempferol as biomarkers for dietary intake. *American Journal of Clinical Nutrition*: 68, 60–5.

Elliott, A. J., S. A. Scheiber, C. Thomas, R. S. Pardini (1992) Inhibition of glutathione reductase by flavonoids. A structure-activity study. *Biochemical Pharmacology*: 44, 1603–8.

Frankel, E. N., A. L. Waterhouse, J. E. Kinsella (1993a) Inhibition of human LDL oxidation by resveratrol. *Lancet*: **341**, 1103–4.

Frankel, E. N., J. Kanner, J. B. German, E. Parks, J. E. Kinsella (1993b) Inhibition of oxidation of human low density lipoprotein by phenolic substances in red wine. *Lancet*: **341**, 454–7.

Frankel E. N., A. L. Waterhouse, P. L. Teissedre (1995) Principal phenolic phytochemicals in selected Californian wines and their antioxidant activity in inhibiting oxidation of human low-density lipoproteins. *Journal of Agricultural and Food Chemistry*: **43**, 890–3.

Fuhrman, B. and M. Aviram (1996) White wine reduces the susceptibility of low density lipoprotein to oxidation. *American Journal of Clinical Nutrition*: **63**, 403–4.

Fuhrman, B. and M. Aviram (2001a) Flavonoids protect LDL from oxidation and attenuate atherosclerosis. *Current Opinion in Lipidology*: 12, 41–8.

Fuhrman, B. and M. Aviram (2001b) Anti-atherogenicity of nutritional antioxidants. *International Drugs*: 4, 82–92.

Fuhrman, B. and M. Aviram (2001c) Polyphenols and flavonoids protect LDL against atherogenic modifications, in *Handbook of Antioxidants: Biochemical, Nutritional and Clinical Aspects*, (eds E. Cadenas and L. Packer). 2nd edition, Marcel Dekker, New York, pp. 303–36.

Fuhrman, B., J. Oiknine, M. Aviram (1994) Iron induces lipid peroxidation in cultured macrophages, increases their ability to oxidatively modify LDL and affects their secretory properties. *Atherosclerosis*: **111**, 65–78.

Fuhrman, B., A. Lavy, M. Aviram, (1995) Consumption of red wine with meals reduces the susceptibility of human plasma and LDL to undergo lipid peroxidation *American Journal of Clinical Nutrition*: **61**, 549–54.

Fuhrman, B., J. Oiknine, S. Keidar, M. Kaplan, M. Aviram (1997) Increased uptake of low density lipoprotein (LDL) by oxidized macrophages is the result of enhanced LDL receptor activity and of progressive LDL oxidation. *Free Radical Biology and Medicine*: **23**, 34–46.

Fuhrman, B., N. Volkova, M. Aviram (2001) White wine with red wine-like properties: increased extraction of grape skin's-polyphenols improves the antioxidant capacity of the derived white wine. *Journal of Agricultural and Food Chemistry*: **49**, 3164–8.

Gerrity, R. G. (1981) The role of monocytes in atherogenesis. *American Journal of Pathology*: **103**, 181–90.

Goldstein, J. L. and M. S. Brown (1990) Regulation of the mevalonate pathway. *Nature*: **343**, 425–30.

Hayek, T., B. Fuhrman, J. Vaya, M. Rosenblat, P. Belinky, R. Coleman, A. Elis, M. Aviram (1997) Reduced progression of atherosclerosis in the apolipoprotein E deficient mice following consumption of red wine, or its polyphenols quercetin or catechin, is associated with reduced susceptibility of LDL to oxidation and aggregation. *Arteriosclerosis, Thrombosis and Vascular Biology*: **17**, 2744–52.

Heinecke, J. W., H. Rosen, A. Chait (1984) Iron and copper promote modification of low density lipoprotein by human arterial smooth muscle cells in culture. *Journal of Clinical Investigation*: **74**, 1980–4.

Hertog, M. G. L., P. O. H. Hollman, M. B. Katan, D. Kromhout (1993a) Intake of potentially anticarcinogenic flavonoids and their determinants in adults in The Netherlands. *Nutrition and Cancer*: **20**, 21–9.

Hertog, M. G. L., P. C. H. Hollman, B. van de Putte (1993b) Content of potentially anticarcinogenic flavonoids of tea infusions, wines, and fruit juices. *Journal of Agricultural and Food Chemistry*: **41**, 1242–6.

Hertog, M. G., D. Kromhout, C. Aravanis, H. Blackburn, R. Buzina, F. Fidanza, S. Giampaoli, A. Jansen, A. Menotti, S. Nedeljkovic, M. Pekkarinen, B. S. Simic, H. Toshima, E. J. M. Feskens, P. C. H. Hollman, M. B. Katan (1995) Flavonoid intake and long-term risk of coronary heart disease and cancer in the seven countries study. *Archives of Internal Medicine*: **155**, 381–6.

Herttuala, S. Y. (1998) Is oxidized low density lipoprotein present *in vivo*? *Current Opinion in Lipidology*: **9**, 337–44.

Hollman, P. C. H. (1997) Bioavailability of flavonoids. *European Journal of Clinical Nutrition*: **51** (suppl. 1), S66–S69.

Howard, A., C. Mridula, D. I. Thurnham, J. J. Strain, B. Fuhrman, M. Aviram (2002) Red wine consumption and inhibition of LDL oxidation. *Medical Hypotheses*: **59**, 101–4.

Hsiech, R., B. German, J. Kinsella (1988) Relative inhibitory potencies of flavonoids on 12-lipoxygenase of fish gill. *Lipids*: **23**, 322–6.

Jialal, I. and S. Devaraj (1996) The role of oxidized low density lipoprotein in atherogenesis. *Journal of Nutrition*: **126** (suppl.), 1053S–1057S.

Kaplan, M. and M. Aviram (1999) Oxidized low density lipoprotein: Atherogenic and proinflammatory characteristics during macrophage foam cell formation. An inhibitory role for nutritional antioxidants and serum paraoxonase. *Clinical Chemistry and Laboratory Medicine*: **37**, 777–87.

Kerry, N. L. and M. Abbey (1997) Red wine and fractionated phenolic compounds prepared from red wine inhibit low density lipoprotein oxidation *in vitro*. *Atherosclerosis*: **135**, 93–102.

Khan, N. B. V., S. Parthasarathy, R. W. Alexander (1995) Modified LDL and its constituents augment cytokine-activated vascular cell adhesion molecule-1 gene expression in human vascular endothelial cells. *Journal of Clinical Investigation*: **95**, 1262–70.

Kim, J. A., M. C. Territo, E. Wayner, T. M. Carlos, F. Parhami, C. W. Smith, M. E. Haberland, A. M. Fogelman, J. A. Berliner (1994) Partial characterization of leukocyte binding molecules on endothelial cells induced by minimally oxidized LDL. *Arteriosclerosis and Thrombosis*: **14**, 427–33.

La Du, B. N., S. Adkins, C. L. Kuo, D. Lipsig (1993) Studies on human serum paraoxonase/arylesterase. *Chemico-biological Interactions*: **87**, 25–34.

La Du, B. N., M. Aviram, S. Billecke, M. Navab, S. Primo-Parmo, R. C. Sorenson, T. J. Standiford (1999) On the physiological role(s) of the paraoxonases. *Chemico-biological Interactions*: **119/120**, 379–88.

Lairon, D. and M. J. Amiot (1999) Flavonoids in food and natural antioxidants in wine. *Current Opinion in Lipidology*: **10**, 23–8.

Lamuela-Raventos, R. M. and M. C. de la Torre-Boronat (1999) Beneficial effects of white wines. *Drugs under Experimental and Clinical Research*: **25**, 121–4.

Lanninghamfoster, L., C. Chen, D. S. Chance, G. Loo (1995) Grape extract inhibits lipid peroxidation of human low density lipoprotein. *Biological and Pharmaceutical Bulletin*: **18**, 1347–51.

Leibovitz, B. E. and J. A. Mueller (1993) Bioflavonoids and polyphenols: medical application. *Journal of Optimal Nutrition*: **2**, 17–35.

Luiz da Silva, E., T. Tsushida, J. Terao (1998) Inhibition of mammalian 15-lipoxygenase-dependent lipid peroxidation in low density lipoprotein by quercetin and quercetin monoglucosides. *Archives of Biochemistry and Biophysics*: **349**, 313–20.

Mackness, M. I., B. Mackness, P. N. Durrington, P. W. Connelly, R. A. Hegele (1996) Paraoxonases: biochemistry, genetics and relationship to plasma lipoproteins. *Current Opinion in Lipidology*: **7**, 69–76.

Manach, C., C. Morand, O. Texier, M. I. Favier, G. Agullo, C. Demigne, F. Regerat, C. Remesy (1995) Quercetin metabolites in plasma of rats fed diets containing rutin or quercetin. *Journal of Nutrition*: **125**, 1911–22.

McDonald, M. S., M. Hughes, J. Burns, M. E. J. Lean, D. Matthews, A. Crozier (1998) Survey of free and conjugated myricetin and quercetin content of red wines of different geographical origin. *Journal of Agricultural and Food Chemistry*: **46**, 368–75.

Nardini, M., M. D'Aquino, G. Tomassi, V. Gentili, M. Di Felice, C. Scaccini (1995) Inhibition of human low density lipoprotein oxidation by caffeic acid and other hydroxycinnamic acid derivatives. *Free Radical Biology and Medicine*: **19**, 541–52.

Nigdikar, S. V., N. Williams, B. A. Griffin, A. H. Howard (1998) Consumption of red wine polyphenols reduces the susceptibility of low density lipoproteins to oxidation *in vivo*. *American Journal of Clinical Nutrition*: **68**, 258–65.

Paganga, G., N. Miller, C. A. Rice-Evans (1999) The polyphenolic content of fruit and vegetables and their antioxidant activities. What does a serving constitute? *Free Radical Research*: **30**, 153–62.

Parthasarathy, S. and S. M. Rankin (1992) The role of oxidized LDL in atherogenesis. *Progress in Lipid Research*: **31**, 127–43.

Parthasarathy, S., D. J. Printz, D. Boyd, L. Joy, D. Steinberg (1986) Macrophage oxidation of low density lipoprotein generates a modified form recognized by the scavenger receptor. *Arteriosclerosis*: **6**, 505–10.

Parthasarathy, S., N. Santanam, N. Auge (1998) Oxidized low-density lipoprotein, a two-faced janus in coronary artery disease? *Biochemical Pharmacology*: **56**, 279–84.

Penn, M. S. and G. M. Chisolm (1994) Oxidized lipoproteins, altered cell function and atherosclerosis. *Atherosclerosis*: **108**, S21–S29.

Price, S. F., P. J. Breen, M. Valladao, B. T. Watson (1995) Cluster sun exposure and quercetin in Pinot noir grapes and wine. *American Journal of Enology and Viticulture*: **46**, 187–94.

Rangaswamy, S., M. S. Penn, G. M. Saidel, G. M. Chisolm (1997) Exogenous oxidized low density lipoprotein injures and alters the barrier function of endothelium in rats *in vivo*. *Circulation Research*: **80**, 37–44.

Rao, A. V. and M. T. Yatcilla (2000) Bioabsorption and *in vivo* antioxidant properties of grape extract BioVin: A human intervention study. *Journal of Medicine and Food*: **3**, 15–22.

Renaud, S. and M. de Lorgeril (1992) Wine alcohol, platelets and the French paradox for coronary heart disease. *Lancet*: **339**, 1523–6.

Rice-Evans, C. A., N. J. Miller, P. G. Bolwell, P. M. Bramley, J. B. Pridham (1995) The relative antioxidant activities of plant-derived polyphenolic flavonoids. *Free Radical Research*: **22**, 375–83.

Rice-Evans, C. A., N. J. Miller, G. Paganga (1996) Structure-antioxidant activity relationships of flavonoids and phenolic acids. *Free Radical Biology and Medicine*: **20**, 933–56.

Rifici, V. A., E. M. Stephan, S. H. Schneider, A. K. Khachadurian (1999) Red wine inhibits the cell-mediated oxidation of LDL and HDL. *Journal of the American College of Nutrition*: **18**, 137–43.

Rosenblat, M., P. Belinky, J. Vaya, R. Levy, T. Hayek, R. Coleman, S. Merchav, M. Aviram, (1999) Macrophage enrichment with the isoflavan glabridin inhibits NADPH oxidase-induced cell mediated oxidation of low density lipoprotein. *Journal of Biological Chemistry*: **274**, 13790–9.

Salah, N., N. J. Miller, G. Paganga, L. Tijburg, G. P. Bolwell, C. Rice-Evans (1995) Polyphenolic flavanols as scavengers of aqueous phase radicals and as chain-breaking antioxidants. *Archives of Biochemistry and Biophysics*: **322**, 339–46.

Schaffner, T., K. Taylor, E. J. Bartucci, K. Fischer-Dzoga, J. H. Beenson, S. Glagov, R. Wissler (1980) Arterial foam cells with distinctive immuno-morphologic and histochemical features of macrophages. *American Journal of Pathology*: **100**, 57–80.

Serafini, M., G. Maiani, A. Ferro-Luzzi (1998) Alcohol-free red wine enhances plasma antioxidant capacity in humans. *Journal of Nutrition*: **128**, 1003–7.

Soleas, G. J., E. P. Diamandis, D. M. Goldberg (1997) Wine as a biological fluid: history, production, and role in disease prevention. *Journal of Clinical and Laboratory Analysis*: **11**, 287–313.

Steinberg, D. (1997) Low density lipoprotein oxidation and its pathobiological significance. *Journal of Biological Chemistry*: **272**, 20963–6.

Steinberg, D., S. Parthasarathy, T. E. Carew, J. C. Khoo, J. L. Witztum (1989) Beyond cholesterol: modifications of low-density lipoprotein that increase its atherogenicity. *New England Journal of Medicine*: **320**, 915–24.

Tubaro, F., P. Rapuzzi, F. Ursini (1999) Kinetic analysis of antioxidant capacity of wine. *Biofactors*: **9**, 37–47.

van Acker, S. A. B. E., D. J. Van-den Berg, M. N. J. L. Tromp, D. H. Griffioen, W. P. van Bennekom, W. J. F. Van der Vijgh, A. Bast (1996) Structural aspects of antioxidant activity of flavonoids. *Free Radical Biology and Medicine*: **20**, 331–42.

Vinson, J. A. and B. A. Hontz (1995) Phenol antioxidant index: Comparative antioxidant effectiveness of red and white wines. *Journal of Agricultural and Food Chemistry*: **43**, 401–3.

Vinson, J. A., Y. A. Dabbagh, M. M. Serry, J. Janj (1995) Plant flavonoids, especially tea flavonols, are powerful antioxidants using an *in vitro* model for heart disease. *Journal of Agricultural and Food Chemistry*: **45**, 2800–2.

Witztum, J. L. and D. Steinberg (1984) Modification of low density lipoprotein by endothelial cells involves lipid peroxidation and degradation of low density lipoprotein phospholipids. *Proceedings of the National Academy of Sciences of the United States of America*: **81**, 3883–7.

Witztum, J. L. and D. Steinberg (1991) Role of oxidized low density lipoprotein in atherogenesis. *Journal of Clinical Investigation*: **88**, 1785–92.

# 8 Resveratrol: biochemistry, cell biology and the potential role in disease prevention

*D. M. Goldberg and G. J. Soleas*

## Introduction

Since the discovery of *trans*-resveratrol (3,5,4′-trihydroxystilbene) as a constituent of wine by Siemann and Creasy (1992), the possibility that this compound, almost unique to red wine among constituents of the human diet, may in large measure account for the putative health benefits of this beverage beyond its mere content of vulgar ethanol, excited the imagination of the scientific and medical communities, initiating a ferment of research and enquiry that continues to this day. Indeed, ripples of these activities from time to time flow into the pages of the lay press, so that resveratrol has become a molecule impacting the consciousness of many well-informed members of the lay public. In March, 1997 we published a major review incorporating 183 references forming the bulk of the world literature on resveratrol up to that time (Soleas *et al.* 1997a). Our 'bottom line' was that the future of resveratrol did not look particularly promising given the reality that, despite its miraculous performances in the culture dish and the test tube, the intact bodies of mice and men proved to be an inhospitable milieu, robbing it of its presumed powers.

These views were reiterated in a further review presented at a symposium in September 1999, but only published in May 2001 (Soleas *et al.* 2001a). To us, it seemed to be a compound 'for whom the bell tolled', but others did not share this gloomy prognosis. In fact, so much new work on this topic has been published since September 1999 that a re-appraisal of the situation should be welcomed.

Resveratrol exists as *trans* and *cis* isomers. Very little is known about the latter. When the nature of resveratrol is not specified, the reader should assume that the text refers to the *trans* isomer.

Background information concerning its chemical nature, natural occurrence and biosynthesis was extensively presented in our earlier reviews (Soleas *et al.* 1997a, 2001a). The only further preliminary task is to draw attention to other reviews that describe its functions in plant biology (Daniel *et al.* 1999); human health, with special reference to cardiovascular disease (Constant 1997; Lin and Tsai 1999; Wu *et al.* 2001) and cancer (Lin and Tsai 1999; Gusman *et al.*

2001; Pervaiz 2001; Yang *et al.* 2001); its role in preventing inflammation (Surh *et al.* 2001); and its general biological effects (Fremont 2000).

## Inflammation and atherosclerosis

These disease processes involve mechanisms common to both, notably activation of polymorphonuclear leucoytes (PMN) with release of cytokines, and synthesis of pro-inflammatory eicosanoids. Oxygen free-radicals and immune responses also play important roles in their initiation and propagation. By contrast, lipid abnormalities predispose to atherosclerosis but not inflammation as such, while infectious agents are common precursors of inflammation. However, these distinctions are becoming increasingly blurred as our knowledge expands to reveal that in many respects atherosclerosis behaves as an inflammatory disease (Ross 1999), and that micro-organisms may play a role in its aetiology (Nicholson and Hajjar 1998).

### *Cyclo-oxygenase activity and eicosanoid synthesis*

Resveratrol has been shown to modulate a number of metabolic and enzymatic pathways that are central to the inflammatory response, and has been reported to inhibit carrageenin-induced injury in the mouse (Jang *et al.* 1997) and rat (Gentilli *et al.* 2001). These authors attributed the observed protection to down-regulation of cyclo-oxygenase (COX) 1, a constitutively expressed enzyme responsible for the biosynthesis of prostaglandins and thromboxanes, but they were unable to detect resveratrol-induced changes in the COX 1 or COX 2 content of mouse skin cells stimulated by phorbol ester (Jang and Pezzuto 1998, 1999).

Subbaramaiah *et al.* (1998, 1999; Subbaramaiah and Dannenberg 2001) described inhibition by resveratrol of the inducible enzyme COX 2 that is increased following stimulation by mitogens such as phorbol ester. This inhibition operated at several levels: direct enzymatic activity, protein and mRNA synthesis, a cyclic AMP (adenosine monophosphate) response element, protein kinase C, and AP-1-mediated gene expression. Inhibition of prostaglandin synthesis was an important consequence of resveratrol treatment. Their experiments were performed with human mammary and oral epithelial cells, and with resveratrol concentrations in the range 2.5–30 μmol/l. Suppression of the eukaryotic nuclear transcription factor-kappa B (NF-κB) may, at least in part, account for this down-regulation of COX-2 expression (Surh *et al.* 2001). Resveratrol, in common with other resorcin-type structures, suppressed COX-2 promoter activity in DLD-1 human colon cancer cells, with and without prior stimulation by TGF α (Mutoh *et al.* 2000). It appears, therefore, that the alterations in COX gene expression by resveratrol are not identical in different experimental models of tissue damage, and that they may manifest tissue specificity dependent upon the response evoked by the compound on the expression of c-fos.

In actual fact, COX acts in conjunction with a peroxidase to comprise the prostaglandin H synthase (PGHS) multienzyme system. Employing a different *in vitro* model, Johnson and Maddipati (1998) found that resveratrol inhibits PGHS-1, but causes a two-fold increase in the activity of PGHS-2. Inhibition of the peroxidase activity of the former took place with an $IC_{50}$ (the dose or concentration required for 50% inhibition) of 15 μmol/l; the peroxidase activity of the latter was inhibited with an $IC_{50}$ of 200 μmol/l. These data are inconsistent with a number of contemporaneous reports. Resveratrol appears to inhibit the COX activity of PGHS-2 purified from sheep seminal vesicles although with a potency that is only 25–30% of the inhibitory capacity manifested by the resveratrol polymer α-viniferin (Lee *et al.* 1998; Shin *et al.* 1998a). It also inhibits a COX-like enzyme, tentatively identified as a COX-2, from the invertebrate *Ciona intestinalis* (Knight *et al.* 1999). The overexpression of COX-2 and increased protaglandin synthesis induced in murine resident peritoneal macrophages in response to stimulation by lipopolysaccharide (LPS) and phorbol esters was significantly decreased by resveratrol (Martinez and Moreno 2000), which also significantly impaired COX-2 induction of human smooth muscle cells by oxidized low-density lipoprotein (LDL) (Mietus-Snyder *et al.* 2000). The production of prostaglandin-$E_2$ by an osteoblastic cell line was inhibited by resveratrol (Mizutani *et al.* 1998), which also reduced its synthesis as well as platelet-derived growth factor (PDGF)-stimulated phospholipase $A_2$ activity and arachidonate release in 3T6 fibroblasts (Moreno 2000). Leucotriene $B_4$ as well as prostaglandin-$E_2$ were also decreased in human erythroleukaemia K562 cells (MacCarrone *et al.* 1999).

The inhibition by resveratrol of COX and PGSH *in vitro* leads to a marked reduction in eicosanoid production in affected cells. This was first demonstrated for rat peritoneal polymorphonuclear leucoytes (Kimura *et al.* 1985) in which the cyclo-oxygenase pathway (evaluated by the synthesis of hydroxyheptadecatrienoate (HHT) and thromboxane $B_2$) was blocked ($IC_{50}$ around 0.5–1 μmol/l), as was the 5-lipoxygenase pathway (assessed by 5-HETE (5-hydroxyeicosatetraenoate) production, $IC_{50}$ 2.72 μmol/l). It was subsequently shown that resveratrol inhibited the 5-lipoxygenase and 15-lipoxygenase pathways in washed neutrophils from healthy human subjects with $IC_{50}$ concentrations of 22.4 μmol/l and 8.7 μmol/l, respectively (Goldberg *et al.* 1997), in line with the earlier observations of Kimura and colleagues (1995) who reported that resveratrol prevented the formation of an array of 5-lipoxygenase products in human leucoytes ($IC_{50}$ values 1.37–8.90 μmol/l).

The synthesis of eicosanoids by platelets is also blocked by resveratrol. A series of papers by Chinese investigators described its inhibition of thromboxane $B_2$ production from arachidonate in rabbit platelets (Shan 1988; Shan *et al.* 1990; Chung *et al.* 1992). Similar inhibition of thromboxane $B_2$ synthesis in human platelets was reported, as well as a modest reduction in the activity of the platelet 12-lipoxygenase pathway leading to the production of pro-atherogenic hepoxillins (Pace-Asciak *et al.* 1995).

## *Thrombus formation*

Aggregation of platelets is a prelude to thrombus formation, the mechanism that precipitates coronary artery occlusion in the majority of patients who go on to develop acute myocardial infarction. This phenomenon is prevented by resveratrol in rabbit platelets (Shan *et al.* 1990; Chung *et al.* 1992) and in human platelets (Pace-Asciak *et al.* 1995); in the latter, the $IC_{50}$ is around 10 μmol/l with ADP (adenosine 5′-diphosphate) or thrombin as agonist. Bertelli and colleagues (1995, 1996a) reported an $IC_{50}$ for *trans*-resveratrol with human platelets of 15 nmol/l and a slightly lower value for *cis*-resveratrol when collagen was employed as agonist. These are orders of magnitude less than $IC_{50}$ values for various biological effects reported by virtually all other investigators. Moreover, no information was provided about the nature of the *cis*-resveratrol whose synthesis has never been reported, apart from the name of the institution from which it was obtained. Confirmation of the antiplatelet aggregation activity of resveratrol as well as of its glucoside has been provided (Orsini *et al.* 1997a).

One of the mechanisms thought to underlie the platelet-aggregating activity of thrombin is its propensity to increase intracellular calcium-ion concentrations, an effect that was blocked by resveratrol in thrombin-stimulated as well as unstimulated human platelets, with a relatively low $IC_{50}$ of 0.5 μmol/l (Dobrydneva *et al.* 1999). Platelet aggregation is accompanied by the generation of significant amounts of free radicals and reactive oxygen species. These effects were blocked in pig platelets stimulated with either endotoxin (LPS) or by thrombin, but only at very high resveratrol concentrations (Olas *et al.* 1999, 2001a). The two agonists displayed very different interactions with resveratrol on the release of adenine nucleotides and secretory proteins, both being inhibited by resveratrol in thrombin-stimulated platelets and enhanced in LPS-stimulated platelets (Olas 2001b).

Besides platelet aggregation, fibrin formation is a major component of thrombosis. Protection against this event is afforded through the action of the proteolytic enzyme plasmin, generated from its precursor plasminogen, which digests the fibrin clot. The conversion of plasminogen to plasmin is mediated through at least two classes of plasminogen activators synthesized and secreted by endothelial cells. Alcohol is a potent stimulant of plasminogen activator expression in these cells (see Goldberg and Soleas 2001 for review); it has now been shown that resveratrol as well as a number of other wine polyphenols upregulates gene transcription of both plasminogen activators in human vascular endothelial cells in the concentration range 0.001–10 μmol/l (Abou-Agag *et al.* 2001). Taken together, the inhibitory *in vitro* effects of resveratrol upon platelet aggregation and its stimulation of fibrinolysis are consistent with potential antithrombotic activity *in vivo* that, among wine drinkers, would be enhanced by the alcohol content of their preferred beverage.

### *Cell adhesion*

Contact between circulatory blood cells, especially polymorphonuclear leucoytes and monocytes, and the vascular endothelium is a necessary prelude to the entry of these cells into the underlying intimal layers where the latter can undergo transformation into macrophages that then take up lipids, particularly oxidized LDL, to become the 'foam cells' characteristic of the early atherosclerotic lesion known as the 'fatty streak'. This contact is facilitated by an array of adhesion molecules whose expression is upregulated by endothelial damage and cytokine signalling typically observed in atherosclerosis (Cavenagh *et al.* 1998; Chia 1998). Resveratrol, in concentrations ranging from 100 nmol/l to 1 μmol/l blocked the expression of at least two of these adhesion molecules in tumour necrosis factor (TNF) α-stimulated human umbilical vein endothelial cells (Ferrero *et al.* 1998). It also reduced the expression of the $\beta_2$ integrin cell adhesion molecule MAC-1 on the surface of activated human polymorphonuclear leucoytes (Rotondo *et al.* 1998). Both of these functions could have important implications for the putative antiatherosclerotic role of resveratrol.

### *Lysosomal enzymes*

The release of lysosomal enzymes from activated PMN leucoytes causes degranulation and contributes to inflammatory damage in adjacent tissues. Resveratrol prevents this secretion in cells stimulated by the calcium ionophore A23187, but the $IC_{50}$ has been reported to be around 0.1–1 mmol/l by one group (Kimura *et al.* 1995) and around 30 μmol/l by other investigators (Rotondo *et al.* 1998). The liberation of β-hexosaminidase from cultured RBL-2H3 cells was also inhibited by resveratrol with an $IC_{50}$ of 14 μmol/l (Cheong *et al.* 1999).

### *Cytokine production*

Formation of fibrous tissue to replace necrotic cells is a fundamental and irreversible part of the chronic inflammatory process. In the liver, stellate cells subserve this function. Kawada *et al.* (1998) demonstrated inhibition by resveratrol of rat hepatic stellate cell proliferation using an *in vitro* model. Simultaneously, the following functions were decreased: activity of mitogen-activated protein (MAP) kinase; concentration of the cell cycle protein cyclin D1; production of nitric oxide and of the pro-inflammatory cytokine TNF-α. These effects would be expected to reduce both acute and chronic inflammatory reactions in the liver, but the resveratrol concentrations used were quite high (10–100 μmol/l). A possible protective effect of resveratrol against immunological hepatocyte damage assessed by release of the enzyme alanine aminotransferase into the medium has recently been demonstrated in experiments with cultured hepatocytes (Chen *et al.* 1999).

An important inflammatory pathway involves activation of the nuclear transcription factor NF-κB which promotes the synthesis of several cytokines, including TNFα, and nitric oxide (NO); the former can then lead to the release of pro-inflammatory tissue factor (TF). Resveratrol seems to disrupt this pathway, although there is some disagreement about how these effects are accomplished. Tsai *et al.* (1999) described inhibition of NO generation accompanied by down-regulation of NF-κB in a macrophage cell line stimulated by LPS at a resveratrol concentration of 30 μmol/l. Similar findings were reported for human monocyte and macrophage cell lines at the same resveratrol concentration following stimulation with either TNF or LPS (Holmes-McNary and Baldwin 2000). Wadsworth and Koop (1999) reported that, in concentrations in the range of 50–100 μmol/l, resveratrol did not inhibit LPS-induced activation of NF-κB in the same cell line. It did reduce LPS-induced NO release but enhanced LPS-induced production of TNFα. Oxidized lipoproteins activate NF-κB binding to the promoter region of target genes in PC12 cells, a rat phaeochromocytoma cell line, but resveratrol protects the cells against this activation and apoptotic cell death that follows (Draczynska-Lusiak *et al.* 1998a). All three groups measured NF-κB activation by the same technique (electrophoretic mobility shift assay). The importance of the inhibition of NF-κB activation in accounting for the ability of resveratrol to down-regulate COX-2 and inducible NO synthase has been emphasized in a recent review (Surh *et al.* 2001).

A further set of investigators found that resveratrol displayed a dose-dependent inhibition of TF expression in endothelial cells stimulated by a variety of agonists, including TNFα and LPS (Pendurthi *et al.* 1999). They also showed that resveratrol inhibited LPS-induced expression of TNFα and IL-$1_{\beta}$ in endothelial cells and monocytes. However, these phenomena could not be attributed to activation of transcription factors (including NF-κB) necessary for induction of the TF promoter in these cells. Resveratrol has also been reported to block the dioxin-induced increase of IL-$1_{\beta}$ in an endometrial adenocarcinoma cell line (Casper *et al.* 1999), as well as the release of IL-6 from calcium ionophore-stimulated mouse peritoneal macrophages (Zhong *et al.* 1999). Further, production of phorbol ester-induced TGF-β1 in mouse skin was inhibited by resveratrol (Jang and Pezzuto 1998, 1999), although these authors were unable to detect an increased TNFα content in these cells. To summarize, it does appear that resveratrol can attenuate the production of cytokines by vascular cells and peripheral blood cells, but the mechanisms involved remain to be elucidated, especially the issue of whether these effects are indirectly due to radical scavenging activity or whether they are associated with direct alteration of gene expression.

### *Lipids and lipoproteins*

The possibility that resveratrol may modulate lipid metabolism was first proposed by Arichi *et al.* (1982), who provided resveratrol both orally and intraperitoneally to rats and mice fed a high cholesterol diet, and noted reduced deposition of cholesterol and triglyceride in the livers of these animals, as well as a diminished rate of hepatic triglyceride synthesis. Our group utilized the human hepatoma cell line Hep G2 to study the effects of resveratrol upon lipid and lipoprotein metabolism (Goldberg *et al.* 1995a, 1997). The intracellular content of cholesteryl esters and the rate of secretion of both cholesteryl esters and triglycerides were reduced in a dose-dependent manner over concentrations of 1–50 μmol/l. Under these conditions the intracellular content and rates of secretion of apolipoprotein B (the main protein of very low-density lipoprotein (VLDL) and LDL) and of apolipoprotein AI (the main protein of high-density lipoprotein (HDL)) were also reduced; since the former change would tend to prevent and the latter to augment atherosclerosis, these effects would seem to cancel each other out. When wines of high and low resveratrol content were administered to healthy humans for a period of four weeks, there was no major difference in the plasma lipid and apolipoprotein responses between the two experimental groups (Goldberg *et al.* 1996). An absence of change in serum lipoprotein patterns in the rat following the intraperitoneal injection of large doses of resveratrol has also been reported (Turrens *et al.* 1997).

On balance, it appears that resveratrol does not have a beneficial effect on circulating lipid or lipoprotein concentrations and that its antiatherosclerotic properties *in vivo*, if any, are not attributable to such effects. Indeed, a disturbing report published a few years ago described an *increase* in the area of aortic atherosclerosis visualized in resveratrol-fed hypercholesterolaemic rabbits compared with controls (Wilson *et al.* 1996). The resveratrol was given in a dose of 0.6 mg/kg during the first five days and 1 mg/kg from days 6–60, but this amount (up to 3 mg) was stated to be dissolved in 0.05 ml of ethanol, vastly in excess of its solubility.

## Cell growth, proliferation and cancer

### *General*

When added to cultured Hep G2 cells in concentrations ranging from 1–50 μmol/l, resveratrol did not alter the following functions over time periods of up to seven days: number of cells per plate; cell viability as gauged by trypan blue exclusion and lactate dehydrogenase efflux; incorporation of [$^{14}$C]-leucine into cell proteins. [$^{14}$C]-Thymidine incorporation into DNA was stable for three days and showed a sharp increase at day 7, but only at 50 μmol concentration (Goldberg *et al.* 1995a). Two years later Jang *et al.* (1997) reported that resveratrol (1–25 μmol/l) inhibited the initiation and promotion of hydrocarbon-induced skin cancer in the mouse as well as the progression of breast cancer in the same animal. A potent antimutagenic activity of resveratrol was also demonstrated (Uenobe *et al.* 1997). Subsequently, resveratrol has been shown to behave as an antiproliferative agent in oestrogen-dependent as well as oestrogen-independent human breast epithelial cells (Mgbonyebi *et al.* 1998) and breast cancer cells (Damianaki *et al.* 2000); in a human oral cancer cell line (ElAttar and Virji 1999); in androgen-responsive and androgen-non-responsive human prostate cancer cell lines (Hsieh and Wu 1999, 2000), although inhibition of androgen-receptor function in these cells was independently demonstrated (Mitchell *et al.* 1999); and in the Yoshida AH-130 ascites hepatoma inoculated into rats (Carbo *et al.* 1999).

Many other reports have broadened these findings. For example, resveratrol may be more active upon aggressively metastatic cancer cells than upon those that exhibit a lesser invasive potential (Hsieh *et al.* 1999a). In line with this notion, it inhibited the metastatic properties of hepatoma cells at a concentration of 25 μmol/l, whereas concentrations >50 μmol/l were required to suppress the proliferation of these cells (Kozuki *et al.* 2001). It also attenuated the invasive properties of Hep G2 cells stimulated by hepatocyte growth factor as evaluated using the Boyden chamber assay, and while this was accompanied by decreased cell proliferation, cytotoxicity and apoptosis were not enhanced (De Ledinghen *et al.* 2001). Resveratrol and certain of its glucosides (including polydatin) appear to inhibit the metastasis as well as the growth of tumours transplanted into mice, although the free resveratrol did so at lower doses (Kimura and Okuda 2000, 2001).

Relevant to the above is the observation that resveratrol in concentrations as low as 0.5 μmol/l attenuates the growth of a human osteoblast cell line in response to conditioned media from a wide range of tumour cell lines grown in culture (Ulsperger *et al.* 1999). An important and practical conundrum surrounding the interaction of resveratrol with cancer cells is whether the former is able to suppress the production by the latter of tumour-specific cancer marker proteins. Hsieh and Wu (1999) reported a decrease in the intracellular and secreted prostate-specific antigen (PSA) content of an androgen-sensitive prostate cancer cell line. We have been unable to demonstrate an effect of resveratrol upon the secretion of either PSA or carcinoembryonic antigen (CEA) in four different breast carcinoma cell lines, nor could we demonstrate any changes in cancer-associated p53 gene expression even though the presence of genetic mutations in these cell lines was structurally determined by DNA analysis (Soleas *et al.* 2001b).

These findings suggest that, if it should prove to merit consideration as a chemotherapeutic agent as opposed to a cancer-preventive lifestyle nutrient, its role may be in adjunct therapy for advanced cancers. Of some concern is the apparent potential of resveratrol to suppress the proliferation of normal human skin cells (keratinocytes) even at low (0.25 μmol/l) concentrations, but true cytotoxicity was not observed below concentrations of 40 μmol/l (Molian and Walter 2001).

### *Apoptosis*

One mechanism responsible for its anticancer behaviour seems to be induction by resveratrol of apoptosis. This was first described in resveratrol-treated

HL-60 cells, a human leukaemia cell line, and was mediated by a dose-dependent increase in intracellular caspases as well as CD-95L expression (Clement *et al.* 1998; Surh *et al.* 1999). By contrast, normal human lymphocytes were unaffected. In lymphoblasts expressing wild-type p53, but not in p53-deficient cells, resveratrol suppressed tumour promoter-induced cell transformation accompanied by apoptosis together with transactivation of p53 activity and expression of p53 protein (Huang *et al.* 1999). The dose responses for these apparently related phenomena manifested a similar pattern. However, we observed only minor and inconsistent effects of resveratrol upon p53 expression in a number of human cancer cell lines in which we rigorously identified p53 gene status by DNA sequence analysis (Soleas *et al.* 2001b).

Leukaemic and related cells have been extensively used to probe further the relationship between the anticancer effects of resveratrol and apoptosis. The T-cell-derived leukaemic cell line CEM-C7H2, which is deficient in the tumour-suppressor gene products p53 and p16, underwent S-phase arrest and apoptosis accompanied by massive proteolytic activation of caspase-6 and lesser activation of caspase-2 and caspase-3 in the presence of 20 μmol/l resveratrol (Bernhard *et al.* 2000). It also induced apoptosis in a range of mouse and human leukaemia cell lines at concentrations that did not inhibit normal haematopoietic progenitor cells, suggesting a possible role in the *ex vivo* purging of bone marrow autografts (Gautam *et al.* 2000). In similar fashion, it caused apoptosis in THP-1 human monocytic leukaemia cells, but not in differentiated monocytes and macrophages, with an $IC_{50}$ of 12 μmol/l (Tsan *et al.* 2000). Resveratrol was also shown to induce extensive apoptosis in several cell lines derived from human leukaemia patients by a mechanism involving loss of mitochondrial membrane potential and increased caspase-9 activity independent of CD-95 signalling, with no effect on normal mononuclear cells (Dorrie *et al.* 2001). One report described prevention by resveratrol of apoptosis in human erythroleukaemia K562 cells, accompanied by inhibition of eicosanoid synthesis (MacCarrone *et al.* 1999).

Androgen-responsive human prostate cancer cells also undergo apoptosis in response to resveratrol (Hsieh and Wu 1999). High doses caused apoptosis accompanied by increased caspase-3 activity (as well as enhanced expression of cyclins A and E) and decreased levels of cyclin D kinase (cdk)2 and cdk6 in the human colon cancer cell line CaCo-2 (Wolter *et al.* 2001). Growth suppression due to apoptosis was noted at high concentrations of resveratrol in both oestrogen receptor-positive and oestrogen receptor-negative human breast cancer cell lines, whereas cell proliferation occurred at low concentrations only in the oestrogen receptor-positive cells (Nakagawa *et al.* 2001). In a human epidermoid keratinocytic cell line, it caused concentration-dependent $G_1$-phase arrest followed by apoptosis, accompanied by down-regulation of the cyclin–cdk machinery (cyclins D1, D2 and E; cdk2, cdk4 and cdk6) over the range 1–50 μmol/l (Ahmad *et al.* 2001). It also induced apoptosis in a virally transformed normal human cell line together with increased p53 expression, but not in the untransformed cells (Lu *et al.* 2001). Induction of apoptosis with p53 activation followed exposure to resveratrol in a mouse epidermal cell line (She *et al.* 2001), whereas resveratrol-induced apoptosis in Chinese hamster lung cells was associated with sister chromatid exchange (Matsuoka *et al.* 2001). Reviewing this topic, Pervaiz (2001) laid strong emphasis on mitochondrial release of cytochrome$_c$ and activation of caspases 3 and 9 among the mechanisms responsible for resveratrol-associated apoptosis. Another mitochondrial pathway controlled by Bcl-2 has been invoked as possibly accounting for apoptosis induced by resveratrol (Tinhofer *et al.* 2001).

Paradoxically, resveratrol was reported to block apoptosis induced by oxidized lipoproteins and by hydrogen peroxide in PC-12 cells, an outcome that is preceded by NF-κB activation and binding to DNA (Draczynska-Lusiak *et al.* 1998a; Jang and Surh 2001). It also manifested an anti-apoptotic effect in a human neuroblastoma cell line in which apoptosis was induced by exposure to the drug paclitaxel (Nicolini *et al.* 2001).

### *Modulation of cell cycle*

Arrest of cell division is an alternative mechanism for resveratrol-induced inhibition of cell growth, and may be accompanied by enhanced differentiation or similar phenotypic changes in growth-arrested cells. This was first demonstrated for human leukaemia cell line HL-60 cells, which, at a concentration of 30 μmol/l, became arrested at S-phase concomitant with a significant increase in cyclins A and E and cdc 2 in the inactive phosphorylated forms (Mgbonyebi *et al.* 1998). However, rat hepatic stellate cells did not show any increase in these cyclins in response to resveratrol, although cyclin D1 content was reduced (Kawada *et al.* 1998). Differentiation of the HL-60 cells towards a myelomonocytic phenotype occurred synchronously with arrest of cell division at S-phase (Ragione *et al.* 1998).

These findings are consistent with an earlier report that resveratrol inhibits ribonucleotide reductase ($IC_{50}$ around 4 μmol/l), the enzyme that provides proliferating cells with the deoxyribonucleotides required for DNA synthesis during early S-phase (Fontecave *et al.* 1998). Indeed, its potency on a molar basis was orders of magnitude greater than that of hydroxyurea, another inhibitor of the same enzyme that has been used therapeutically as an anticancer and anti-HIV (human immunodeficiency virus) agent. Resveratrol also leads to an increase in the proportion of androgen-non-responsive human prostate cancer cell lines in S-phase, but this does not occur with androgen-responsive cells (Hsieh and Wu 1999). It will be recalled that the latter undergo apoptosis in response to resveratrol, whereas the former do not. Upregulation of the cyclin-dependent kinase inhibitor p21 also occurs in these cells (Mitchell *et al.* 1999). Arrest in S-phase was noted in human breast carcinoma cell lines exposed to resveratrol (Hsieh *et al.* 1999a). Resveratrol inhibits another important enzyme involved in DNA synthesis,

DNA polymerase (Sun *et al.* 1998; Stivala *et al.* 2001), and causes cleavage of DNA in the presence of $Cu^{2+}$ ions (Fukuhara and Miyata 1998) and $Fe^{3+}$ (Miura *et al.* 2000). At variance with the above findings is a report that resveratrol in low concentrations ($10^{-9}$–$10^{-7}$mol/l) dose-dependently increased DNA synthesis, proliferation and differentiation of osteoblastic MC 3T3-E1 cells (Johnson and Maddipati 1998). This action was blocked by the antio-estrogenic compound tamoxifen. The interaction of resveratrol with DNA will be explored more fully in a later section.

More recent reports have consolidated the evidence favouring cell-cycle arrest as an important action of resveratrol in cancer cells. The CEM-C7H2 lymphocytic leukaemia cell line, which is deficient in functional p53 and p16, underwent G0–G1-phase arrest (Bernhard *et al.* 2000), while S-phase arrest was reported in rat and human hepatoma cell lines, the effect being enhanced by ethanol (Delmas *et al.* 2000). Loss of p21 expression occurred in the normal colonic mucosa of rats treated with the carcinogen azoxymethane and given resveratrol in their drinking water (Tessitore *et al.* 2000); early neoplastic changes were also depressed in these animals. G1-phase arrest was noted in resveratrol-treated human epidermoid carcinoma cells with marked changes in the cyclin system as described above (Ahmad *et al.* 2001). Of special interest is a report describing S-phase arrest in human histiocytic lymphoma (U937) cells that was reversed by removal of the resveratrol, being accompanied by changes in cyclins E and A (Park *et al.* 2001). Resveratrol (50–100 μmol/l) suppressed proliferation of cultured smooth muscle cells by an apoptosis-independent mechanism resulting in arrest at G1–S-phase (Zou *et al.* 1999). Finally, human colon cancer (CaCo-2) cells and related HCT-116 cells were arrested in S-phase, accompanied by decreased levels of cyclin D1 and cdk4, but with no change in cyclo-oxygenases (Wolter *et al.* 2001).

Some variability in the precise phase of the cell-cycle that is blocked and in the related changes in the cyclin system in response to resveratrol may depend upon the particular cells under study, but in the former respect S-phase transition appears to be the most frequent target.

### *Aryl hydrocarbon receptors*

In addition to its putative antimutagenic, pro-apoptotic and DNA anti-synthetic properties, resveratrol may target another biological system involved in carcinogenesis. Aryl hydrocarbons such as dioxin and dimethylbenzan-thracene are taken up by a specific cytosolic receptor (AHR) in susceptible cells. The complex translocates to the nucleus where it dimerizes with another protein and initiates the transcription of a number of genes involved in carcinogenesis. The best characterized of these is the *CYP1A1* gene that encodes a cytochrome $P_{450}$-dependent microsomal enzyme; the *CYP1A1* gene product hydroxylates aryl hydrocarbons to genotoxic metabolites that bind DNA with consequent mutational events. Ciolino and colleagues (1998) reported that resveratrol in concentrations of between 0.5 and 20 μmol/l inhibited the induction of CYP1A1 protein, mRNA and enzyme activity by the halogenated dioxin derivative TCDD (2,3,7,8-tetrachlorobenzo-p-dioxin) in Hep G2 cells in a dose-dependent manner. It also prevented the TCDD-induced transformation of cytosolic AHR to its nuclear DNA-binding form and blocked its binding to promoter sequences that regulate CYP1A1 transcription. It did not modulate the binding of TCDD to cytosolic AHR. They extended these observations using benzo [a] pyrene as the inducer in both Hep G2 and human mammary carcinoma MCF-7 cells (Ciolino and Yeh 1999). Similar findings with respect to cytochrome P4501A1 were described by Chun *et al.* (1999) using a model system incorporating human liver microsomes; weaker inhibition of CYP1A2 was also demonstrated.

These results were partially confirmed by Casper *et al.* (1999) in a human breast cancer cell line, but with some differences. They found that resveratrol displaced labelled dioxin from AHR with an $IC_{50}$ of 6 μmol/l. Neither nuclear translocation nor DNA-binding of AHR were altered by resveratrol, but its transcriptional activity for CYP1A1 was blocked. The most novel and exciting aspect of this report was the finding that resveratrol, when given in doses of 1 and 5 mg/kg to rats, inhibited the expression of CYP1A1 in lung and kidney induced by benzpyrene and dimethylbenzanthracene. The likely tissue concentrations of resveratrol were three orders of magnitude less than those required for CYP1A1 inhibition in the *in vitro* experiments, suggesting that resveratrol may be converted *in vivo* to a metabolite one thousand-fold more active than the parent compound.

The same group described the ability of resveratrol in low concentrations to block the antiosteogenetic effects of TCDD in chicken and rat *in vitro* models of bone formation (Singh *et al.* 2000). It should be added that resveratrol in concentrations <1 μmol/l inhibited a range of cytochrome $P_{450}$-linked enzymes in hamster liver microsomes *in vitro* (Teel and Huynh 1998), but it also induced the mRNA for CYP1A1 in cultured Hela cells derived from a human cervical carcinoma (Frotschl *et al.* 1998). Other cytochromes whose catalytic activity or gene expression have been reported to be inhibited by resveratrol include CYP3A4, CYP3A5, and CYP1B1 (Chan and Delucchi 2000; Chang *et al.* 2000; Chang and Yeung 2001; Mollerup *et al.* 2001). Finally, modulation of c-fos gene expression has been postulated to be an important target for the anticancer activity of resveratrol, at least in the mouse skin carcinogenesis model (Jang and Pezzuto 1998, 1999).

### *Other mechanisms*

Angiogenesis is an important process in maintaining the blood supply to tumours and permitting their proliferation. Resveratrol and its glucosides have been reported to inhibit capillary formation by vascular endothelial cells (an essential component of angiogenesis), thereby restricting tumour metastasis in animal models (Kimura and Okuda 2000, 2001; Brakenhielm *et al.* 2001; Igura *et al.* 2001).

Receptor-mediated signalling through activation of protein kinases (PK) is another mechanism deemed to be of importance in cancer development. Resveratrol may interfere with this pathway as it inhibits the catalytic activity of PKC$\alpha$ (Garcia-Garcia *et al.* 1999). On the other hand, it does not affect the autophosphorylation of PKC while inhibiting that of the related phorbol ester-responsive isoenzyme PKD (Stewart *et al.* 2000). Extracellular-regulated kinases (ERK) are believed to play a negative role in cancer growth by phosphorylation-activation of p53, but their response to resveratrol is controversial: one group of investigators reported their upregulation (She *et al.* 2001), another group observed no effect (De Ledinghen *et al.* 2001), and a third found inhibition of ERK2 as well as of PKC (Yu *et al.* 2001). It has recently been proposed that the effects of resveratrol in blocking phorbol ester-mediated activities are generated, at least in part, by inhibition of mitogen-activated protein kinase (MAPK) pathways (Schneider *et al.* 2000).

Polyamines are growth-promoting factors present in high concentrations in cancer and other proliferating cells, where the rate-limiting enzyme in their synthesis, ornithine decarboxylase (ODC), is upregulated. Their concentrations, as well as the activity of ODC, were reduced in CaCo-2 cells exposed to 25 μmol/l resveratrol (Nielsen *et al.* 2000). Resveratrol at 17–50 μmol/l also increased gap–junctional intercellular communications in rat liver epithelial cells, and reversed their down-regulation by the tumour promoters phorbol ester and dichlorodiphenyltrichloroethane (DDT) (Kong *et al.* 2001). Finally, modulation of c-fos gene expression has been postulated to be an important target for the anticancer activity of resveratrol, at least in the mouse skin carcinogenesis model (Subbaramiah *et al.* 1998; Jang and Pezzuto 1999).

## Antioxidant activity

The ability of *trans*-resveratrol to function as an antioxidant was first demonstrated by Frankel *et al.* (1993). On a molar basis it was less effective than a number of flavonoids in preventing the copper-mediated oxidation of human LDL, but it was much more potent than $\alpha$-tocopherol. It was the second most potent of eight food additives in preventing lipid peroxidation and scavenging hypochlorous acid (HOCl) (Murcia and Martinez-Tome 2001). Zou *et al.* (1999) confirmed the reduced production of thiobarbituric acid-reactive substances (TBARS) and increased lag phase when human LDL from normal subjects was oxidized in the presence of $Cu^{2+}$. Frankel *et al.* (1995) examined the relative contribution of individual wine phenolics to the inhibition of LDL oxidation, based upon their concentrations in the wines utilized, and concluded that resveratrol did not correlate with this activity. Subsequently, Soleas *et al.* (1997b) reported that resveratrol contributed significantly to the total antioxidant activity of wine as evaluated using the Randox *in vitro* assay.

Belguendouz *et al.* (1997) carried out an extensive examination of the inhibition by resveratrol of porcine LDL oxidation in the presence of the free radical generator 2,2′-azobis (2-amidinopropane dihydrochloride) (AAPH) or copper ions. The slope of the propagation phase and the prolongation of the lag phase were much greater with the latter than with the former. Formation of TBARS was completely inhibited by up to 200 min in the copper-mediated system by 1 μmol/l resveratrol, more effective than trolox or the flavonoids tested. The relevant mechanisms appeared to be a combination of copper chelation and free radical scavenging; surprisingly, resveratrol was unable to chelate iron. In a subsequent report it proved to be more effective than flavonoids as a chelator of copper and less effective as a free-radical scavenger (Fremont *et al.* 1999), but later authors found it to be very potent in scavenging superoxide and peroxyl radicals, and in blocking lipid peroxidation (Martinez and Moreno 2000; Miura *et al.* 2000; Tadolini *et al.* 2000; Stivala *et al.* 2001; Stojanovic *et al.* 2001). Resveratrol added to plasma was distributed between the lipoprotein classes according to their lipid content in the order VLDL > LDL > HDL (Belguendouz *et al.* 1998). The authors' suggestion that resveratrol may be effective in a lipid as well as in an aqueous environment was supported by experiments showing that it blocked the formation of TBARS by AAPH in phospholipid liposomes. Not only does resveratrol inhibit oxidation of LDL, it also blocks the uptake of oxidized LDL by the scavenger receptor in human smooth muscle cells (Mietus-Snyder *et al.* 2000). Using a different system (inhibition of cytochrome C oxidation by hydroxyl radicals generated by photolysis of $H_2O_2$), Turrens *et al.* (1997) found that *trans*-resveratrol manifested antioxidant activity, the $EC_{50}$ (effective) concentration being 33 μmol/l. These findings were extended by Fauconneau *et al.* (1997), who reported that *trans*-resveratrol protected rat liver microsomes against $Fe^{2+}$-mediated lipid peroxidation and human LDL against $Cu^{2+}$-mediated lipid peroxidation. The $EC_{50}$ concentrations were 3.0 μmol/l and 2.6 μmol/l, respectively: in the same range as anthocyanins but around two-fold higher than catechins and the stilbene astringinin. The activity of resveratrol in scavenging the stable free radical 1,1-diphenyl-2-picryl hydrazyl (DPPH) ($EC_{50}$ = 74 μmol/l) was much higher than that of the other previously mentioned compounds.

Resveratrol appears to be a potent antioxidant in a number of other biological systems. It was more effective than vitamins C or E in preventing oxidative damage and death in a rat phaeochromocytoma cell line (Chanvitayapongs *et al.* 1997). It protected the same cells against damage induced by oxidized lipoproteins (Draczynska-Lusiak *et al.* 1998b), and also proved to be a powerful inhibitor of reactive oxygen species production in murine macrophages, human monocytes and human neutrophils, although the $IC_{50}$ values for these effects ranged from 17 to 23 μmol/l (Jang *et al.* 1999). Jang and Pezzuto (1998, 1999) have attributed, at least in part, its anticancer activity in the mouse skin carcinogenesis model to its antioxidant properties, since phorbol ester-mediated increases in myeloperoxidase, superoxide dismutase and $H_2O_2$ production were restored to control levels by treatment with resveratrol. Related stilbenes also have antioxidant properties. *Cis*-resveratrol and resveratrol glucosides demonstrated protection against lipid peroxidation in mouse liver microsomes and human LDL, although the $IC_{50}$ values were an order of magnitude greater than that of *trans*-resveratrol (Fauconneau *et al.* 1997; Waffo-Teguo *et al.* 1998). Oxyresveratrol is a potent inhibitor of dopa oxidase activity (Shin *et al.* 1998b).

A number of biological effects of resveratrol have been attributed to its antioxidant and radical-scavenging properties. These include the ability to protect cultured rat adrenal phaeochromocytoma cells against apoptotic cell death induced by oxidized LDL and VLDL (Draczynska-Lusiak *et al.* 1998c), inhibition of proliferation of cancer cells stimulated by oxidative DNA damage (Sgambato *et al.* 2001), and prevention of oxidative DNA damage in the kidney of rats treated with potassium bromate (Cadenas and Barja 1999). The latter was thought to involve suppression of NF-κB activation, which is induced by TNF and blocked by resveratrol (Manna *et al.* 2000). The adhesion of platelets to fibrinogen was also inhibited by resveratrol more potently than by other antioxidants tested (Zbikowska and Olas 2000), although concentrations in the mmol/l range were employed by these authors. The inhibitory effects of resveratrol upon a series of breast cancer cell lines were associated with reduction in the production and cytotoxicity of peroxide and reactive oxygen species (Damianaki *et al.* 2000). Its inhibition of hepatoma cell invasion was also attributed to its antioxidant properties (Kozuki *et al.* 2001). Of a number of antioxidants studied, only resveratrol was able to suppress IL-6 (interleukin-6) production induced by reactive oxygen species in glial cell cultures subjected to hypoxia/hypoglycaemia followed by reoxygenation (Wang *et al.* 2001). It also protected the kidneys of intact rats (Giovannini *et al.* 2001) and isolated rat hearts (Ray *et al.* 1999) from ischaemia/reperfusion injury, and decreased oxidative DNA damage in spontaneously hypertensive rats (Mizutani *et al.* 2001).

The interaction of resveratrol, oxygen and metal ions on DNA is a controversial topic. In the presence of $Cu^{2+}$ ions it caused cleavage of DNA (Fukuhara *et al.* 1998; Ahmad *et al.* 2000). DNA strand breaks were stimulated during the interaction of resveratrol with ADP-$Fe^{3+}$ in the presence of $H_2O_2$, suggesting that in these specific conditions it exercises a pro-oxidant effect on DNA damage (Miura *et al.* 2000). However, in the presence of physiological concentrations of ascorbic acid and glutathione, resveratrol protected DNA from copper-mediated damage by acting as a radical-scavenging antioxidant (Burkitt and Duncan 2000). Support for the DNA-protective role of resveratrol has recently been provided by experiments demonstrating its inhibition of DNA damage in response to $Cr^{3+}$ and $H_2O_2$, with an $IC_{50}$ as low as 0.1 mmol/l (Burkhardt *et al.* 2001).

This optimistic account of resveratrol and its antioxidant activity is soured by an important negative report. Turrens *et al.* (1997) injected synthetic resveratrol intraperitoneally into rats over a 21-day period at doses of 20 and

40 mg/kg body weight (equivalent to the ingestion in a 70 kg human of approximately 700 and 1400 l of red wine, respectively). The formation of TBARS from plasma proteins (including lipoproteins) chromatographed to remove water-soluble antioxidants in the presence of $Cu^{2+}$ at two concentrations (2 µmol/l and 10 µmol/l) was not significantly reduced in the resveratrol-treated rats; neither was any change detected in the serum lipoprotein profile. As in the study described by Wilson *et al.* (1996), *in vivo* findings do not necessarily support the results to be expected on the basis of the behaviour of resveratrol *in vitro*, raising the important issues of absorption and bioavailability that will be dealt with in the final section of this review.

## Vascular relaxation and nitric oxide production

NO is produced in a wide range of cells. At least two different enzymes, nitric oxide synthases (NOS), are involved in its synthesis. One, iNOS, is inducible in response to inflammatory stimulants such as LPS in macrophages and other cells involved in inflammatory reactions. In this scenario, NO is pro-oxidant and potentially noxious. Its production by iNOS is inhibited by resveratrol in rat hepatic stellate cells (Kawada *et al.* 1998), and macrophages (Tsai *et al.* 1999; Wadsworth and Koop 1999; Matsuda *et al.* 2000). This inhibition is synergistically enhanced by ethanol in a concentration-dependent manner over the range 0.1–0.75%, in which range ethanol alone is without appreciable effect (Man-Ying *et al.* 2000). Very low concentrations of resveratrol (5–25 nmol/l) can attenuate the death of rat hippocampal neuronal cells following exposure to nitric oxide free-radical donors; this protection is almost certainly due to its antioxidant properties since no changes in intracellular enzymes (including iNOS) were detected (Bastianetto *et al.* 2000).

The second, cNOS, is a constitutive enzyme in vascular endothelial cells. NO produced in this location prevents adherence of platelets to the endothelial surface and diffuses distally to promote relaxation of the smooth muscle layer, in part attributable to its antagonism of the vasoconstricting agent endothelin. Using rat aortic rings, Fitzpatrick *et al.* (1993) showed that red wine extracts were able to abolish vasoconstrictive events; a number of wine constituents (including quercetin) could reproduce this phenomenon, but resveratrol was inactive in this regard. Subsequently, Chen and Pace-Asciak (1996) were able to demonstrate quite effective vasorelaxation by resveratrol in a similar system, but could not offer any explanation for the discordance between these and the previous results. A recent report (Jager and Nguyen-Duong 1999) lends strong support to the findings of Chen and Pace-Asciak (1996) by demonstrating dose-dependent inhibition by resveratrol of histamine and fluoride-induced contractions in isolated porcine coronary arteries at very low $IC_{50}$ concentrations of <1 nmol/l.

Among its spectrum of activities upon bovine pulmonary artery endothelial cells, resveratrol enhanced cNOS concomitantly with increased levels of p53 protein and elevated concentrations of the cyclin-dependent kinase inhibitor p21, resulting in accumulation of cells in S- and $G_2$–M-phases (Hsieh *et al.* 1999b). The authors hypothesized that this inhibition of cell proliferation may reduce damage to the vascular endothelium, thereby attenuating the development of atherosclerosis, but the logic underlying this speculation is not very compelling. A more plausible relationship has been reported between the ability of resveratrol to lower arrhythmias and mortality in anaesthetized rats after ischaemia/reperfusion, and the increase in NO concentrations of the carotid blood (Hung *et al.* 2000). Increased NO production has also been cited as the reason for the ability of resveratrol to reduce from 50% to 10% the renal damage caused by ischaemia/reperfusion injury in intact rats (Giovannini *et al.* 2001).

A recent report suggests that the vasorelaxation induced by resveratrol is dependent on the type of artery (Naderali *et al.* 2000), being greater in resistance arteries (e.g. mesenteric) than upon conductance arteries (e.g. uterine). The effects on mesenteric arteries were the same in lean rats as in obese rats manifesting endothelial dysfunction (Naderali *et al.* 2001); in the former group they were dependent upon NO production and endothelial integrity, whereas in the latter group they were independent of both and seemed to occur by virtue of a different mechanism. Resveratrol is also able to block the proliferation, DNA synthesis and prolyl hydroxylase activity of vascular smooth muscle cells induced by advanced glycosylation end-products (AGEs), components responsible for the vascular complications of diabetes mellitus, including atherosclerosis (Mizutani *et al.* 2000).

Relevant to the vascular pathology of atherosclerosis and the cardio-protective properties of resveratrol is the observation that it inhibited the migration and proliferation of smooth muscle cells by a mechanism that does not involve apoptosis but results in arrest at $G_1$–S phase (Zou *et al.* 1999). Intimal hyperplasia of the denuded iliac artery in rabbits was also prevented by intragastric administration of resveratrol (Zou *et al.* 2000). Intravenous injection of resveratrol in doses of 1 μg and 0.1 μg/kg body weight in rats exposed to occlusion/reperfusion of the middle cerebral artery significantly reduced the total volume of infarcted cerebral tissue (Huang *et al.* 2001). Finally, the ability of bovine pulmonary artery endothelial cells to resist simulated arterial flow was enhanced by resveratrol synchronously with an increase in phosphorylated ERK1/2 and NOS (Bruder *et al.* 2001). All of the above effects are consistent with the likelihood that resveratrol can prevent or attenuate many of the events leading to vascular degeneration and atherosclerosis; it is particularly noteworthy that several of these experiments were performed in whole animals.

## Oestrogenic activity

Following the knowledge that pinosylvin, a stilbene containing one less OH group than resveratrol, is oestrogenic in human breast cancer cell lines (Mellanen *et al.* 1996), Gehm *et al.* (1997) were the first to report the interaction of resveratrol with oestrogen and its receptor. At concentrations of 3–10 μmol/l, resveratrol inhibited the binding of labelled oestradiol to the oestrogen receptor in human breast cancer cells, and activated transcription of oestrogen-responsive reporter genes transfected into these cells. This transcriptional activation was oestrogen receptor-dependent, required an oestrogen response element in the reporter gene, and was inhibited by specific oestrogen antagonists. Depending on the cell type, it produced activation greater than, equal to or less than that of oestradiol. It also increased the expression of native oestrogen-regulated genes and stimulated the proliferation of an oestrogen-dependent human breast carcinoma cell line.

A subsequent report described resveratrol as lacking the ability to bind to oestrogen receptor in a pituitary cell line (Stahl *et al.* 1998). Unlike other phyto-oestrogens tested in the same system, resveratrol did not stimulate the growth of these cells, but like them it was able to stimulate their prolactin secretion in a dose- and time-dependent manner. When tested in concentrations ranging from approximately 2.5–200 μmol/l, resveratrol inhibited the growth and proliferation of human breast cancer cell lines irrespective of their oestrogen responsiveness (Mgbonyebi *et al.* 1998). These results were extended by Lu and Serrero (1999) who also grew oestrogen receptor-positive breast cancer cells (MCF-7) in the presence of resveratrol. At concentrations of 1 μmol/l and above, the latter antagonized the growth-promoting effect of 17-beta-oestradiol (1 nmol/l) upon these cells, as well as its stimulation of progesterone receptor gene expression. At higher concentrations (100 μmol/l), resveratrol blocked the expression of mRNA for TGFα and insulin-like growth factor-I (ILGF-I) receptor but increased mRNA for $TGF_{\beta 2}$ in MCF-7 cells. The increase in DNA synthesis and stimulation by resveratrol of an osteoblastic cell line was inhibited by the antioestrogen tamoxifen (Mizutani *et al.* 1998), which also diminished its ability to block the deleterious actions of AGEs upon vascular smooth muscle cells previously described (Mizutani *et al.* 2000). Its cytostatic effect upon a human endometrial carcinoma cell line was attributed to both oestrogenic and antioestrogenic effects (Bhat and Pezzuto 2001).

Both isomers of resveratrol were tested for oestrogenic activities over a range of concentrations in two breast cancer cell lines, including MCF-7. High concentrations blocked proliferation induced by oestradiol, medium concentrations enhanced oestradiol activity, and low concentrations were without effect (Basly *et al.* 2000). At low concentrations it caused cell proliferation in two oestrogen receptor-positive human breast cancer cell lines but not in a receptor-negative cell line (Nakagawa *et al.* 2001); at higher concentrations growth was suppressed in all three. *trans*-Resveratrol manifested differential effects upon $ER_{\alpha}$ and $ER_{\beta}$ receptors in MCF-7 cells and rat uterine cells (Bowers *et al.* 2001). It binds both receptors with equal affinity that is 7000-fold lower than that of oestradiol, but it exhibits antagonistic activity towards the latter only with oestrogen receptor α ($OR_{\alpha}$). Resveratrol-liganded $OR_{\beta}$ was found to have higher transcriptional activity than oestradiol-liganded $OR_{\beta}$, suggesting that tissues in which the latter is the predominant receptor will be more sensitive to its oestrogen-agonist effects. Of 72 phytochemical flavonoids tested with BT-474 human breast cancer cells at concentrations of $10^{-8}$–$10^{-5}$ mol, resveratrol ranked eighth for oestrogenic activity (assessed by production of the oestrogen-regulated protein pS2), but with only 22% of the potency of genistein and <50% of the potency of luteolin and daidzein (Rozenberg Zand *et al.* 2000). It inhibited sulphation of oestradiol (thereby enhancing its effect) in human mammary epithelial cells with an $IC_{50}$ of 1 μmol/l, but quercetin was 10-fold more potent (Otake *et al.* 2000).

It is interesting to note that resveratrol was reported to down-regulate the androgen receptor in a human prostate cancer cell line; decreased expression of PSA and p21 at the protein and mRNA levels were among the consequences observed (Mitchell *et al.* 1999). Although the lower PSA levels induced by resveratrol were confirmed, subsequent authors could find no change in androgen receptor-content based upon immunoblot assays (Hsieh and Wu 2000).

Its role as a phyto-oestrogen has been invoked to account, at least in part, for its protection against atherosclerosis (Kopp 1998), but *in vivo* experiments cast doubt upon the extrapolation of these *in vitro* antioestrogenic effects to whole animals. When given to rats by oral gavage or subcutaneous injection, resveratrol in doses ranging up to the equivalent in a 70 kg human of 2800 l of wine daily did not show any activity in an assay using uterine oestrogen receptors (Ashby *et al.* 1999). In fact, *in vitro* studies carried out by this group and reported in the same paper described the affinity of resveratrol for rat uterine oestrogen receptors as five orders of magnitude lower than that of oestradiol or diethylstilboestrol. Similar conclusions were reached with experiments utilizing oestrogen receptor-transfected yeast cells and cos-1 cells. Turner *et al.* (1999) carried out an extensive investigation on the effect of orally-administered resveratrol (graded doses equivalent to 0.5–500 ml of red wine per day for 6 days to weanling rats) upon oestrogen target tissues assessed by growth rate, body weight, serum cholesterol and radial bone growth, all of which were enhanced by equivalent doses of oestradiol. Resveratrol did not alter any of these parameters. Finally, massive doses of up to 500 mg/kg subcutaneously, which raised plasma concentrations to 1–2 μmol, had only very weak effects upon the rat uterus with respect to $OR\alpha$-protein and progesterone receptor-protein compared with doses of oestradiol lower by a magnitude of $2 \times 10^5$ (Freyberger *et al.* 2001).

Before accepting these negative conclusions, two caveats are worthy of consideration. First, it is conceivable that the affinity of resveratrol for oestrogen receptors is subject to some tissue specificity, being more potent for breast than for uterus; moreover, since Ashby *et al.* (1999) had found that the affinity of resveratrol for uterine oestrogen receptors was five orders of magnitude less than that of oestradiol, it is surprising that they employed similar doses of both agonists in their whole animal experiments. Second, the parameters selected by Turner *et al.* (1999) may have been appropriate for oestradiol, but conceivably not for resveratrol, whose widespread biological actions, including those on cell growth and DNA synthesis, may have masked its possible role as a phyto-oestrogen as judged by these criteria. Finally, resveratrol at the same concentration can act as an oestrogen or antioestrogen in the same cells depending upon whether they have the wild-type or a mutant oestrogen receptor (Yoon *et al.* 2001).

## Other biological effects

Two papers have drawn attention to possible antiviral activity attributable to resveratrol. The first described inhibition of the replication of *Herpes simplex* virus (HSV) types 1 and 2 in a dose-dependent and reversible manner by targeting an early event in the virus replication cycle; it was most effective when resveratrol was added within 1 h of infection, less effective if added 6 h later, and ineffective at 9 h post-infection. Cell-cycle delay at the S–$G_2$M interphase was noted as the likely site of action (Docherty *et al.* 1999). It also synergistically enhanced the anti-HIV activity of a number of nucleoside analogues in combating infection in peripheral white blood cells (Heredia *et al.* 2000). A third noted the ability of resveratrol selectively to inhibit two bacteria responsible for gonorrhoea and meningitis (Docherty *et al.* 2001).

Three reports are worth citing, given the interest in the possibility that antioxidant flavonoids may prevent dementia and improve cerebral function. The first described induction by resveratrol of phosphorylation of several protein kinases (including MAPK and ERK) in both differentiated and undifferentiated human neuroblastoma cells (Miloso *et al.* 1999). The second involved an intriguing contrast between the effects of resveratrol in two rat models, one *in vivo* and the other *in vitro*. In the former, it protected against the damage to the hippocampus caused by injection of the toxin kainic acid, but it had no beneficial effect on cultured hippocampal slices (Virgili and Contestabile 2000). The third indicated that intravenous injection of resveratrol to rats prior to occlusion of the artery reduced the size of the subsequent brain infarct (Huang *et al.* 2001).

A number of miscellaneous effects have very recently been described. Rats that developed proteinuria, hypoalbuminaemia and hyperlipidaemia as a consequence of experimental immune-induced nephritis had these manifestations, as well as liver and kidney enlargement reduced by oral resveratrol in a dose of 50 mg/kg body weight (Nihei *et al.* 2001); the blood concentrations of triglycerides fell much more dramatically than those of cholesterol. Resveratrol was 10-fold more potent than hydroxyurea (an established inducer of erythroid differentiation) in increasing haemoglobin production by the human erythroleukaemic K562 cell line, a process strongly linked to increased expression of p21 (Rodrigue *et al.* 2001).

Inhibition of phosphatidate and diacylglycerol synthesis in human neutrophils has been demonstrated, possibly because of suppression of phospholipase

D activity, and may contribute to the anti-inflammatory effects of resveratrol (Tou and Urbizo 2001). Finally, it appears to block the intracellular uptake of glucose and dehydroascorbic acid by a competitive process (Park 2001), although the biological consequences of this phenomenon have not been clarified.

The above observations obviously require confirmation, and it will be some time before they can be integrated with the more established mechanisms to provide a comprehensive picture of how resveratrol accomplishes its biological effects and precisely what these effects are; but several are consistent with the anti-tumour and anti-inflammatory functions described in the earlier sections of this review.

## Absorption and bioavailability

The equivocal results of human and whole animal experiments designed to reproduce the *in vitro* actions of resveratrol give urgency to the question of whether it is efficiently absorbed and, if so, whether its metabolic and excretory patterns are consistent with tissue concentrations adequate to achieve desirable effects. The prerequisite to such investigations was the development of assays sensitive enough to allow the measurement of resveratrol in blood, as well as its distribution between plasma or serum and formed elements of the blood. Our group was the first to describe investigations relevant to these issues (Goldberg *et al.* 1995b; Tham *et al.* 1995). Using a high-performance liquid chromatography (HPLC) method, we found that *trans*-resveratrol added to whole human blood was >90% recoverable and partitioned as follows: serum 54.8%; erythrocytes 36.0%; leucoytes and platelets 6.1%. However, taking protein content into account, the latter fraction was the most highly enriched in resveratrol.

A subsequent report, incorrectly claiming to be the first method developed to measure resveratrol in animal and human samples, presented results that were not quite consistent with these findings (Blache *et al.* 1997). Washed human and rat erythrocytes, rat platelets and human LDL incorporated approximately 50%, 10% and 17.5% respectively of *trans*-resveratrol when incubated at room temperature for 15–30 min. Two HPLC methods for the measurement of plasma resveratrol were described, with detection limits of 20 μg/l (Juan *et al.* 1999) and 5 μg/l (Zhu *et al.* 1999), respectively. Only the former was applied in whole animals; 15 min after unspecified administration of 2 mg/kg to rats, the plasma resveratrol concentration was stated to be 175 μg/l, almost 10% of the dose given (Juan *et al.* 1999), a result completely at variance with all other literature reports on this topic (Soleas *et al.* 2001c).

Bertelli and colleagues developed an HPLC method to determine resveratrol concentrations in rat serum, and claimed a detection limit of 1 ng/ml. They found that after 4 ml of red wine containing 26 μg of resveratrol given to the animals by gavage, the blood resveratrol concentration peaked around

60 min at about 15 ng/ml, or $5.8 \times 10^{-2}$ percent of the dose administered, and returned to baseline by 4 h (Bertelli *et al.* 1996b). In a further experiment, they gave 13 μg of total *trans*- and *cis*-resveratrol per day to a second group of rats for 15 days, at which time the concentrations of resveratrol in plasma (7.6 ng/ml), urine (66 ng/ml), heart (3 ng/ml), liver (54 ng/ml) and kidneys (44 ng/ml) were well below those required for pharmacological activity based upon *in vitro* studies (Bertelli *et al.* 1996c). A kinetic analysis revealed that relative tissue bioavailability, calculated as 'area under the curve' (AUC) for tissues expressed as a percentage of AUC for plasma, accorded with the following ratios: heart, 24; liver, 218; kidneys, 295. However, they subsequently claimed that by the administration of red wine they were able to achieve resveratrol concentrations compatible with the inhibition of platelet aggregation (Bertelli *et al.* 1998a).

Our group gave 298 nCi of [$H^3$]-resveratrol added to 10% ethanol, white grape juice or vegetable homogenate (V-8) by stomach tube to male Wistar rats (average weight 300 g), following which the animals were held in metabolic cages for collection of urine and faeces independently (Soleas *et al.* 2001d). After 24 h, they were sacrificed with collection of blood, various organs, and the contents of colon and bladder, which were added to stool and urine, respectively. Only traces of radioactivity were detected in the blood after 24 h, or in groups of rats sacrificed at 30 min intervals over the first 2 h. Urine and bladder accounted for 50–60%, and stool and colon for 11–14% of the radioactivity after 24 h (Table 8.1). There were no significant differences between the three beverages. Only traces of radioactivity were detected in spleen, liver, kidney, or the cellular elements of the blood. Using ethanol as the vehicle, competition experiments were performed with cold resveratrol as well as unlabelled catechin and quercetin (two flavonoid polyphenols present in red wine that share some similar structural features with resveratrol). None of these compounds altered the amount of radioactivity in stool or urine after 24 h. We concluded, based upon urine measurements, that around 50–60% of the *trans*-resveratrol entering the rat intestine is absorbed, probably by bulk fluid transfer rather than by receptor-mediated mechanisms, and that its clearance from the bloodstream is very rapid. We speculated that this percentage may be closer to 90 if we assume that all of the radioactivity not recovered in the stool was actually absorbed. Between

*Table 8.1* Recovery of labelled resveratrol 24 hours after gastric administration in rats*

| *Matrix* | *Total (%)* | *Stool (%)* | *Urine (%)* |
|---|---|---|---|
| Grape juice | 61.1 ± 4.9 | 10.8 ± 0.9 | 50.3 ± 5.7 |
| V-8 juice | 73.6 ± 3.5 | 14.1 ± 1.0 | 59.5 ± 4.5 |
| Ethanol | 62.2 ± 2.3 | 13.2 ± 1.4 | 49.0 ± 3.1 |

* Mean of 8 experiments ± SEM.

25% and 40% of tracer could not be accounted for and may have been deposited in adipose tissue and brain in view of its lipophilic nature. The presence of alcohol does not seem to be necessary for effective absorption to take place.

Next, we developed a sensitive method to assay resveratrol in whole blood, serum and urine employing liquid extraction, derivatization and gas–liquid chromatography followed by mass spectrometry (GC–MS). With a retention time of 10 min, the limits of detection (LOD) and quantitation (LOQ) were 0.1 μg/l and1.0 μg/l, respectively, with very satisfactory recovery and imprecision in all three matrices from human and rat (Soleas *et al.* 2001c). Preliminary experiments in two healthy subjects given 25 mg of resveratrol in white wine by mouth revealed a peak at 30 min in both whole blood and serum, with little difference in concentration between the two at any time-point over the first 180 min when the values were declining towards baseline levels. Based on the AUC it appeared that 10–15% of the administered dose was absorbed, but a much smaller fraction (<1%) was recovered in the urine over the first 24 h.

It was difficult to reconcile these findings with the results of the tritiated-resveratrol experiments in rats (Soleas *et al.* 2001d), but a plausible explanation soon presented itself. De Santi and colleagues (2000a, b, c), in a series of papers, described the sulphation and glucuronidation of resveratrol in human liver and duodenum, processes that were inhibited by natural flavonoids in a non-competitive manner. These observations were consistent with a large body of recent work demonstrating that flavonoids, at least in the rat, are conjugated to glucuronides and sulphates by intestinal cells prior to release into the bloodstream, in which very little of the free flavonoids could be detected (Crespy *et al.* 1999, 2001; Donovan *et al.* 2001).

We therefore developed an ultrasensitive assay that could independently measure free resveratrol and the sum of its glucuronide and sulphate conjugates (Soleas *et al.* 2001e). The latter exceeded the former 15–50-fold in the blood and plasma at all time periods after the oral administration of resveratrol to humans. As with free resveratrol in our earlier experiments (Soleas *et al.* 2001c), the peak concentration occurred at 30 min and by 4 h had fallen to 15% of the peak value, although being still 10-fold the zero-time concentration. The urinary conjugated resveratrol concentrations exceeded those of the free by a similar order of magnitude over a 24 h period after an oral dose. Table 8.2 presents data for the plasma concentrations up to 60 min, and the 24 h urine output of free and conjugated resveratrol after an oral dose of 25 mg. Absorption from grape juice and vegetable homogenate was just as effective as from white wine (Soleas and Goldberg, submitted for publication).

Two reports published since this work was undertaken emphasize the importance of conjugation in determining the bioavailability of resveratrol. The first described and quantitated vascular uptake of resveratrol from the small intestinal lumen of the rat (Andlauer *et al.* 2000). Of the dose given, 20.5% was recovered, 16.8% as the glucuronide and 0.3% as the sulphate.

*Table 8.2* Concentrations of *trans*-resveratrol in blood serum and urine after oral consumption (25 mg) in three different matrices by healthy males*

| *Matrix* | *V-8 juice* | *White wine* | *Grape juice* |
|---|---|---|---|
| *Serum* (*μg/l*) | | | |
| 0 min | 2.8 | 3.7 | 2.2 |
| 30 min | 471 | 416 | 424 |
| 60 min | 250 | 245 | 344 |
| *Urine* | | | |
| 24 h output (mg) | 4.24 | 4.19 | 4.01 |
| Recovery (%) | 16.0 | 18.3 | 16.2 |

* Each data point is the mean of four individual subjects.

In the second, the authors demonstrated the glucuronidation of both *trans*- and *cis*-resveratrol by human liver microsomes (Aumont *et al.* 2001). The latter isomer was conjugated at a faster rate, and different isoforms of uridine diphosphate-glucuronyltransferase showed different specificities for the two isomers as well as for the 3-0 and 4′-0 hydroxyl positions of each isomer.

While it is uncertain if, like in the rat, the major site of resveratrol conjugation in humans is the intestine, this seems to be highly likely in view of the small proportion of free resveratrol measurable in the circulation at all times after an oral dose. The relative importance of the liver and intestine for resveratrol metabolism is an important issue that still needs to be resolved. A greater imperative is a solution to the conundrum of whether these conjugates are biologically active, and, if so, how this relates qualitatively and quantitatively to the activities of free resveratrol.

## Conclusion

Since our first review on resveratrol (Soleas *et al.* 1997a), many more biological effects of the compound have been demonstrated *in vitro*. Further, the molecular and biochemical basis for several of these effects has been elucidated. Anyone faced with having to enunciate the biological actions of resveratrol could be forgiven for replying in a parody of one of John F. Kennedy's more celebrated battle cries: 'ask not what resveratrol can do; ask rather, what resveratrol cannot do'. One of the paradoxes, as true today as it was several years ago, is the difficulty of reproducing these biological activities in whole animals or humans. A concern at that time was the issue of whether resveratrol can actually undergo intestinal absorption. This now seems to have been affirmatively resolved. However, its excretion appears to be fairly rapid; even when quite large amounts are given, its concentrations in blood and tissues fall well below the levels required for most biological activities, and suggestions that long-term administration may lead to cumulatively higher concentrations are not convincing. Given the levels in naturally-produced red wines, it seems unlikely that biologically useful concentrations will be achieved from this source alone, although one report did suggest that resveratrol may be converted *in vivo* to metabolites with much greater activity (Casper *et al.* 1999).

We have clearly demonstrated the nature and extent of the metabolic conversion of resveratrol in humans. There is a real risk that this phenomenon will consign most of the papers forming the basis of this review to the dust heap of science. One can legitimately ask: what biological relevance can there possibly be in a molecule that enters the *milieu intérieur* of humans and animals only as a conjugate? Indeed, the situation may be even worse than we imagine, since our method uses an enzyme cocktail capable of measuring only sulphates and glucuronides. There may be many other forms that would decrease even further the ratio of free to conjugated resveratrol. One conclusion is that we should be highly sceptical of, and perhaps even ignore completely, all *in vitro* work to date based upon the use of free resveratrol. A second is that further *in vitro* studies with this compound should be abandoned. A third, and the most compelling, is the need to define the biological properties of resveratrol glucuronide and resveratrol sulphate. The investigations described in this review provide an excellent starting point for this endeavour in terms of pinpointing the most likely activities and suggesting useful experimental protocols whereby they can be evaluated and tested. We do not have high expectations in this regard, since glucuronidation and sulphation are the consequences of evolutionary efforts to find the means of inactivating drugs and xenobiotics, not to increase or maintain their potency.

Three papers have appeared that testify to the efforts underway for the large-scale production of resveratrol and its glucosides, presumably for pharmaceutical purposes. Orsini *et al.* (1997b) have reported the synthesis of a range of resveratrol derivatives, including polydatin (piceid), by means of Wittig reactions followed by glucosylation under phase-transfer catalysis. Polydatin has also been generated on a preparative scale through the microbial transformation of resveratrol by a strain of *Bacillus cereus* (Cichewicz and Kouzi 1998). Resveratrol is much more stable than other polyphenols in grape skins and pomace after fermentation when stored at room temperature for relatively lengthy periods (Bertelli *et al.* 1998b), raising the expectation that these materials will be excellent sources for its extraction and purification on a commercial scale.

Our finding that resveratrol appears to be absorbed as effectively in matrices that do not contain alcohol as in those that do lends credibility to the notion that it could feasibly be provided as a capsule or elixir, joining the rank of natural health preparations that are increasingly invading our retail outlets. Alternatively, winemakers may wake up to the realization that enriching their products with resveratrol may give them a market advantage, provided that there are no adverse organoleptic characteristics as a result. The future of resveratrol as a non-patentable natural product will depend upon the delicate interplay of scientific and commercial forces, but this balance could change dramatically if synthetic patentable analogues with greater potency are developed.

## Acknowledgements

We thank Mrs Sheila Acorn and Mrs Patricia Machado for their help in preparing this manuscript. The personal work cited in this paper has been generously supported by the National Research Council of Canada (IRAP) and the Wine Institute, San Francisco.

## References

Abou-Agag, L. H., Aikens, M. L., Tabengwa, E. M., Benza, R. L., Shows, S. R., Grenett, H. E., Booyse, F. M. (2001) Polyphenolics increase t-PA and u-PA gene transcription in cultured human endothelial cells. *Alcoholism, Clinical and Experimental Research*: **25**, 155–62.

Ahmad, A., Farhan, A. S., Singh, S., Hadi, S. M. (2000) DNA breakage by resveratrol and Cu(II): reaction mechanism and bacteriophage inactivation. *Cancer Letters*: **154**, 29–37.

Ahmad, N., Adhami, V. M., Afaq, F., Feyes, D. K., Mukhtar, H. (2001) Resveratrol causes WAF-1/p21 $G_1$-phase arrest of cell cycle and induction of apoptosis in human epidermoid carcinoma A431 cells. *Clinical Cancer Research*: **7**, 1466–73.

Andlauer, W., Kolb, J., Siebert, K., Furst, P. (2000) Assessment of resveratrol bioavailability in the perfused small intestine of the rat. *Drugs in Experimental and Clinical Research*: **26**, 47–55.

Arichi, H., Kimura, Y., Okuda, H., Baba, K., Kozawa, M., Arichi, S. (1982) Effects of stilbene components of the roots of *Polygonum cuspidatum* Sieb. et Zucc. on lipid metabolism. *Chemical and Pharmaceutical Bulletin*: **30**, 1766–70.

Ashby, J., Tinwell, H., Pennie, W., Brooks, A. N., Lefevre, P. A., Beresford, N., Sumpter, J. P. (1999) Partial and weak oestrogenicity of the red wine constituent resveratrol: consideration of its superagonist activity in MCF-7 cells and its suggested cardiovascular protective effects. *Journal of Applied Toxicology*: **19**, 39–45.

Aumont, V., Krisa, S., Battaglia, E., Netter, P., Richard, T., Merillon, J. M., Magdalou, J., Sabolovic, N. (2001) Regioselective and stereospecific glucuronidation of *trans*- and *cis*-resveratrol in human. *Archives of Biochemistry and Biophysics*: **393**, 281–90.

Basly, J. P., Marre-Fournier, F., Le Bail, J. C., Habrioux, G., Chulia, A. J. (2000) Oestrogenic/antioestrogenic and scavenging properties of (E)- and (Z)-resveratrol. *Life Sciences*: **66**, 769–77.

Bastianetto, S., Zheng, W. H., Quirion, R. (2000) Neuroprotective abilities of resveratrol and other red wine constituents against nitric oxide-related toxicity in cultured hippocampal neurons. *British Journal of Pharmacology*: **131**, 711–20.

Belguendouz, L., Fremont, L., Linard, A. (1997) Resveratrol inhibits metal ion-dependent and independent peroxidation of porcine low-density lipoproteins. *Biochemical Pharmacology*: **53**, 1347–55.

Belguendouz, L., Fremont, L., Gozzelino, M. T. (1998) Interaction of *trans*-resveratrol with plasma lipoproteins. *Biochemical Pharmacology*: **55**, 811–16.

Bernhard, D., Tinhofer, I., Tonko, M., Hubi, H., Ausserlechner, M. J., Greil, R., Kofler, R., Csordas, A. (2000) Resveratrol causes arrest in the S-phase prior to fas-independent apoptosis in CEM-C7H2 acute leukemia cells. *Cell Death and Differentiation*: 7, 834–42.

Bertelli, A. A., Giovannini, L., Giannessi, D., Migliori, M., Bernini, W., Fregoni, M., Bertelli, A. (1995) Antiplatelet activity of synthetic and natural resveratrol in red wine. *International Journal of Tissue Reactions*: 17, 1–3.

Bertelli, A. A., Giovannini, L., Bernini, W., Migliori, M., Fregoni, M., Bavaresco, L., Bertelli, A. (1996a) Antiplatelet activity of *cis*-resveratrol. *Drugs in Experimental and Clinical Research*: 22, 61–3.

Bertelli, A. A., Giovannini, L., Stradi, R., Bertelli, A., Tillement, J. P. (1996b) Plasma, urine and tissue of *trans*- and *cis*-resveratrol (3,4′,5-trihydroxystilbene) after short-term or prolonged administration of red wine to rats. *International Journal of Tissue Reactions*: 18, 67–71.

Bertelli, A. A., Giovannini, L., Stradi, R., Urien, S., Tillement, J. P., Bertelli, A. (1996c) Kinetics of *trans*- and *cis*-resveratrol (3,4′,5-trihydroxystilbene) after red wine oral administration in rats. *International Journal of Clinical Pharmacology Research*: 16, 77–81.

Bertelli, A., Bertelli, A. A., Gozzini, A., Giovannini, L. (1998a) Plasma and tissue resveratrol concentrations and pharmacological activity. *Drugs in Experimental and Clinical Research*: 24, 133–8.

Bertelli, A. A., Gozzini, A., Stradi, R., Stella, S., Bertelli, A. (1998b) Stability of resveratrol over time and in the various stages of grape transformation. *Drugs in Experimental and Clinical Research*: 24, 207–11.

Bhat, K. P. and Pezzuto, J. M. (2001) Resveratrol exhibits cytostatic and anti-oestrogenic properties with human endometrial adenocarcinoma (Ishikawa) cells. *Cancer Research*: 61, 6137–44.

Blache, D., Rustan, I., Durand, P., Lesgards, G., Loreau, N. (1997) Gas chromatographic analysis of resveratrol in plasma, lipoproteins and cells after in vitro incubations. *Journal of Chromatography B*: 702, 103–10.

Bowers, J. L., Tyulmenkov, V. V., Jernigan, S. C., Klinge, C. M. (2000) Resveratrol acts as a mixed agonist/antagonist for oestrogen receptors alpha and beta. *Endocrinology*: 141, 3657–67.

Brakenhielm, E., Cao, R., Cao, Y. (2001) Suppression of angiogenesis, tumor growth, and wound healing by resveratrol, a natural compound in red wine and grapes. *FASEB Journal*: 15, 1798–800.

Bruder, J. L., Hsieh, T. T., Lerea, K. M., Olson, S. C., Wu, J. M. (2001) Induced cytoskeletal changes in bovine pulmonary artery endothelial cells by resveratrol and the accompanying modified responses to arterial shear stress. *BMC Cell Biology*: 2, 1.

Burkhardt, S., Reiter, R. J., Tan, D., Hardeland, R., Cabrera, J., Karbownik, M. (2001) DNA oxidatively damaged by chromium (III) and H(2)O(2) is protected by the antioxidants melatonin, N(1)-acetyl-N-(2)-formyl-5-methoxykynuramine, resveratrol and uric acid. *International Journal of Biochemistry and Cell Biology*: 33, 775–83.

Burkitt, M. J. and Duncan, J. (2000) Effects of trans-resveratrol on copper-dependent hydroxyl-radical formation and DNA damage: evidence for hydroxyl-radical scavenging and a novel, glutathione-sparing mechanism of action. *Archives of Biochemistry and Biophysics*: 381, 253–63.

Cadenas, S. and Barja, G. (1999) Resveratrol, melatonin, vitamin E, and PBN protect against renal oxidative DNA damage induced by the kidney carcinogen KBrO3. *Free Radical Biology and Medicine*: **26**, 1531–7.

Carbo, N., Costelli, P., Baccino, F. M., Lopez-Soriano, F. J., Argiles, J. M. (1999) Resveratrol, a natural product present in wine, decreases tumour growth in a rat tumour model. *Biochemical and Biophysical Research Communications*: **254**, 739–43.

Casper, R. F., Quesne, M., Rogers, I. M., Shirota, T., Jolivet, A., Milgrom, E., Savouret, J. F. (1999) Resveratrol has antagonist activity on the aryl hydrocarbon receptor: Implications for prevention of dioxin toxicity. *Molecular Pharmacology*: **56**, 784–90.

Cavenagh, J. D., Cahill, M. R., Kelsey, S. M. (1998) Adhesion molecules in clinical medicine. *Critical Reviews in Clinical and Laboratory Sciences*: **34**, 415–59.

Chan, W. K. and Delucchi, A. B. (2000) Resveratrol, a red wine constituent, is a mechanism-based inactivator of cytochrome P450 3A4. *Life Sciences*: **67**, 103–12.

Chang, T. K. and Yeung, R. K. (2001) Effect of trans-resveratrol on 7-benzyloxy-4-trifluoromethylcoumarin O-dealkylation catalyzed by human recombinant CYP3A4 and CYP3A5. *Canadian Journal of Physiology and Pharmacology*: **79**, 220–6.

Chang, T. K., Lee, W. B., Ko, H. H. (2000) Trans-resveratrol modulates the catalytic activity and mRNA expression of the procarcinogen-activating human cytochrome P450 1B1. *Canadian Journal of Physiology and Pharmacology*: **78**, 874–81.

Chanvitayapongs, S., Draczynska-Lusiak, B., Sun, A. Y. (1997) Amelioration of oxidative stress by antioxidants and resveratrol in PC12 cells. *NeuroReport*: **8**, 1499–1502.

Chen, C. K. and Pace-Asciak, C. R. (1996) Vasorelaxing activity of resveratrol and quercetin in isolated rat aorta. *General Pharmacology*: **27**, 363–6.

Chen, T., Li, J., Cao, J., Xu, Q., Komatsu, K., Namba, T. (1999) A new flavanone isolated from rhizoma Smilacis glabrae and the structural requirements of its derivatives for preventing immunological hepatocyte damage. *Planta Medica*: **65**, 56–9.

Cheong, H., Ryu, S. Y., Kim, K. M. (1999) Anti-allergic action of resveratrol and related hydroxystilbenes. *Planta Medica*: **65**, 266–8.

Chia, M. C. (1998) The role of adhesion molecules in atherosclerosis. *Critical Reviews in Clinical and Laboratory Sciences*: **35**, 573–602.

Chun, Y. J., Kim, M. Y., Guengerich, F. P. (1999) Resveratrol is a selective human cytochrome P450 1A1 inhibitor. *Biochemical and Biophysical Research Communications*: **262**, 20–4.

Chung, M.-I., Teng, C.-M., Cheng, K.-L., Ko, F.-N., Lin, C.-N. (1992) An antiplatelet principle of *Veratrum formosanum. Planta Medica*: **58**, 274–6.

Cichewicz, R. H. and Kouzi, S. A. (1998) Biotransformation of resveratrol to piceid by Bacillus cereus. *Journal of Natural Products*: **61**, 1313–14.

Ciolino, H. P. and Yeh, G. C. (1999) Inhibition of aryl hydrocarbon-induced cytochrome P-450 1A1 enzyme activity and CYP1A1 expression by resveratrol. *Molecular Pharmacology*: **56**, 760–7.

Ciolino, H. P. and Yeh, G. C. (2001) The effects of resveratrol on CYP1A1 expression and aryl hydrocarbon receptor function in vitro. *Advances in Experimental Medicine and Biology*: **492**, 183–93.

Ciolino, H. P., Daschner, P. J., Yeh, G. C. (1998) Resveratrol inhibits transcription of CYP1A1 in vitro by preventing activation of the aryl hydrocarbon receptor. *Cancer Research*: **58**, 5707–12.

Clement, M. V., Hirpara, J. L., Chawdhury, S. H., Pervaiz, S. (1998) Chemopreventive agent resveratrol, a natural product derived from grapes, triggers CD95 signaling-dependent apoptosis in human tumor cells. *Blood*: **92**, 996–1002.

Constant, J. (1997) Alcohol, ischemic heart disease, and the French paradox. *Coronary Artery Disease*: **8**, 645–9.

Crespy, V., Morand, C., Manach, C., Besson, C., Demigne, C., Remesy, C. (1999) Part of quercetin absorbed in the small intestine is conjugated and further secreted in the intestinal lumen. *American Journal of Physiology*: 277, G120–6.

Crespy, V., Morand, C., Besson, C., Manach, C., Demigne, C., Remesy, C. (2001) Comparison of the intestinal absorption of quercetin, phloretin and their glucosides in rats. *Journal of Nutrition*: **131**, 2109–14.

Damianaki, A., Bakogeorgou, E., Kampa, M., Notas, G., Hatzoglou, A., Panagiotou, S., Gemetzi, C., Kouroumalis, E., Martin, P. M., Castanas, E. (2000) Potent inhibitory action of red wine polyphenols on human breast cancer cells. *Journal of Cellular Biochemistry*: **78**, 429–41.

Daniel, O., Meier, M. S., Schlatter, J., Frischknecht, P. (1999) Selected phenolic compounds in cultivated plants: Ecologic functions, health implications, and modulation of pesticides. *Environmental Health Perspectives*: **107** (suppl. 1), 109–14.

De Ledinghen, V., Monvoisin, A., Neaud, V., Krisa, S., Payrastre, B., Bedin, C., Desmouliere, A., Bioulac-Sage, P., Rosenbaum, J. (2001) Trans-resveratrol, a grapevine-derived polyphenol, blocks hepatocyte growth factor-induced invasion of hepatocellular carcinoma cells. *International Journal of Oncology*: **19**, 83–8.

Delmas, D., Jannin, B., Malki, M. C., Latruffe, N. (2000) Inhibitory effect of resveratrol on the proliferation of human and rat hepatic derived cell lines. *Oncology Reports*: 7, 847–52.

De Santi, C. A., Pietrabissa, A., Spisni, R., Mosca, F., Pacifici, G. M. (2000a) Sulphation of resveratrol, a natural product present in grapes and wine, in the human liver and duodenum. *Xenobiotica*: **30**, 609–17.

De Santi, C. A., Pietrabissa, A., Spisni, R., Mosca, F., Pacifici, G. M. (2000b) Sulphation of resveratrol, a natural compound present in wine, and its inhibition by natural flavonoids. *Xenobiotica*: **30**, 857–66.

De Santi, C. A., Pietrabissa, A., Mosca, F., Pacifici, G. M. (2000c) Glucuronidation of resveratrol, a natural product present in grape and wine, in the human liver. *Xenobiotica*: **30**, 1047–54.

Dobrydneva, Y., Williams, R. L., Blackmore, P. F. (1999) Trans-resveratrol inhibits calcium influx in thrombin-stimulated human platelets. *British Journal of Pharmacology*: **128**, 149–57.

Docherty, J. J., Fu, M. M., Stiffler, B. S., Limperos, R. J., Pokabla, C. M., DeLucia, A. L. (1999) Resveratrol inhibition of herpes simplex virus replication. *Antiviral Research*: **43**, 145–55.

Docherty, J. J., Fu, M. M., Tsai, M. (2001) Resveratrol selectively inhibits Neisseria gonorrhoeae and Neisseria meningitidis. *Journal of Antimicrobial Chemotherapy*: 47, 243–4.

Donovan, J. L., Crespy, V., Manach, C., Morand, C., Besson, C., Scalbert, A., Remesy, C. (2001) Catechin is metabolized by both the small intestine and liver of rats. *Journal of Nutrition*: **131**, 1753–7.

Dorrie, J., Gerauer, H., Wachter, Y., Zunino, S. J. (2001) Resveratrol induces extensive apoptosis by depolarizing mitochondrial membranes and activating caspase-9 in acute lymphoblastic leukemia cells. *Cancer Research:* **61**, 4731–9.

Draczynska-Lusiak, B., Chen, Y. M., Sun, A. Y. (1998a) Oxidized lipoproteins activate NF-kappaB binding activity and apoptosis in PC12 cells. *NeuroReport*: **9**, 527–32.

Draczynska-Lusiak, B., Doung, A., Sun, A. Y. (1998b) Oxidized lipoproteins may play a role in neuronal cell death in Alzheimer disease. *Molecular and Chemical Neuropathology*: **33**, 139–48.

Draczynska-Lusiak, B., Chen, Y. M., Sun, A. Y. (1998c) Oxidized lipoproteins activate NF-kB binding activity and apoptosis in PC12 cells. *NeuroReport*: **9**, 527–32.

ElAttar, T. M. and Virji, A. S. (1999) Modulating effect of resveratrol and quercetin on oral cancer cell growth and proliferation. *Anticancer Drugs*: **10**, 187–93.

Fauconneau, B., Waffo-Teguo, P., Huguet, F., Barrier, L., Decendit, A., Merillon, J. M. (1997) Comparative study of radical scavenger and antioxidant properties of phenolic compounds from Vitis vinifera cell cultures using in vitro tests. *Life Sciences*: **61**, 2103–10.

Ferrero, M. E., Bertelli, A. E., Fulgenzi, A., Pellegatta, F., Corsi, M. M., Bonfrate, M., Ferrara, F., De Catarina, R., Giovannini, L., Bertelli, A. (1998) Activity *in vitro* of resveratrol on granulocyte and monocyte adhesion to endothelium. *American Journal of Clinical Nutrition*: **68**, 1208–14.

Fitzpatrick, D. F., Hirschfield, S. L., Coffey, R. G. (1993) Endothelium-dependent vasorelaxing activity of wine and other grape products. *American Journal of Physiology*: **265**, H774–H778.

Fontecave, M., Lepoivre, M., Elleingand, E., Gerez, C., Guittet, O. (1998) Resveratrol, a remarkable inhibitor of ribonucleotide reductase. *FEBS Letters*: **421**, 277–9.

Frankel, E. N., Waterhouse, A. L., Kinsella, J. E. (1993) Inhibition of human LDL oxidation by resveratrol. *Lancet*: **341**, 1103–4.

Frankel, E. N., Waterhouse, A. L., Teissedre, P. L. (1995) Principal phenolic phytochemicals in selected California wines and their antioxidant activity in inhibiting oxidation of human low-density lipoproteins. *Journal of Agricultural and Food Chemistry*: **43**, 890–4.

Fremont, L. (2000) Biological effects of resveratrol. *Life Sciences*: **66**, 663–73.

Fremont, L., Belguendouz, L., Delpal, S. (1999) Antioxidant activity of resveratrol and alcohol-free wine polyphenols related to LDL oxidation and polyunsaturated fatty acids. *Life Sciences*: **64**, 2511–21.

Freyberger, A., Hartmann, E., Hildebrand, H., Krotlinger, F. (2001) Differential response of immature rat uterine tissue to ethinyloestradiol and the red wine constituent resveratrol. *Archives of Toxicology*: **74**, 709–15.

Frotschl, R., Chichmanov, L., Kleeberg, U., Hildebrandt, A. G., Roots, I., Brockmoller, J. (1998) Prediction of aryl hydrocarbon receptor-mediated enzyme induction of drugs and chemicals by mRNA quantification. *Chemical Research in Toxicology*: **11**, 1447–52.

Fukuhara, K. and Miyata, N. (1998) Resveratrol as a new type of DNA-cleaving agent. *Bioorganic and Medicinal Chemistry Letters*: **8**, 3187–92.

Garcia-Garcia, J., Micol, V., de Godos, A., Gomez-Fernandez, J. C. (1999) The cancer chemopreventive agent resveratrol is incorporated into model membranes and inhibits protein kinase C alpha activity. *Archives of Biochemistry and Biophysics*: **372**, 382–8.

Gautam, S. C., Xu, Y. X., Dumaguin, M., Janakiraman, N., Chapman, R. A. (2000) Resveratrol selectively inhibits leukemia cells: a prospective agent for ex vivo bone marrow purging. *Bone Marrow Transplantation*: **25**, 639–45.

Gehm, B. D., McAndrews, J. M., Chien, P. Y., Jameson, J. L. (1997) Resveratrol, a polyphenolic compound found in grapes and wine, is an agonist for the estrogen receptor. *Proceedings of the National Academy of Sciences of the United States of America*: **94**, 14138–43.

Gentilli, M., Mazoit, J. X., Bouaziz, H., Fletcher, D., Casper, R. F., Benhamou, D., Savouret, J. F. (2001) Resveratrol decreases hyperalgesia induced by carrageenan in the rat hind paw. *Life Sciences*: **68**, 1317–21.

Giovannini, L., Migliori, M., Longoni, B. M., Das, D. K., Bertelli, A. A., Panichi, V., Filippi, C., Bertelli, A. (2001) Resveratrol, a polyphenol found in wine, reduces ischemia reperfusion injury in rat kidneys. *Journal of Cardiovascular Pharmacology*: **37**, 262–70.

Goldberg, D. M. and Soleas, G. J. (2001) Beverage alcohol consumption as a negative risk factor for coronary heart disease: Biochemical mechanisms, in *Alcohol in Health and Disease* (eds D. P. Agarwal and H. K. Seitz). Marcel Dekker, New York, pp. 547–72.

Goldberg, D. M., Hahn, S. E., Parkes, J. G. (1995a) Beyond alcohol: Beverage consumption and cardiovascular mortality. *Clinica Chimica Acta*: **237**, 155–87.

Goldberg, D. M., Tham, L., Diamandis, E. P., Karumanchiri, A., Soleas, G. J. (1995b) The assay of resveratrol and its distribution in human blood. *Clinical Chemistry*: **41**, S115 (abstract).

Goldberg, D. M., Garovic-Kocic, V., Diamandis, E. P., Pace-Asciak, C. R. (1996) Wine: does the colour count? *Clinica Chimica Acta*: **246**, 183–93.

Goldberg, D. M., Soleas, G. J., Hahn, S. E., Diamandis, E. P., Karumanchiri, A. (1997) Identification and assay of trihydroxystilbenes in wine and their biological properties, in *Wine composition and health benefits* (ed. T. Watkins). American Chemical Society, Washington, DC, pp. 24–43.

Gusman, J., Malonne, H., Atassi, G. (2001) A reappraisal of the potential chemopreventive and chemotherapeutic properties of resveratrol. *Carcinogenesis*: **22**, 1111–17.

Heredia, A., Davis, C., Redfield, R. (2000) Synergistic inhibition of HIV-1 in activated and resting peripheral blood mononuclear cells, monocyte-derived macrophages, and selected drug-resistant isolates with nucleoside analogues combined with a natural product, resveratrol. *Journal of Acquired Immune Deficiency Syndrome*: **25**, 246–55.

Holian, O. and Walter, R. J. (2001) Resveratrol inhibits the proliferation of normal human keratinocytes in vitro. *Journal of Cellular Biochemistry*: **36**, 55–62.

Holmes-McNary, M. and Baldwin, Jr, A. S. (2000) Chemopreventive properties of *trans*-resveratrol are associated with inhibition of activation of the IkappaB kinase. *Cancer Research*: **60**, 3477–83.

Hsieh, T. C. and Wu, J. M. (1999) Differential effects on growth, cell cycle arrest, and induction of apoptosis by resveratrol in human prostate cancer cell lines. *Experimental Cell Research*: **249**, 109–15.

Hsieh, T. C. and Wu, J. M. (2000) Grape-derived chemopreventive agent resveratrol decreases prostate-specific antigen (PSA) expression in LNCaP cells by an androgen receptor (AR)-independent mechanism. *Anticancer Research*: **20**, 225–8.

Hsieh, T. C., Burfeind, P., Laud, K., Backer, J. M., Traganos, F., Darzynkiewicz, Z., Wu, J. M. (1999a) Cell cycle effects and control of gene expression by resveratrol in human breast carcinoma cell lines with different metastatic potentials. *International Journal of Oncology*: **15**, 245–52.

Hsieh, T.-C., Juan, G., Darzynkiewicz, Z., Wu, J. M. (1999b) Resveratrol increases nitric oxide synthase, induces accumulation of p53 and p21$^{WAF1/CIP1}$, and suppresses cultured bovine pulmonary artery endothelial cell proliferation by perturbing progression through S and $G_2$. *Cancer Research*: **59**, 2596–2601.

Huang, C., Ma, M. Y., Goranson, A., Dong, Z. (1999) Resveratrol suppresses cell transformation and induces apoptosis through a p53-dependent pathway. *Carcinogenesis*: **20**, 237–42.

Huang, S. S., Tsai, M. C., Chih, C. L., Hung, L. M., Tsai, S. K. (2001) Resveratrol reduction of infarct size in Long-Evans rats subjected to focal cerebral ischemia. *Life Sciences*: **69**, 1057–65.

Hung, L. M., Chen, J. K., Huang, S. S., Lee, R. S., Su, M. J. (2000) Cardioprotective effect of resveratrol, a natural antioxidant derived from grapes. *Cardiovascular Research*: **47**, 549–55.

Igura, K., Ohta, T., Juroda, Y., Kaji, K. (2001) Resveratrol and quercetin inhibit angiogenesis in vitro. *Cancer Letters*: **171**, 11–16.

Jager, U. and Nguyen-Duong, H. (1999) Relaxant effect of *trans*-resveratrol on isolated porcine coronary arteries. *Arzneimittelforschung*: **49**, 207–11.

Jang, D. S., Kang, B. S., Ryu, S. Y., Chang, I. M., Min, K. R., Kim, Y. (1999) Inhibitory effects of resveratrol analogs on unopsonized zymosan-induced oxygen radical production. *Biochemical Pharmacology*: **57**, 705–12.

Jang, J. and Surh, Y. (2001) Protective effects of resveratrol on hydrogen peroxide-induced apoptosis in rat pheochromocytoma (PC12) cells. *Mutation Research*: **496**, 181–90.

Jang, M. and Pezzuto, J. M. (1998) Effects of resveratrol on 12-O-tetradecanoylphorbol-13-acetate-induced oxidative events and gene expression in mouse skin. *Cancer Letters*: **134**, 81–9.

Jang, M. and Pezzuto, J. M. (1999) Cancer chemopreventive activity of resveratrol. *Drugs in Experimental and Clinical Research*: **25**, 65–77.

Jang, M., Cai, L., Udeani, G. O., Slowing, K. V., Thomas, C. F., Beecher, C. W., Fong, H. H., Farnsworth, N. R., Kinghorn, A. D., Mehta, R. G., Moon, R. C., Pezzuto, J. M. (1997) Cancer chemopreventive activity of resveratrol, a natural product derived from grapes. *Science:* **275**, 218–20.

Johnson, J. L. and Maddipati, K. R. (1998) Paradoxical effects of resveratrol on the two prostaglandin H synthases. *Prostaglandins and Other Lipid Mediators:* **56**, 131–43.

Juan, M. E., Lamuela-Raventos, R. M., de la Torre-Boronat, M. C., Planas, J. M. (1999) Determination of *trans*-resveratrol in plasma by HPLC. *Analytical Chemistry*: **71**, 747–50.

Kawada, N., Seki, S., Inoue, M., Kuroki, T. (1998) Effect of antioxidants, resveratrol, quercetin, and N-acetylcysteine, on the functions of cultured rat hepatic stellate cells and Kupffer cells. *Hepatology*: **27**, 1265–74.

Kimura, Y. and Okuda, H. (2000) Effects of naturally occurring stilbene glucosides from medicinal plants and wine, on tumour growth and lung metastasis in Lewis lung carcinoma-bearing mice. *Journal of Pharmacy and Pharmacology*: **52**, 1287–95.

Kimura, Y. and Okuda, H. (2001) Resveratrol isolated from Polygonum cuspidatum root prevents tumor growth and metastasis to lung and tumor-induced neovascularization in Lewis lung carcinoma-bearing mice. *Journal of Nutrition*: **131**, 1344–9.

Kimura, Y., Okuda, H., Arichi, S. (1985) Effects of stilbenes on arachidonate metabolism in leucoytes. *Biochimica Biophysica Acta*: **834**, 275–8.

Kimura, Y., Okuda, H., Kubo, M. (1995) Effects of stilbenes isolated from medicinal plants on arachidonate metabolism and degranulation in human polymorphonuclear leukocytes. *Journal of Ethnopharmacology*: 45, 131–9.

Knight, J., Taylor, G. W., Wright, P., Clare, A. S., Rowley, A. F. (1999) Eicosanoid biosynthesis in an advanced deuterostomate invertebrate, the sea squirt (*Ciona intestinalis*). *Biochimica Biophysica Acta*: 1436, 467–78.

Kong, A. T., Yu, R., Hebbar, V., Chen, C., Owuor, E., Hu, R., Ee, R., Mandlekar, S. (2001) Signal transduction events elicited by cancer prevention compounds. *Mutation Research*: 480–1, 231–41.

Kopp, P. (1998) Resveratrol, a phytooestrogen found in red wine. A possible explanation for the conundrum of the 'French paradox'? *European Journal of Endocrinology*: 138, 619–20.

Kozuki, Y., Miura, Y., Yagasaki, K. (2001) Resveratrol suppresses hepatoma cell invasion independently of its anti-proliferative action. *Cancer Letters*: 167, 151–6.

Lee, S. H., Shin, N. H., Kang, S. H., Park, J. S., Chung, S. R., Min, K. R., Kim, Y. (1998) Alpha-viniferin: a prostaglandin H2 synthase inhibitor from root of Carex humilis. *Planta Medica*: 64, 204–7.

Lin, J. K. and Tsai, S. H. (1999) Chemoprevention of cancer and cardiovascular disease by resveratrol. *Proceedings of the National Science Council, Republic of China, Part B, Life Sciences*: 23, 99–106.

Lu, J., Ho, C. H., Ghai, G., Chen, K. Y. (2001) Resveratrol analog, 3,4,5,4′-tetrahydroxystillbene, differentially induces pro-apoptotic p53/Bax gene expression and inhibits the growth of transformed cells but not their normal counterparts. *Carcinogenesis*: 22, 321–8.

Lu, R. and Serrero, G. (1999) Resveratrol, a natural product derived from grape, exhibits antioestrogenic activity and inhibits the growth of human breast cancer cells. *Journal of Cell Physiology*: 179, 297–304.

MacCarrone, M., Lorenzon, T., Guerrieri, P., Agro, A. F. (1999) Resveratrol prevents apoptosis in K562 cells by inhibiting lipoxygenase and cyclooxygenase activity. *European Journal of Biochemistry*: 265, 27–34.

Manna, S. K., Mukhopadhyay, A., Aggarwal, B. B. (2000) Resveratrol suppresses TNF-induced activation of nuclear transcription factors NF-kappa B, activator protein-1, and apoptosis: potential role of reactive oxygen intermediates and lipid peroxidation. *Journal of Immunology*: 164, 6509–19.

Man-Ying, C. M., Mattiacci, J. A., Hwang, H. S., Shah, A., Fong, D. (2000) Synergy between ethanol and grape polyphenols, quercetin, and resveratrol, in the inhibition of the inducible nitric oxide synthase pathway. *Biochemical Pharmacology*: 60, 1539–48.

Martinez, J. and Moreno, J. J. (2000) Effect of resveratrol, a natural polyphenolic compound, on reactive oxygen species and prostaglandin production. *Biochemical Pharmacology*: 59, 865–70.

Matsuda, H., Kageura, T. R., Morikawa, T., Toguchida, I., Harima, S., Yoshikawa, M. (2000) Effects of stilbene constituents from rhubarb on nitric oxide production in lipopolysaccharide-activated macrophages. *Bioorganic and Medicinal Chemistry Letters*: 10, 323–7.

Matsuoka, A., Furuta, A., Ozaki, M., Fukuhara, K., Miyata, N. (2001) Resveratrol, a naturally occurring polyphenol, induces sister chromatid exchanges in a Chinese hamster lung (CHL) cell line. *Mutation Research*: 494, 107–13.

Mellanen, P., Petanen, T., Lehtimaki, J., Makela, S., Bylund, G., Holmbom, B., Mannila, E., Oikari, S. A., Santti, R. (1996) Wood-derived oestrogens: studies in vitro with breast cancer cell lines and in vivo in trout. *Toxicology and Applied Pharmacology*: **136**, 381–8.

Mgbonyebi, O. P., Russo, J., Russo, I. H. (1998) Antiproliferative effect of synthetic resveratrol on human breast epithelial cells. *International Journal of Oncology*: **12**, 865–9.

Mietus-Snyder, M., Gowri, M. S., Pitas, R. E. (2000) Class A scavenger receptor up-regulation in smooth muscle cells by oxidized low density lipoprotein. Enhancement by calcium flux and concurrent cyclooxygenase-2 up-regulation. *Journal of Biological Chemistry*: **275**, 17661–70.

Miloso, M., Bertelli, A. A., Nicolini, G., Tredici. G. (1999) Resveratrol-induced activation of the mitogen-activated protein kinases, ERK1 and ERK2, in human neuroblastoma SH-SY5Y cells. *Neuroscience Letters*: **264**, 141–4.

Mitchell, S. H., Zhu, W., Young, C. Y. (1999) Resveratrol inhibits the expression and function of the androgen receptor in LNCaP prostate cancer cells. *Cancer Research*: **59**, 5892–5.

Miura, T., Muraoka, S., Ikeda, N., Watanabe, M., Fujimoto, Y. (2000) Antioxidative and prooxidative action of stilbene derivatives. *Pharmacology and Toxicology*: **86**, 203–8.

Mizutani, K., Ikeda, K., Kawai, Y., Yamori, Y. (1998) Resveratrol stimulates the proliferation and differentiation of osteoblastic MC3T3-E1 cells. *Biochemical and Biophysical Research Communications*: **253**, 859–63.

Mizutani, K., Ikeda, K., Yamori, Y. (2000) Resveratrol inhibits AGEs-induced proliferation and collagen synthesis activity in vascular smooth muscle cells from stroke-prone spontaneously hypertensive rats. *Biochemical and Biophysical Research Communications*: **274**, 61–7.

Mizutani, K., Ikeda, K., Kawai, Y., Yamori, Y. (2001) Protective effect of resveratrol on oxidative damage in male and female stroke-prone spontaneously hypertensive rats. *Clinical and Experimental Pharmacology*: **28**, 55–9.

Mollerup, S., Ovrebo, S., Haugen, A. (2001) Lung carcinogenesis: resveratrol modulates the expression of genes involved in the metabolism of PAH in human bronchial epithelial cells. *International Journal of Cancer*: **92**, 18–25.

Moreno, J. J. (2000) Resveratrol modulates arachidonic acid release, prostaglandin synthesis, and 3T6 fibroblast growth. *Journal of Pharmacology and Experimental Therapeutics*: **294**, 333–8.

Murcia, M. A., Martinez-Tome, M. (2001) Antioxidant activity of resveratrol compared with common food additives. *Journal of Food Protection*: **64**, 379–84.

Mutoh, M., Takahashi, M., Fukuda, K., Matsushima-Hibiya, Y., Mutoh, H., Sugimura, T., Wakabayashi, K. (2000) Suppression of cyclooxygenase-2 promoter-dependent transcriptional activity in colon cancer cells by chemopreventive agents with resorcin-type structure. *Carcinogenesis*: **21**, 959–63.

Naderali, E. K., Doyle, P. J., Williams, G. (2000) Resveratrol induces vasorelaxation of mesenteric and uterine arteries from female guinea-pigs. *Clinical Science*: **98**, 537–43.

Naderali, E. K., Smith, S. L., Doyle, P. J., Williams, G. (2001) The mechanism of resveratrol-induced vasorelaxation differs in the mesenteric resistance arteries of lean and obese rats. *Clinical Science*: **100**, 55–60.

Nakagawa, H., Kiyozuka, Y., Uemura, Y., Senzaki, H., Shikata, N., Hioki, K., Tsubura, A. (2001) Resveratrol inhibits human breast cancer cell growth and may mitigate the effect of linoleic acid, a potent breast cancer cell stimulator. *Journal of Cancer Research and Clinical Oncology*: **127**, 258–64.

Nicholson, A. C. and Hajjar, D. P. (1998) Herpes viruses in atherosclerosis and thrombosis: etiologic agents or ubiquitous bystanders? *Arteriosclerosis, Thrombosis and Vascular Biology*: **18**, 339–48.

Nicolini, G., Rigolio, R., Miloso, M., Bertelli, A. A., Tredici, G. (2001) Anti-apoptotic effect of trans-resveratrol on paclitaxel-induced apoptosis in the human neuroblastoma SH-SY5Y cell line. *Neuroscience Letters*: **302**, 41–4.

Nielsen, M., Ruch, R. J., Vang, O. (2000) Resveratrol reverses tumor-promoter-induced inhibition of gap-junctional intercellular communication. *Biochemical and Biophysical Research Communications*: **275**, 804–9.

Nihei, T., Miura, Y., Yagasaki, K. (2001) Inhibitory effect of resveratrol on proteinuria, hypoalbuminemia and hyperlipidemia in nephritic rats. *Life Sciences*: **68**, 2845–52.

Olas, B., Zbikowska, H. M., Wachowicz, B., Krajewski, T., Buczynski, A., Magnuszewska, A. (1999) Inhibitory effect of resveratrol on free radical generation in blood platelets. *Acta Biochimica Polonica*: **46**, 961–6.

Olas, B., Wachowicz, B., Saluk-Juszczak, J., Zielinski, T., Kaca, W., Buczynski, A. (2001a) Antioxidant activity of resveratrol in endotoxin-stimulated blood platelets. *Cell Biology and Toxicology*: **17**, 117–25.

Olas, B., Wachowicz, B., Szewczuk, J., Saluk-Juszczak, J., Kaca, W. (2001b) The effect of resveratrol on the platelet secretory process induced by endotoxin and thrombin. *Microbios*: **105**, 7–13.

Orsini, F., Pelizzoni, F., Verotta, L., Aburjai, T., Rogers, C. B. (1997a) Isolation, synthesis, and antiplatelet aggregation activity of resveratrol 3-O-beta-D-glucopyranoside and related compounds. *Journal of Natural Products*: **60**, 1082–7.

Orsini, F., Pelizzoni, F., Bellini, B., Miglierini, G. (1997b) Synthesis of biologically active polyphenolic glycosides (combretastatin and resveratrol series). *Carbohydrate Research*: **301**, 95–109.

Otake, Y., Nolan, A. L., Walle, U. K., Walle, T. (2000) Quercetin and resveratrol potently reduce oestrogen sulfotransferase activity in normal human mammary epithelial cells. *Journal of Steroid Biochemistry and Molecular Biology*: **5**, 265–70.

Pace-Asciak, C. R., Hahn, S., Diamandis, E. P., Soleas, G., Goldberg, D. M. (1995) The red wine phenolics *trans*-resveratrol and quercetin block human platelet aggregation and eicosanoid synthesis: Implications for protection against coronary heart disease. *Clinica Chimica Acta*: **235**, 207–19.

Park, J. B. (2001) Inhibition of glucose and dehydroascorbic acid uptakes by resveratrol in human transformed myelocytic cells. *Journal of Natural Products*: **64**, 381–4.

Park, J. W., Choi, Y. J., Jang, M. A., Lee, Y. S., Jun, D. Y., Suh, S. I., Baek, W. K., Suh, M. H., Jin, I. N., Kwon, T. K. (2001) Chemopreventive agent resveratrol, a natural product derived from grapes, reversibly inhibits progression through S and G2 phases of the cell cycle in U937 cells. *Cancer Letters*: **163**, 43–9.

Pendurthi, U. R., Williams, J. T., Rao, L. V. (1999) Resveratrol, a polyphenolic compound found in wine, inhibits tissue factor expression in vascular cells: A possible mechanism for the cardiovascular benefits associated with moderate consumption of wine. *Arteriosclerosis, Thrombosis and Vascular Biology*: **19**, 419–26.

Pervaiz, S. (2001) Resveratrol–from the bottle to the bedside? *Leukemia and Lymphoma*: **40**, 491–8.

Ragione, F. D., Cucciolla, V., Borriello, A., Pietra, V. D., Racciopi, L., Soldati, G., Manna, C., Galletti, P., Zappia, V. (1998) Resveratrol arrests the cell division cycle at S/G2 phase transition. *Biochemical and Biophysical Research Communications*: **250**, 53–8.

Ray, P. S., Maulik, G., Cordis, G. A., Bertelli, A. A., Bertelli, A., Das, D. K. (1999) The red wine antioxidant resveratrol protects isolated rat hearts from ischemia reperfusion injury. *Free Radical Biology and Medicine*: **27**, 160–9.

Rodrigue, C. M., Bachir, D., Smith-Ravin, J., Romeo, P. H., Galacteros, F., Garel, M. C. (2001) Resveratrol, a natural dietary phytoalexin, possesses similar properties to hydroxyurea towards erythroid differentiation. *British Journal of Haematology*: **113**, 500–7.

Rosenberg Zand, R. S., Jenkins, D. J. A., Diamandis, E. P. (2000) Steroid hormone activity of flavonoids and related compounds. *Breast Cancer Research and Treatment*: **62**, 35–49.

Ross, R. (1999) Atherosclerosis – An inflammatory disease. *New England Journal of Medicine*: **340**, 115–26.

Rotondo, S., Rajtar, G., Manarini, S., Celardo, A., Rotillo, D., De Gaetano, G., Evangelista, V., Ceretti, C. (1998) Effect of *trans*-resveratrol, a natural polyphenolic compound, on human polymorphonuclear leucocyte function. *British Journal of Pharmacology*: **123**, 1691–9.

Schneider, Y., Vincent, F., Duranton, B., Badolo, L., Gosse, F., Bergmann, C., Seiler, N., Raul, F. (2000) Anti-proliferative effect of resveratrol, a natural component of grapes and wine, on human colonic cancer cells. *Cancer Letters*: **158**, 85–91.

Sgambato, A., Ardito, R., Faraglia, B., Boninsegna, A., Wolf, F. I., Cittadini, A. (2001) Resveratrol, a natural phenolic compound, inhibits cell proliferation and prevents oxidative DNA damage. *Mutation Research*: **496**, 171–80.

Shan, C. W. (1988) Effects of polydatin on platelet aggregation of rabbits. *Acta Pharmacologica Sinica*: **23**, 394–6.

Shan, C. W., Yang, S. Q., He, H. D., Shao, S. L., Zhang, P. W. (1990) Influences of 3,4,5-trihydroxy stilbene-3-β-mono-D-glucoside on rabbits' platelet aggregation and thromboxane $B_2$ production *in vitro*. *Acta Pharmacologica Sinica*: **11**, 527–30.

She, Q. B., Bode, A. M., Ma, W. Y., Chen, N. Y., Dong, Z. (2001) Resveratrol-induced activation of p53 and apoptosis is mediated by extracellular-signal-regulated protein kinases and p38 kinase. *Cancer Research*: **61**, 1604–10.

Shin, N. H., Ryu, S. Y., Lee, H., Min, K. R., Kim, Y. (1998a) Inhibitory effects of hydroxystilbenes on cyclooxygenase from sheep seminal vesicles. *Planta Medica*: **64**, 283–4.

Shin, N. H., Ryu, S. Y., Choi, E. J., Kang, S. H., Chang, I. M., Min, K. R., Kim, Y. (1998b) Oxyresveratrol as the potent inhibitor on dopa oxidase activity of mushroom tyrosinase. *Biochemical and Biophysical Research Communications*: **243**, 801–3.

Siemann, E. H. and Creasy, L. L. (1992) Concentration of the phytoalexin resveratrol in wine. *American Journal of Enology and Viticulture*: **43**, 49–52.

Singh, S. U., Casper, R. F., Fritz, P. C., Sukhu, B., Ganss, B., Girard, Jr, B., Savouret, J. F., Tenenbaum, H. C. (2000) Inhibition of dioxin effects on bone formation in vitro by a newly described aryl hydrocarbon receptor antagonist, resveratrol. *Journal of Endocrinology*: 167, 183–95.

Soleas, G. J., Diamandis, E. P., Goldberg, D. M. (1997a) Resveratrol: A molecule whose time has come? And gone? *Clinical Biochemistry*: 30, 91–113.

Soleas, G. J., Tomlinson, G., Diamandis, E. P., Goldberg, D. M. (1997b) Relative contributions of polyphenolic constituents to the antioxidant status of wines: Development of a predictive model. *Journal of Agricultural and Food Chemistry*: 45, 3995–4003.

Soleas, G. J., Diamandis, E. P., Goldberg, D. M. (2001a) The world of resveratrol. *Advances in Experimental Medicine and Biology*: 492, 159–82.

Soleas, G. J., Goldberg, D. M., Grass, L., Levesque, M., Diamandis, E. P. (2001b) Do wine polyphenols modulate p53 gene expression in human cancer cell lines? *Clinical Biochemistry*: 34, 415–20.

Soleas, G. J., Yan, J., Goldberg, D. M. (2001c) Measurement of *trans*-resveratrol, (+)-catechin, and quercetin in rat and human blood and urine by gas chromatography with mass selective detection. *Methods in Enzymology*: 335, 130–45.

Soleas, G. J., Angelini, M., Grass, L., Diamandis, E. P., Goldberg, D. M. (2001d) Absorption of *trans*-resveratrol in rats. *Methods in Enzymology*: 335, 145–54.

Soleas, G. J., Yan, J., Goldberg, D. M. (2001e) Ultrasensitive assay for three polyphenols (catechin, quercetin and resveratrol) and their conjugates in biological fluids utilizing gas chromatography with mass selective detection. *Journal of Chromatography: B*: 757, 161–72.

Stahl, S., Chun, T. Y., Gray, W. G. (1998) Phytooestrogens act as oestrogen agonists in an oestrogen-responsive pituitary cell line. *Toxicology and Applied Pharmacology*: 152, 41–8.

Stewart, J. R., Christman, K. L., O'Brian, C. A. (2000) Effects of resveratrol on the autophosphorylation of phorbol ester-responsive protein kinases. Inhibition of protein kinase d but not protein kinase c isozyme autophosphorylation. *Biochemical Pharmacology*: 60, 1355–9.

Stivala, L. A., Savio, M., Carafoli, F., Perucca, P., Bianchi, L., Maga, G., Forti, L., Pagnoni, U. M., Albini, A., Prosperi, E., Vannini, V. (2001) Specific structural determinants are responsible for the antioxidant activity and the cell cycle effects of resveratrol. *Journal of Biological Chemistry*: 276, 22586–94.

Stojanovic, S., Sprinz, H., Brede, O. (2001) Efficiency and mechanism of the antioxidant action of trans-resveratrol and its analogues in the radical liposome oxidation. *Archives of Biochemistry and Biophysics*: 391, 79–89.

Subbaramaiah, K. and Dannenberg, A. J. (2001) Resveratrol inhibits the expression of cyclooxygenase-2 in mammary epithelial cells. *Advances in Experimental Medicine and Biology*: 492, 147–57.

Subbaramaiah, K., Chung, W. J., Michaluart, P., Telang, N., Tanabe, T., Inoue, H., Jang, M., Pezzuto, J. M., Dannenberg, A. J. (1998) Resveratrol inhibits cyclooxygenase-2 transcription and activity in phorbol ester-treated human mammary epithelial cells. *Journal of Biological Chemistry*: 273, 21875–82.

Subbaramaiah, K., Michaluart, P., Chung, W. J., Tanabe, T., Telang, N., Dannenberg, A.J. (1999) Resveratrol inhibits cyclooxygenase-2 transcription in human mammary epithelial cells. *Annals of the New York Academy of Sciences*: 880, 214–23.

Sun, N. J., Woo, S. H., Cassady, J. M., Snapka, R. M. (1998) DNA polymerase and topoisomerase II inhibitors from Psoralea corylifolia. *Journal of Natural Products*: 61, 362–6.

Surh, Y. J., Hurh, Y. J., Kang, J. Y., Lee, E., Kong, G., Lee, S. J. (1999) Resveratrol, an antioxidant present in red wine, induces apoptosis in human promyelocytic leukemia (HL-60) cells. *Cancer Letters*: **140**, 1–10.

Surh, Y., Chun, K., Cha, H., Han, S. S., Keum, Y., Park, K., Lee, S. S. (2001) Molecular mechanisms underlying chemopreventive activities of anti-inflammatory phytochemicals: down-regulation of COX-2 and iNOS through suppression of NF-kappaB activation. *Mutation Research*: **480–1**, 243–68.

Tadolini, B., Juliano, C., Piu, L., Franconi, F., Cabrini, L. (2000) Resveratrol inhibition of lipid peroxidation. *Free Radical Research*: **33**, 105–14.

Teel, R. W. and Huynh, H. (1998) Modulation by phytochemicals of cytochrome P450-linked enzyme activity. *Cancer Letters*: **133**, 135–41.

Tessitore, L., Davit, A., Sarotto, I., Caderni, G. (2000) Resveratrol depresses the growth of colorectal aberrant crypt foci by affecting bax and p21(CIP) expression. *Carcinogenesis*: **21**, 1619–22.

Tham, L., Goldberg, D. M., Diamandis, P., Karumanchiri, A., Soleas, G. J. (1995) Extraction of resveratrol from human blood. *Clinical Biochemistry*: **28**, 339 (abstract).

Tinhofer, I., Bernhard, D., Senfter, M., Anether, G., Loeffler, M., Kroemer, G., Kofler, R., Csordas, A., Greil, R. (2001) Resveratrol, a tumor-suppressive compound from grapes, induces apoptosis via a novel mitochondrial pathway controlled by Bcl-2. *FASEB Journal*: **15**, 1613–15.

Tou, J. and Urbizo, C. (2001) Resveratrol inhibits the formation of phosphatidic acid and diglyceride in chemotactic peptide- or phorbol ester-stimulated human neutrophils. *Cellular Signalling*: **13**, 191–7.

Tsai, S. H., Lin-Shiau, S. Y., Lin, J. K. (1999) Suppression of nitric oxide synthase and the down-regulation of the activation of NFkappaB in macrophages by resveratrol. *British Journal of Pharmacology*: **126**, 673–80.

Tsan, M. F., White, J. E., Maheshwari, J. G., Bremner, T. A., Sacco, J. (2000) Resveratrol induces Fas signalling-independent apoptosis in THP-1 human monocytic leukaemia cells. *British Journal of Haematology*: **109**, 405–12.

Turner, R. T., Evans, G. L., Zhang, M., Maran, A., Sibonga, J. D. (1999) Is resveratrol an oestrogen agonist in growing rats? *Endocrinology*: **140**, 50–4.

Turrens, J. F., Lariccia, J., Nair, M. G. (1997) Resveratrol has no effect on lipoprotein profile and does not prevent peroxidation of serum lipids in normal rats. *Free Radical Research*: **27**, 557–62.

Uenobe, F., Nakamura, S.-I., Miyazawa, M. (1997) Antimutagenic effect of resveratrol against Trp-P-1. *Mutation Research*: **373**, 197–200.

Ulsperger, E., Hamilton, G., Raderer, M., Baumgartner, G., Hejna, M., Hoffmann, O., Mallinger, J. (1999) Resveratrol pretreatment desensitizes AHTO-7 human osteoblasts to growth stimulation in response to carcinoma cell supernatants. *International Journal of Oncology*: **15**, 955–9.

Virgili, M. and Contestabile, A. (2000) Partial neuroprotection of in vivo excitotoxic brain damage by chronic administration of the red wine antioxidant agent, trans-resveratrol in rats. *Neuroscience Letters*: **281**, 123–6.

Wadsworth, T. L. and Koop, D. R. (1999) Effects of the wine polyphenolics quercetin and resveratrol on pro-inflammatory cytokine expression in RAW 264.7 macrophages. *Biochemical Pharmacology*: **57**, 941–9.

Waffo-Teguo, P., Fauconneau, B., Deffieux, G., Huguet, F., Vercauteren, J., Merillon, J. M. (1998) Isolation, identification, and antioxidant activity of three stilbene glucosides newly extracted from Vitis vinifera cell cultures. *Journal of Natural Products*: **61**, 655–7.

Wang, M. J., Huang, H. M., Hsieh, S. J., Jeng, K. C., Kuo, J. S. (2000) Resveratrol inhibits interleukin-6 production in cortical mixed glial cells under hypoxia/ hypoglycemia followed by reoxygenation. *Journal of Neuroimmunology*: **112**, 28–34.

Wilson, T., Knight, T. J., Beitz, D. C., Lewis, D. S., Engen, R. L. (1996) Resveratrol promotes atherosclerosis in hypercholesterolemic rabbits. *Life Sciences*: **59**, 15–21.

Wolter, F., Akoglu, B., Clausnitzer, A., Stein, J. (2001) Downregulation of the cyclin d1/cdk4 complex occurs during resveratrol-induced cell cycle arrest in colon cancer cell lines. *Journal of Nutrition*: **131**, 2197–203.

Wu, J. M., Wang, Z. R., Hsieh, T. C., Bruder, J. L., Zou, J. G., Huang, Y. Z. (2001) Mechanism of cardioprotection by resveratrol, a phenolic antioxidant present in red wine. *International Journal of Molecular Medicine*: **8**, 3–17.

Yang, C. S., Landau, J. M., Huang, M. T., Newmark, H. L. (2001) Inhibition of carcinogenesis by dietary polyphenolic compounds. *Annual Review of Nutrition*: **21**, 381–406.

Yoon, K., Pallaroni, L., Stoner, M., Gaido, K., Safe, S. (2001) Differential activation of wild-type and variant forms of oestrogen receptor alpha by synthetic and natural oestrogenic compounds using a promoter containing three oestrogen-responsive elements. *Journal of Steroid Biochemistry and Molecular Biology*: **78**, 25–32.

Yu, R., Hebbar, V., Kim, D. W., Mandlekar, S., Pezzuto, J. M., Kong, A. N. (2001) Resveratrol inhibits phorbol ester and UV-induced activator protein 1 activation by interfering with mitogen-activated protein kinase pathways. *Molecular Pharmacology*: **60**, 217–24.

Zbikowska, H. M. and Olas, B. (2000) Antioxidants with carcinostatic activity (resveratrol, vitamin E and selenium) in modulation of blood platelet adhesion. *Journal of Physiology and Pharmacology*: **51**, 513–20.

Zhong, M., Cheng, G. F., Wang, W. J., Guo, Y., Zhu, X. Y., Zhang, J. T. (1999) Inhibitory effect of resveratrol on interleukin 6 release by stimulated peritoneal macrophages of mice. *Phytomedicine*: **6**, 79–84.

Zhu, Z., Klironomos, G., Vachereau, A., Neirinck, L., Goodman, D. W. (1999) Determination of *trans*-resveratrol in human plasma by high-performance liquid chromatography. *Journal of Chromatography B*: **724**, 389–92.

Zou, J., Huang, Y., Chen, Q., Wang, N., Cao, K., Hsieh, T. C., Wu, J. M. (1999) Suppression of mitogenesis and regulation of cell cycle traverse by resveratrol in cultured smooth muscle cells. *International Journal of Oncology*: **15**, 647–51.

Zou, J., Huang, Y., Cao, K., Yang, G., Yin, H., Len, J., Hsieh, T. C., Wu, J. M. (2000) Effect of resveratrol on intimal hyperplasia after endothelial denudation in an experimental rabbit model. *Life Sciences*: **68**, 153–63.

Zou, J. G., Huang, Y. Z., Chen, Q., Wei, E. H., Hsieh, T. C., Wu, J. M. (2001) Resveratrol inhibits copper ion-induced and azo compound-initiated oxidative modification of human low density lipoprotein. *Biochemistry and Molecular Biology International*: 47, 1089–96.

# 9 Grape-derived wine flavonoids and stilbenes

*G. L. Creasy and L. L. Creasy*

## Introduction

Flavonoids and stilbenes are significant constituents of grapes, affecting how they develop into fruit that will make quality wine. The flavonoids represent a very large group of phenolics that contribute to colour, taste and 'mouth-feel', while the stilbenes are a much more specific phenolic group of compounds that, although contributing little in terms of the sensory perception of grapes and wine, can add significantly to the ability of the grape to reach harvest and to the perceived health benefits of moderate wine consumption: the former through combating disease on the vine, and the latter through combating disease in humans. Both fall into a much broader category of antioxidants, which are a subject of much discussion in health circles.

This chapter will explore the contribution of flavonoids and stilbenes to the grape and, subsequently, to the wine. In order to do so, we will have to examine the physiological roles of each group of compounds, the ways in which our management of vines influences their production, and finally identify factors that affect their transfer into wine and continued evolution during wine ageing. A review of what these compounds are and how they are made is in order.

### *Flavonoids*

The flavonoids are vitally important in making wine, affecting how we see in the glass and how we perceive it in our mouths. Primarily this is through appearance – the first, and, for most people, one of the most influential of the senses when we come to evaluate grapes or wine.

Anthocyanins, the most important coloured compounds found in grapes, make up one of three major groups of flavonoids. Another group, the flavonols, make up another important group in that they can have an influence on wine colour without actually being highly coloured themselves. The final group, the flavans, give bitterness and astringency to grapes and wines, most famously through the action of tannins. Together, these three groups of compounds, which are closely related but very different in action, influence grape and wine colour, mouth-feel, flavour and antioxidant capacity.

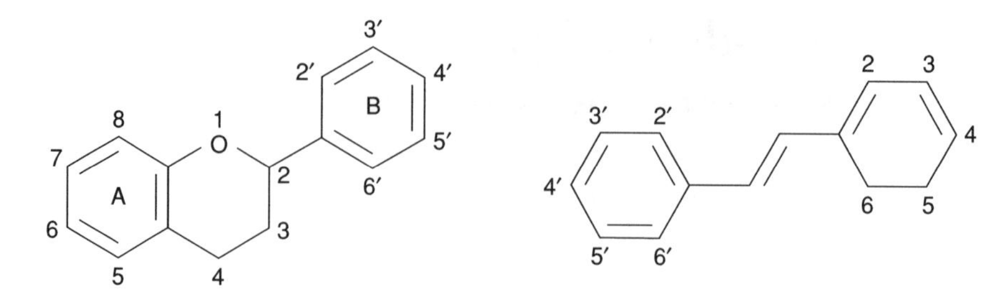


*Figure 9.1* Ring-numbering system for flavonoids (left) and stilbenes (right). A- and B-rings are named as indicated.

In talking about the flavonoids, one should at least mention their counterpart group – the aptly named non-flavonoids. These are simpler phenolic compounds found primarily in the juice of grapes. They are important adjuncts to the flavonoids and can themselves influence the bitterness and aroma of wine. Their effect is much smaller than that of the flavonoids, however, and this chapter will not discuss them in detail, other than in their interaction with the large flavonoids.

Flavonoids are a family of double aromatic-ring molecules that have the basic structure and position-naming convention shown in Figure 9.1. Consistency in numbering of the positions is important in describing the molecules. The spatial arrangement of the molecule is important to their reactivity, which is related to qualities of wine that we can perceive through our eyes (colour), sense of taste (bitterness) and sense of mouth-feel (astringency). Flavonoids are not volatile and therefore do not contribute to wine aroma. Within groups, there are varying levels of substitutions, usually of sugars or non-flavonoids, that alter how they are perceived, as well as how they interact with the other constituents of wine. These associated compounds add to the enormous array of different flavonoid compounds that are found in grapes and wine. More comments about each of the three major groups of flavonoids follows.

### *Anthocyanins*

What makes red wine red? The simple answer is anthocyanins – the principal pigments in certain grapes and wines. The word anthocyanin is derived from the Greek – *anthos*, meaning flower, and *kyanos*, meaning blue – and was first used for these pigments in the 1930s. Anthocyanins are the basis of the colour seen in most plant tissues that appear red or blue, and are usually found only in the skin of the berry. Particular varieties of grapes, known as teinturier types, also have anthocyanins present in the flesh, which means that the juice expressed from them is red, unlike that from ordinary varieties. Similar to green chlorophyll, another important plant pigment, the structure of anthocyanins is such that all visible light energy is absorbed by it, except that in the red to orange range.

In relation to grapes, anthocyanins may have found ecological value in fruit dispersal, as current research indicates that birds are attracted to red grapes before green (Watkins 1999). A happy by-product of this is that with skin contact, anthocyanins are extracted out into juice and fermenting wine – this process will be considered further later in the chapter. The leaves of many plants also contain anthocyanins, although in some their presence is masked by chlorophyll, i.e. they remain hidden until the leaves begin to senesce at the end of the growing season, chlorophyll is degraded and the anthocyanins can show through. Other autumn colouring plants synthesize anthocyanins immediately before leaf fall, causing the blaze of colour.

Structures of the common anthocyanins are shown in Figure 9.2. Substitutions, as indicated, determine anthocyanin type. Malvidin is the principal pigment in grapes, with the balance being made up of delphinidin, peonidin, cyanidin and petunidin. Willstätter and co-workers determined the basic structure of anthocyanidins, and Robinson and colleagues the positions of the sugars by synthesis (see lists of references in Geissman 1962). Willstätter and Zollinger (1915) were the first to isolate malvidin 3-glucoside (oenin) from *Vitis vinifera*.

The aglycones (anthocyanidins) are much more unstable (and able to react with other molecules, to be discussed later) than those molecules with sugar or acyl groups attached to them, and are thus not commonly found in grapes or wine. Anthocyanins undergo hydrolysis (removal of some functional groups) during fermentation (Wightman *et al.* 1997), however, which can result in a loss of colour in the finished wine as those molecules react and are degraded.

The sugar commonly attached (via a β-glycosidic linkage) to the number 3 carbon (Figure 9.1) of grape anthocyanidins is glucose, forming the glucoside form of the pigment. In *V. labrusca* (e.g. Concord) and other native American grape species, sugars attach to two free hydroxyl groups on the molecules, (usually) forming 3,5-diglucosides. *V. labrusca*, *V. rupestris*, *V. riparia*, and the Asian species *V. amurensis* contain both 3-glucosides and 3,5-diglucosides. *V. rotundifolia* (Muscadine grapes) contain only 3,5-diglucosides. *V. vinifera* grapes only have monoglucosides, a finding that has been used as a diagnostic tool for detection of *vinifera* wines adulterated with added French hybrid or pure *labrusca* grapes or wine. Wines with more than 5 mg/l diglycosides are defined as containing some hybrid or American grapes or wine.

Further to glycosylation, anthocyanins may be acylated through the attachment of, most often, *p*-coumaric acid, caffeic acid, or acetic acid to the sugar. This further stabilizes the pigment, but in wine, even while quite young, the acyl groups tend to disassociate from the parent molecule, leaving anthocyanidin monoglucosides.

Pinot noir forms an important exception to acylated pigments in *vinifera* grapes as it does not have any. This, along with lower concentrations of anthocyanins in the skin, may be a contributing factor to its traditionally poor wine colour in comparison with other red varieties such as Cabernet

| A-ring and heterocyclic ring > B-ring ∨ | 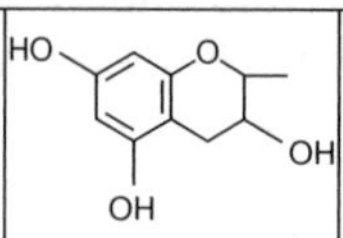 Catechins | 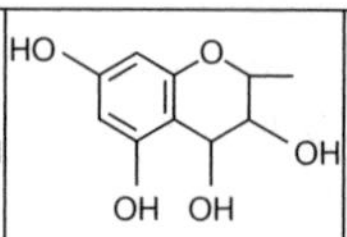 Flavan-3,4-diols | 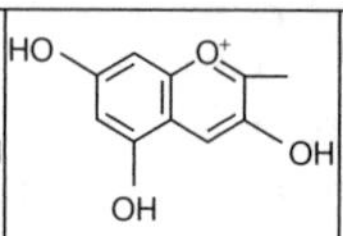 Anthocyanins | 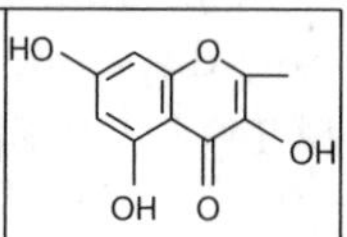 Flavonols |
|---|---|---|---|---|
| 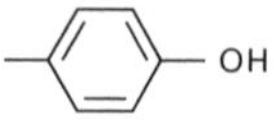 OH | | | | Kaempferol-3-glucoside |
| 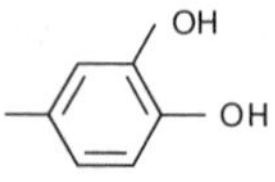 OH, OH | Catechin, epicatechin | Leucocyanidin (as dimers and polymers terminated with a catechin) | Cyanidin-3-glucoside | Quercetin-3-glucoside |
| 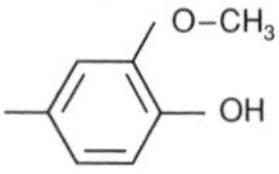 O–$CH_3$, OH | | | Peonidin-3-glucoside | Isorhamnetin-3-glucoside |
| 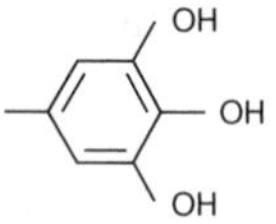 OH, OH, OH | Gallocatechin, epigallocatechin | | Delphinidin-3-glucoside | Myricetin-3-glucoside |
| 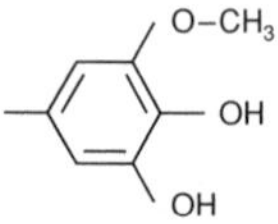 O–$CH_3$, OH, OH | | | Petunidin-3-glucoside | |
| 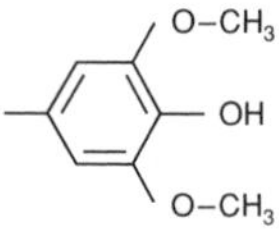 O–$CH_3$, OH, O–$CH_3$ | | | Malvidin-3-glucoside | |

*Figure 9.2* A- and B-ring substitutions for common grape flavonoids.

Sauvignon and Merlot. This non-acylated characteristic can also be used to distinguish pure Pinot noir wines from those with other varieties blended in.

Glycosylation of anthocyanins leads not only to greater pigment stability, but also to a slight change in peak colour absorbance, leading in turn to a shift towards brick (orange) red rather than violet-red. Far more drastic changes to wine colour come about through the sensitivity of anthocyanins to pH changes and sulphite bleaching, as well as interactions with related molecules (e.g. co-pigmentation) and tannins (polymeric pigments).

Co-pigmentation is the association of pigment molecules and other, usually non-coloured, molecules. It has a very significant effect on perceived colour, changing the peak absorbance wavelength and also its magnitude, which can result in colour shifts and more intense colour. In wine, there can be a four- to sixfold increase in colour due to co-pigmentation (Boulton 2001) – meaning that the wine is very much more intensely red (with a colour shift towards purple) than can be accounted for by the amount of anthocyanin alone.

The co-pigmentation co-factors (co-pigments) can take several forms, but from grapes the most important are probably catechin and quercetin (and possibly their derivatives), because they are often found in the highest concentrations relative to anthocyanins. Acylation of the anthocyanin pigments may also enhance co-pigmentation (Boulton 2001).

Wine pH also affects the density and hue of anthocyanins to an important degree. Anthocyanins in wines at acidic pH are red (in their flavylium form), gradually losing their colour as pH increases. Above pH 4, they appear mauve to blue (quinonic base form), and then at neutral pH and above, they are mildly yellow. Even in the pH range of wines (3–4) there can be drastic changes in the apparent colour due to ionization state of the molecules. The percentage of pigment in the red flavylium form decreases from 30% to 8% as pH changes from 3.0 to 3.7 (Glories 1984).

### *Flavonols*

The major flavonol found in grapes is quercetin, although myricetin, kaempferol and isorhamnetin (Cheynier and Rigaud 1986) are also found (Figure 9.2). They are coloured, though not as intensely as anthocyanins, and tend to have a yellow hue. Like the anthocyanins, they are found primarily as 3-monoglucosides in the grapes, although the galactose and glucuronide groups are also found. For example, Pinot noir has equal concentrations of quercetin monoglucoside and quercetin glucuronide in the skins (Price *et al.* 1995). During the winemaking and ageing process, the glycoside can be cleaved off to leave the aglycone form (Price *et al.* 1996). It is, however, possible that the appearance of aglycones could be artefacts arising from processing and analysis. Furthermore, the aglycones are not very soluble in wine, and what there is may be tied up with anthocyanins or other flavonoids, which increases their solubility (Price *et al.* 1995).

Flavonols are also produced and found in the skin of grapes, with virtually none being present in the pulp or seeds. Grape rachis and leaf tissue, however, contain significant amounts of these compounds, which can find their way into the wine either by design (addition of stems to the must) or by accident (plant debris not sorted from the harvested grapes).

Flavonols have several effects on wine. One is their contribution to bitterness, as the detection threshold for these compounds is often exceeded in red wines. The other is as one of the family of flavonoids that can participate in co-pigmentation, as described earlier. This process can be demonstrated by adding a small amount of quercetin to a solution of anthocyanin pigment (such as malvidin-3-glucoside). The result is a fairly dramatic increase in colour density of the solution (Mirabel *et al.* 1999). Scheffeldt and Hrazdina (1978) noted that malvidin-3,5-diglucoside, in combination with rutin (quercetin-3-rutinoside), gave a better co-pigmentation response than did malvidin-3-glucoside. This may perhaps indicate that the *V. labrusca*-derived varieties could lead to more intensely coloured wines, as these are sources of diglucoside anthocyanin pigments.

In young wines, 30–50% of the colour can be attributed to co-pigmentation effects (Boulton 2001). However, this effect can be somewhat short-lived in that, as time passes, the number of anthocyanin molecules able to participate in co-pigmentation decreases (Somers 1998). The ratio of pigment to co-pigment is important to this association, contributing to wine colour significantly.

Additionally, a high concentration of flavonols in wines can lead to a persistent wine haze (Somers and Ziemelis 1985), which has encouraged vigilance in keeping excess leaf and rachis material out of the must due to their high content and ease of extraction of flavonols.

### *Flavans*

The flavans are an integral part of red wine quality, contributing to co-pigmentation, bitterness and phenolic polymers, which are crucial to the astringency of wine, particularly red wine. The past 10 years of research have revealed a great deal about the interactions between these molecules, the compounds that are derived from them, and the impacts they have on our perception of wine. Red wine is again the focus because, like the other flavonoids, flavans (primarily flavan-3-ols and flavan-3,4-diols, see Figure 9.2) are not present in all parts of the berry. Similarly to anthocyanins and flavonols, they are found in the skin, but they are also found in the seeds of the berry, which provides another source during the processing of red wine. Leaves and stems are sources as well, as for the flavonols (Price *et al.* 1996). Therefore, the amount present in grapes can vary greatly depending on whether the seeds are included in the determination and what method of extraction is used (many of the phenolics present in the seeds are not extracted during the winemaking process).

The main flavan-3-ols found in grapes are catechin, epicatechin and epicatechin gallate. Gallocatechin and epigallocatechin are also found in smaller amounts (Figure 9.2). Catechin was first described by Runge (1821), who identified the compound from extracts of an Indian plant, *Acacia catechu*. Flavan-3,4-diols are found in low concentrations, but may, like other flavans, contribute to perceived wine quality. Flavan-3,4-diols are also called proanthocyanidins and leucoanthocyanins – the latter being a term coined in the 1920s to describe non-coloured (leuco meaning white) material found in unripe purple grapes and also in mature white grapes. Robinson and Robinson (1933) suggested the flavan-3,4-diol structure for leucoanthocyanins.

Oxidation of these compounds results in their conversion to the coloured anthocyanins, the origin of the former name (Robinson 1937).

Flavans primarily affect the wine by contributing bitterness. Their polymerization to form tannins and the influence these compounds and their derivatives have on wine bitterness, astringency and colour are major facets of wine quality, particularly for red wines. These factors also have an impact on the antioxidant capacity of the compounds, with the monomers having a higher antioxidant capacity than polymers, and glycosylation of the molecules also decreasing antioxidant capacity (Plumb *et al.* 1998).

The influence of polymeric flavans on the colour of aged wines (when many of the monomeric anthocyanins have disappeared and co-pigmentation plays a more minor role) has been the focus of considerable research in recent years, as technological advances allow greater insight into how they form and interact with other wine constituents.

## *Stilbenes*

Phytoalexins are low-molecular weight compounds that possess antimicrobial properties and are produced upon plant/microbe interaction. Both pathogenic (i.e. those that can produce disease) and non-pathogenic organisms can prompt their appearance. Many phytoalexins are phenolic in nature, and the known examples in grapevines belong to a class called stilbenes.

There has been speculation that the production of stilbenes is antagonistic to that of anthocyanins and other flavonoids (Jeandet *et al.* 1995). However, the hypothesis has not been proven, though this type of notion has been recurring since biosynthetic pathways were found to have branches. Association of other small molecules with stilbenes results in new compounds, sometimes with differing characteristics, just as for the flavonoids – these will be discussed shortly.

Like anthocyanins and flavonols, stilbenes are produced only in the berry skin, but can also be found as constituents of the woody parts of the vine, such as in the canes and wood. In fact, there is a significant body of research on the contribution of stilbenes to preventing heartwood from rotting (such as for eucalyptus and pine). In this regard, the stilbenes are not regarded as phytoalexins.

### *Resveratrol*

Figure 9.3 shows *trans*-resveratrol (trans-3,5,4′-trihydroxystilbene; see Figure 9.1 for the numbering convention for stilbenes), the first stilbene to be synthesized by grapevine tissues. Historically, resveratrol is an active ingredient in a traditional herbal remedy used in the Orient, called *kojo-kon* (Chinese) or *itadori-kon* (Japanese). It was made from the root of *Polygonum cuspidatum* (also known as Japanese knotweed). Roots of *Veratrum grandifolium* and many other plants contain resveratrol and have been used as herbal

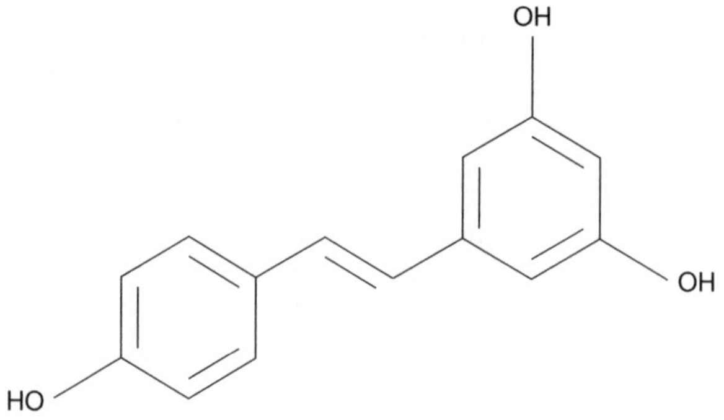

*Figure 9.3 trans*-Resveratrol, the first stilbene to be produced by grapevines.

remedies. However, resveratrol was not identified as the active ingredient of some of these remedies until the 1980s, when extensive biochemical trials were conducted in the Orient. These traditional medicines were used for the treatment of suppurative dermatitis, gonorrhoea, favus, hyperlipaemia, arteriosclerosis, and allergic and inflammatory diseases and athlete's foot (Kubo *et al.* 1981), preceding its identification as a potential antifungal plant phytoalexin (Langcake and Pryce 1976).

Resveratrol and the stilbenes were also studied as chemotaxic markers to distinguish plant species in classic work by Hillis in Australia (Hillis and Ishikura 1968). This was used for tree species such as eucalyptus and pine, but it was also useful for some legumes and lily species.

In the 1970s, Langcake and Pryce (1976) published the first paper acknowledging the presence of stilbenes in grape leaves and wood, based in part on their fluorescence under the influence of ultraviolet (UV) A (350–400 nm) radiation. In this case, resveratrol had been made in healthy leaf cells at the margin of a botrytis infection lesion, and also after exposure to UV-C radiation (Langcake and Pryce 1976). Later, resveratrol was also shown to be made in the berries, though restricted to the skin cells only (Creasy and Coffee 1988).

The link between resveratrol in grapes and wine was started in the laboratory of L. L. Creasy in the early 1990s. Their finding of oriental biochemical research on resveratrol (Arichi *et al.* 1982; Kimura *et al.* 1983) followed an extensive literature search based on a question of whether the disease resistance compound had any toxicity to humans, as Siemann and Creasy (1992) had reported the presence of resveratrol in wine. Because resveratrol is a phytoalexin produced by the berry skin when challenged by micro-organisms, its concentration was found to be much higher in red wine. Siemann and Creasy proposed a link between the resveratrol of red wine and the claimed health benefits of red wine consumption, a hypothesis strengthened by the biochemical research reported in the 1980s by Japanese laboratories on the effect of resveratrol as an active ingredient in circulatory herbal medicines.

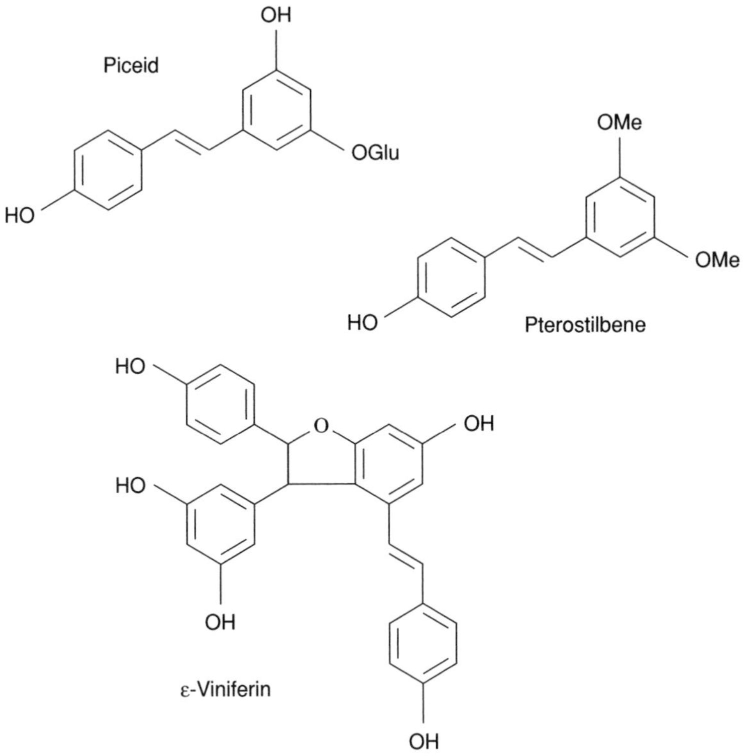

*Figure 9.4* Commonly found resveratrol derivatives in grapevines.

### *Derivatives*

Modifications to resveratrol such as glycosylation, methylation and polymerization, produce piceid, pterostilbene and the viniferins, respectively (Figure 9.4).

The viniferins are primarily found in the woody parts of plants such as some tree species (as previously mentioned) as well as in grapevines, which may well help to prevent the rotting of the permanent parts of the trunk. The exact roles of the other stilbene derivatives have not been determined, though they do have antifungal activity. Concentrations can be quite high, reaching 700 μg/g fresh weight of cane tissue (Langcake and Pryce 1976).

## *Biosynthesis of flavonoids and stilbenes*

Synthesis takes place in the cell cytoplasm, but the flavonoids accumulate in the cell's vacuole and, as far as is known, do not move between cells. They must therefore be manufactured where they are needed. Stilbenes and flavonoids have the same biosynthetic precursors, one molecule of p-coumaryl-CoA and

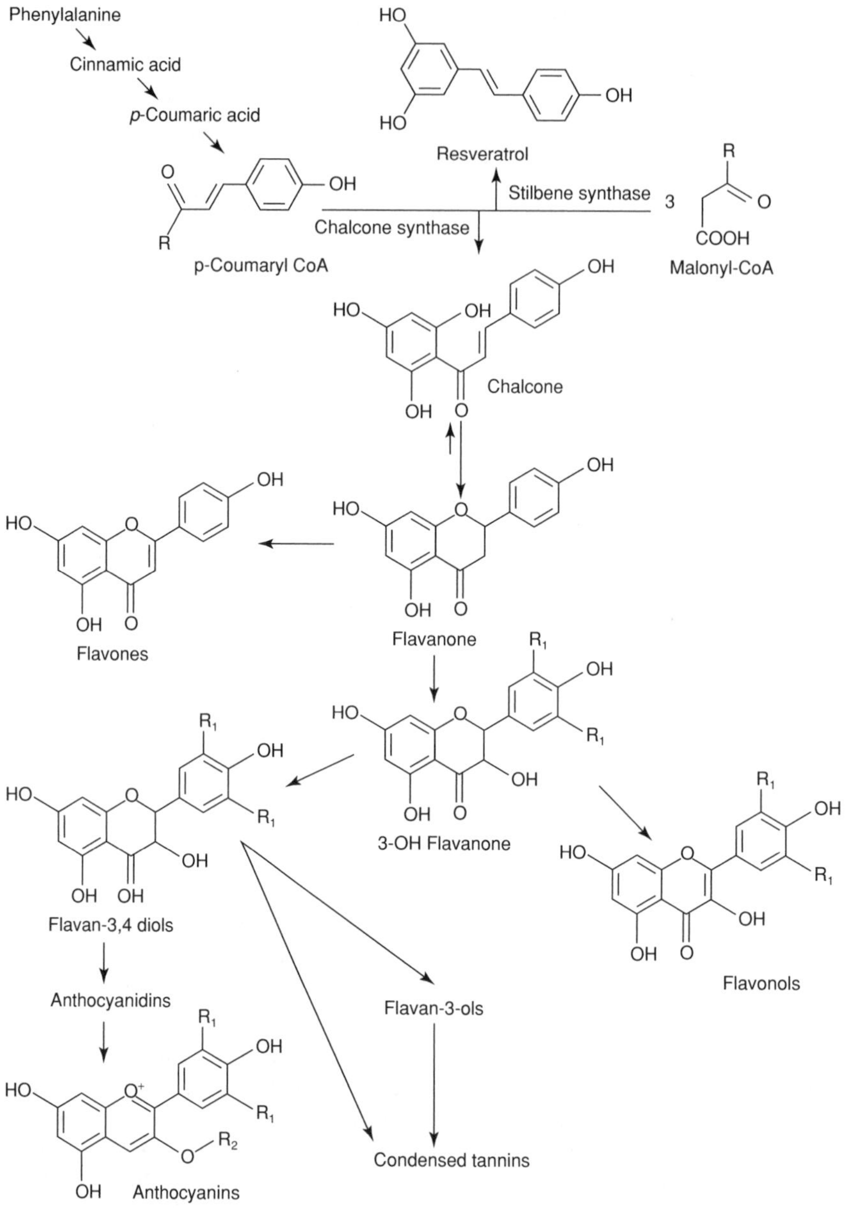

*Figure 9.5* Biosynthesis of flavonoids and resveratrol. R, -S-CoA; $R_1$, -OH or O-Me; $R_2$, -glucose.

3 molecules of malonyl-CoA produce the 14 carbon stilbene resveratrol when catalysed by stilbene synthase, or the 15 carbon naringenin-chalcone when catalysed by chalcone synthase (Figure 9.5). Chalcone synthase is essentially ubiquitous in higher plants, and stilbene synthase is found in widely different

species. It is attractive to consider the possibility that stilbene synthase is a relatively frequent mutation of chalcone synthase, a hypothesis supported by the 80% homogeneity in the structure of the proteins (Schröder *et al.* 1988).

*p*-Coumaric acid is produced from cinnamic acid, which is the first phenolic compound considered to be a 'secondary metabolite' (not part of primary metabolism). Cinnamic acid is produced from the amino acid phenylalanine, an essential amino acid synthesized only by plants. Flavonoids are found in all higher plants and therefore so is chalcone synthase. Chalcone isomerase converts the chalcone to naringenin (a flavanone), which is then hydroxylated to a 3-OH flavanone (also known as a flavanonone) by flavanone 3-hydroxylase. Naringenin is converted to dihydrokaempferol (the 3-OH flavanone) by flavanone 3-hydroxylase. At this stage the B-ring can be hydroxylated from one hydroxyl group to two or three hydroxyl groups, i.e. hydroxylated to dihydroquercetin by flavonoid 3′-hydroxylase or to dihydromyricetin by flavonoid 3′,5′-hydroxylase. The B-ring pattern persists to the final product from this stage. Dihydroflavonols are either converted to flavonols by flavonol synthase, or reduced to flavan-3,4-diols by dihydroflavonol 4-reductase and then further oxidation, dehydration, glycosylation and methylation by anthocyanin synthase, anthocyanin-glucosyltranferase and methyl transferases produce the coloured anthocyanin glucosides.

## Plant physiological functions

Flavonoids and stilbenes are secondary metabolites and therefore contribute to a number of plant processes. Being such a diverse group, the roles that the flavonoids, in particular, play are also varied.

### *Anthocyanins*

Anthocyanins probably developed in fruit in order to aid in the attraction of potential agents for seed dispersal. Birds, for example, are more likely to take fruit that is of a contrasting colour to the foliage around it (Wilson and Whelan 1990). It is unlikely that they came about as compounds to discourage or encourage consumption, as they have no taste on their own. Flavan-3-ols and flavonols contribute significantly to bitterness and astringency in grapes, which may deter some bird species from habituating on grape consumption (Bullard *et al.* 1980).

Anthocyanins also act to cause more absorption of solar radiation, heating fruits and increasing enzyme activity and sink strength. This alters such berry composition factors as acids (particularly causing a reduction in malate) and sugars (increasing both glucose and fructose). In hot climates, sun-exposed red fruit can be 12° C higher than ambient air temperature (Smart and Sinclair 1976). Such fruit is also, therefore, more susceptible to reaching detrimentally high temperatures (greater than 35° C), which can result in the breakdown of anthocyanins and cause enzyme dysfunction. In extreme cases, fruit shrivelling can occur, which complicates yield prediction and can affect winemaking practice. Fruit without anthocyanins is not so susceptible because less sunlight energy is absorbed.

As opposed to hotter grape-growing areas, where fruit can easily reach damagingly high temperatures, in cooler grape-growing areas, such as Oregon, Burgundy and New Zealand, increased fruit temperature as a result of exposure to the sun could result in the beneficial effects of lower acids, higher sugar levels and better colour at harvest compared with shaded fruit.

## *Flavonols*

While anthocyanins absorb radiation quite obviously in the visible range, flavonols are only slightly yellow. However, they do absorb strongly in the UV-B and UV-C range (280 nm). UV radiation is of sufficiently high energy to cause damage to bonds that hold DNA and RNA together, which can interrupt cellular function (Caldwell 1981). Since there is a strong correlation between fruit exposure and the production of flavonols in grape berry skin, some have postulated that they act as a 'sun-screen' to protect from solar radiation damage, though there is no direct scientific evidence to support this theory. The relationship between sunlight reaching the berries and flavonol concentration in the skins is so strong that it has been suggested that the number of quercetin-related compounds in grapes could be used as a measure of fruit exposure, and could also possibly be related to quality (Price 1994).

In the plant species in which they have been investigated, flavonols seem to be produced in the outermost cells of the plant exocarp, which, like anthocyanins, tends to be in the hypodermis (Wellman 1974). This lends (circumstantial) weight to the hypothesis that these compounds function to screen out UV radiation before it can reach and damage other tissue components (Beggs *et al.* 1986).

## *Flavans*

The flavan-3-ols and flavan-3,4-diols are bitter as monomers, but in chains (polymers) their astringency (ability to bind saliva proteins, the effect then being perceived as a sensation on the tongue) generally increases up to a point. Beyond a certain length, the chains are too bulky to interact properly with the proteins, and thus 'lose' astringency. The degree of polymerization of the flavans tends to increase as the berries approach maturity, while the overall flavan-3-ol concentration tends to decrease. This suggests that some of those disappearing monomers are being used to make the polymers in the berry skin and seeds, although this has not been demonstrated directly.

The bitterness and astringency of these compounds may have acted as a deterrent for predators, such as birds, mammals or chewing insects. Within the plant itself, the physiological role of these secondary metabolites is undetermined.

### *Stilbenes*

The antifungal activity of resveratrol and its derivatives has already been mentioned. This is probably the primary physiological role that they play in the vine. It was also mentioned that stilbenes have antifungal activity. Langcake and McCarthy (1979) suggested that there was a good relationship between resistance to botrytis and resveratrol production in leaf tissue. The antifungal role of stilbenes in vine tissues has been the basis of attempts to use them as screening agents in breeding programmes. Pool *et al.* (1981) reported that resveratrol production differed between varieties and clones of a single variety.

As mentioned, resveratrol is the first stilbene to be produced in grapevine tissue. Its production, accumulation and degradation in tissues can occur relatively quickly. As important as the synthesis of the compound is, its degradation is also important as it is very phytotoxic, and if it remained in the cells too long they would be destroyed. The speed with which the vine can accumulate fungitoxic substances like resveratrol may be important in determining the success or failure of an infection attempt (Kuć 1994). Resveratrol is produced through the action of stilbene synthase, then piceid, pterostilbene and the viniferins come later. These derivatives also have antifungal activity, though information about how much is sketchy. Though resveratrol and ε-viniferin have similar activities in preventing spore germination or mycelial growth, since two molecules of resveratrol are necessary to make one molecule of ε-viniferin, there is little benefit to the plant in terms of efficiency of resveratrol use. However, turnover of resveratrol is high once it is produced, whereas ε-viniferin is more stable. Viniferin, because it is persists more than resveratrol, may be a longer-acting antifungal agent in the vine tissue.

The stage of development of vine tissue is also a factor in production. Leaves at the upper and lower extremes of an older shoot, for example, have a lesser capacity to produce resveratrol than leaves that are in the mid-part of the shoot (Pool *et al.* 1981). In older leaves, elicitation using UV radiation may not be as effective due to the accumulation of other constituents, possibly flavonols, lignins or other phenolics that screen the active wavelengths of energy.

A similar relationship holds true for grape berries, with relative potential production (as determined through UV radiation for elicitation) being high in fruit tissue from fruit set to approximately veraison (Figure 9.6). Creasy and Coffee (1988) also found reduced UV elicitation of resveratrol in post-veraison berries.

Botrytis is not the only fungus that can trigger a stilbene response – powdery mildew and downy mildew have also been implicated (Langcake and Pryce 1976; Dercks and Creasy 1989). In Pinot noir grapes grown in New South Wales, Australia, powdery mildew-infected berries showed elevated levels of resveratrol, piceid and pterostilbene (Figure 9.7) compared with

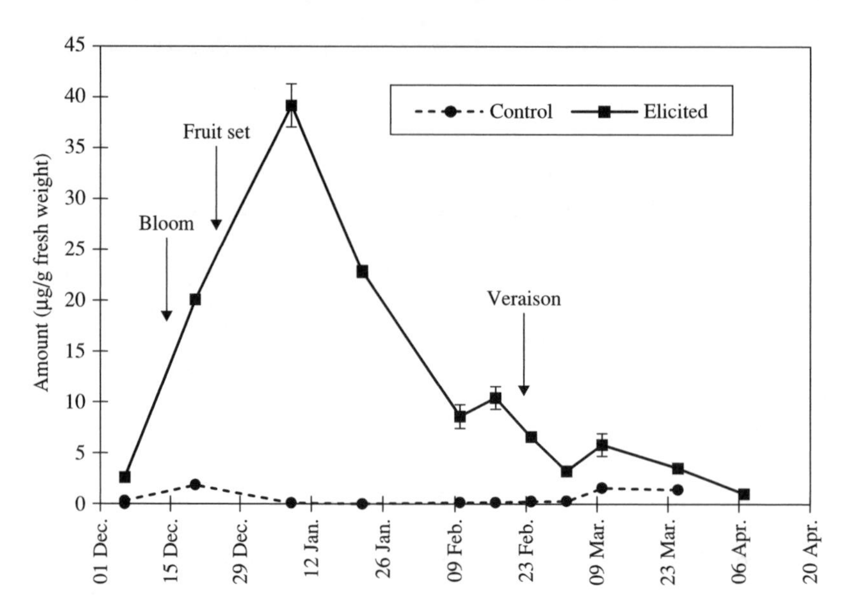


*Figure 9.6* Seasonal resveratrol production potential of Pinot noir berries growing in Tumbarumba, New South Wales, Australia. UV radiation was used to elicit resveratrol production, which was measured after 48 hours of incubation. Bars indicate standard error. Source: Creasy, G. L., Steel, C. C. and Keller, M., unpublished data.

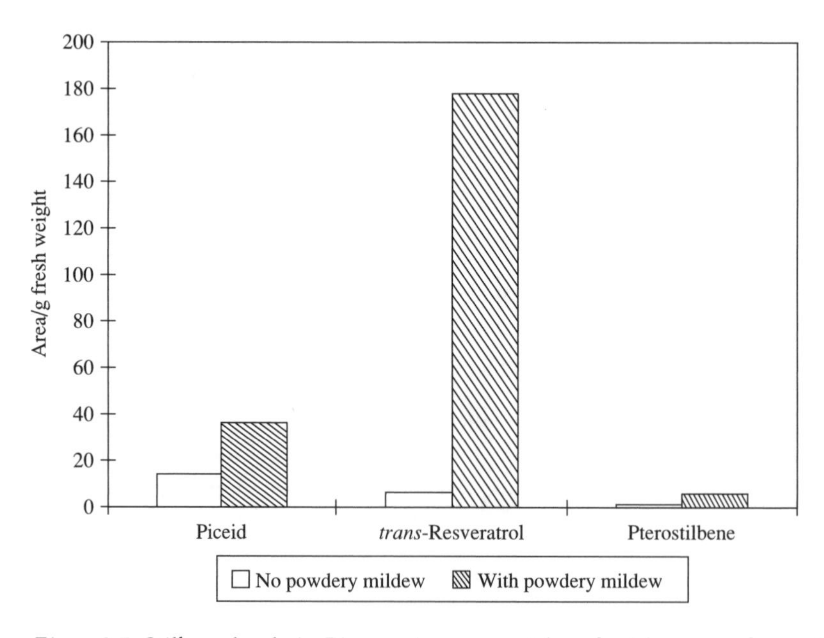


*Figure 9.7* Stilbene levels in Pinot noir grapes with and without powdery mildew infection, expressed as units of high-performance liquid chromatography chromatogram area per gram of fresh weight. Source: Creasy, G. L., Steel, C. C. and Keller, M., unpublished data.

non-infected berries. Resveratrol and viniferins may also be implicated in Grapevine Vein Necrosis (Šubíková 1991). More recently, there have been reports of stilbenes being involved in the vine response to the trunk debilitating diseases of esca (also known as black measles) and *Phaeomoniella* spp. (Petrie vine decline, black goo, etc.) (Alessandro *et al.* 2000; Almafitano *et al.* 2000). While it seems clear that the vine response is not sufficient to prevent infection in pathogenic interactions, it is possible that the presence of stilbenes may modify the vine-fungus result in non-pathogenic interactions in ways that we are not yet aware of.

## Vineyard factors affecting production

### *Flavonoids*

#### *Anthocyanins*

The first and most important factor involved in determining anthocyanin production in grapes is varietal choice. Green grapes lack anthocyanins completely, although there are circumstances where they can be produced. Muscat Gordo Blanco has been found to develop a slight blush on occasion. Cyanidin-3-glucoside was found to be primarily responsible for the colour produced when they have been exposed to excessive sunlight and/or are over-ripe (Gholami and Coombe 1995). Interestingly, the authors linked this anthocyanin production to amounts of monoterpene aroma compounds present, as there seemed to be a linear relationship between the two.

Then there are grapes that produce small amounts of colour, such as Pinot gris, Schönberger or similar. They are easily identifiable as having anthocyanins in them, but it is also clear that the amount is not high. In any case, anthocyanins are not produced in red or blush varieties until the final stage of berry development, which occurs after veraison, or colour change. At this stage, grapes undergo a massive developmental shift, developing colour, but also starting to accumulate sugar and characteristic flavour and aroma compounds, increase in size, and decrease in acidity.

Grapes that are obviously red, such as Cabernet Sauvignon, Syrah, Merlot, etc. generally have the highest concentrations of anthocyanins at harvest, though Pinot noir is another exception to this. Its pigment levels are significantly lower than those found in other red varieties. This is not due to the lack of any one or more pigments, but rather to lower amounts across all pigments.

Anthocyanin production is largely connected to fruit exposure, with the fruit that receives the most light having the highest amount at harvest (Smart *et al.* 1988). At the upper end of the scale, however, too much light can be detrimental to anthocyanin concentration (Price 1994), perhaps because of the associated temperature rises, as noted earlier. Again proving to be not-quite-like-the-others, Pinot noir and its close relative, Pinot gris, showed distinctly less colour in well exposed fruit as opposed to more shaded fruit (Price 1994). One possible explanation for this would be that flavonols are being made in preference to anthocyanins under the high light environment, i.e. compounds that absorb best in the potentially damaging UV range are made in preference to those that reflect light in the visible radiation range, though, as with the proposed stilbene–anthocyanin relationship mentioned earlier, there is no evidence that can be used to support this hypothesis.

The stage of development of the berries at which the light occurs can also have an influence. While Price (1994) found that covering previously exposed Pinot noir clusters with aluminium foil had no visible influence on the development of fruit colour, Creasy *et al.* (1987) covered red table grapes to prevent their direct exposure to light from veraison and found that colour was reduced significantly, though it was dependent on variety. Furthermore, in another experiment on a table grape variety, it was found that 95% shading caused lighter colour, but also brought about changes in the proportions of anthocyanin pigments compared with less shading. Delphinidin-3-glucoside seemed the most affected (Gao and Cahoon 1994). Shading from one month before veraison also caused reductions in berry anthocyanin content, with 90% ambient shading dropping pigments in the skin to one-quarter of the control concentrations for Cabernet Sauvignon (Smart *et al.* 1988).

Seasonal changes in weather are also associated with variable anthocyanin concentrations in the grapes, with a warmer than average season associated with lower anthocyanin production, though concentrations can still be high due to berry shrivelling (Watson *et al.* 1992). Rain occurring as the berries approach harvest will dilute the contents of the berries, which can result in less intense colour in the wine.

In any case, there is a good correlation between grape anthocyanin content and optimal maturity for winemaking purposes, perhaps better than that usually associated with Brix (a measure of soluble solids in grape juice), TA (titrable acidity) and pH (Watson *et al.* 1992). The importance of anthocyanins to the perceived quality of red wine cannot be underestimated.

### *Flavonols*

Flavonols begin to accumulate in the outer cells of the berry as early as fruit set, unlike anthocyanins. Flavonol levels in fruit are remarkably responsive to the vineyard environment, with the general relationship being that increased exposure results in increased production. Price *et al.* (1995) investigated the effect of cluster position on fruit and wine flavonol content in Pinot noir, and found that fruit epidermis exposed to sun had six times the concentration of flavonol glycosides than shaded fruit. Skin from shaded and exposed Chardonnay fruit was even more disparate – a 20-fold increase was seen in the sun-exposed fruit (Price 1994). UV radiation and visible light may be involved in signalling production (Stafford 1990).

Any vineyard practice that increases the exposure of fruit to the sun will effect increased production of flavonols in the fruit (Price *et al.* 1996). This could lead to the ability to distinguish grapes from different exposure regimens, such as are created by different trellises. Quercetin concentration correlates well with measures of canopy density, such as shoot number per metre of row and leaf layer number (Price *et al.* 1996).

Similarly, the greater the proportion of the berry skin that is exposed to the sun, the greater the potential quercetin content. Therefore, large, tight clusters that have a greater proportion of interior berries than smaller or looser clusters will have a lower potential for quercetin in the wine. Price and Watson (1995), found that open-clustered clones of Pinot noir produced grapes with higher quercetin content than clones with more compact clusters.

### *Flavans*

Variety of grape also has an effect on flavan-3-ol content, particularly in the seeds. Thorngate and Singleton (1994) showed that Pinot noir seeds had a higher catechin/epicatechin content than seeds from Cabernet Sauvignon. Since the majority of the catechins are found in the seed of the berry, exposure status of the cluster may not have that much of an influence on content.

Though far more extractable than flavan-3-ols found in the seeds, those found in the skins usually contribute less (approximately half) to the final wine flavan-3-ol content due to the relatively small volume of cells that actually contain them (Thorngate and Singleton 1994). Catechin content in the skin has been found to decrease with increasing exposure (Price 1994). Surprisingly, there is little published information on the effect of vineyard factors on grape flavan content. Given its increasingly defined role in determining wine mouth-feel and the formation of polymeric pigments, and the delicate balance between anthocyanins and flavans necessary for a top quality wine, there is sure to be increased investigation in this area.

### ***Stilbenes***

Other than the influence of cultivar, there are few direct vineyard factors reported to change stilbene levels in grapes. The cultivar, though, can have a large impact. The German-bred variety Castor, given the same elicitation event as other varieties, produces more resveratrol in the leaves, perhaps more quickly (Dercks and Creasy 1989). And between other, more common varieties, there are significant differences reported, with Concord, Cabernet Sauvignon and Catawba having high potential, and White Riesling, Chancellor and Cayuga White having low (Creasy and Coffee 1988). It should be noted that both red and white varieties can have high potential for resveratrol production, but because the stilbenes are located in the grape skin, only wines made with grape skin extraction have high levels of resveratrol.

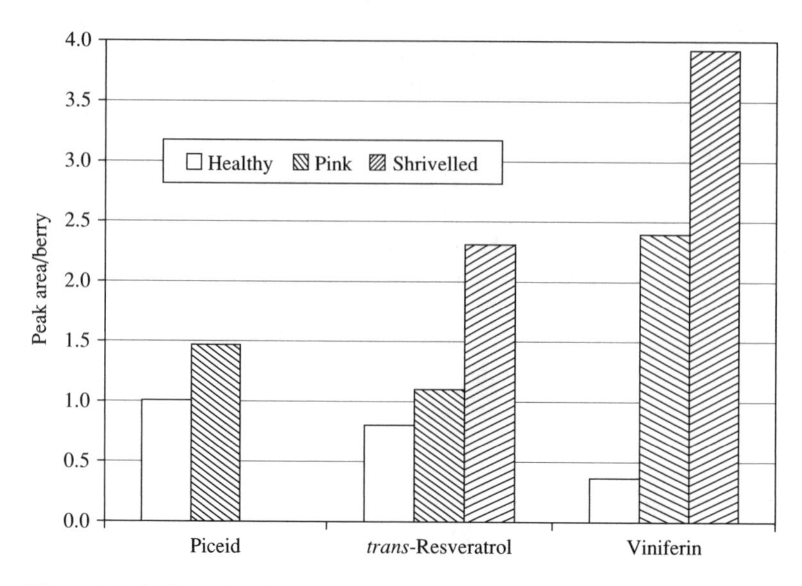


*Figure 9.8* Stilbene levels in Semillon grapes at varying stages of botrytis development, expressed as units of high-performance liquid chromatography chromatogram area per gram of fresh weight. Source: Creasy, G. L., Steel, C. C. and Keller, M., unpublished data.

Indirectly, there can be many factors in the vineyard that influence stilbene content in the grapes. Many would be related to affecting the disease pressure that the vines are under, particularly for botrytis. Therefore, dense canopies, poor fungicide spray programmes, humid growing seasons, late season rains, etc. can increase the incidence and severity of disease in the vineyard. It is thought that this results in a stilbene response in the grapes.

As mentioned earlier, pathogens such as downy mildew, powdery mildew and botrytis can effect resveratrol production in grape tissues. The stage of infection of botrytis in grape berries has a significant influence on stilbene production as well (Figure 9.8).

There are some suggestions that the ability of grapevines to manufacture stilbenes is related to carbohydrate supply. Experiments with detached shoot segments showed that the presence of leaves increased the ability of the shoots to manufacture resveratrol in the flower clusters (Keller *et al.* 2000). While stilbene production is thought to be very much a localized response, it is possible that precursors to production, or some other co-factors are necessarily supplied from nearby cells or organs. If this is the case, then ensuring that vines are healthy and not suffering from low carbohydrate status, as can occur with severe defoliation, particularly in cooler climates where leaf fall happens very soon after harvest, could boost the vine's ability to manufacture stilbenes.

Since UV irradiation at the Earth's surface has increased dramatically in some areas of the globe in recent years, could this have an influence in elicitation of stilbenes in grape plants, much as we use UV in the laboratory to trigger production? Certainly, the amount of potentially damaging radiation is on the increase, but so far only levels of UV-B radiation (280–320 nm) are rising, not the UV-C (100–280 nm) levels that trigger a stilbene response.

## Transfer to wine

Wine is the extracted essence of grape. The process of winemaking is to modify and guide the extraction of the grapes and other plant parts to make a particular wine style. In other words, the transfer of materials in the grape to a juice and wine is what makes the wine. John Durham, winemaker at Cape Mentelle Vineyard in Western Australia stated: '. . . if there were any place for artistic interpretation in the winemaking procedure, then it would have to be in the area of phenolic management' (Durham 1997). It is an area we are still learning about, hence its shroud of mystery.

In thinking about the transfer of phenolic (and other) components to wine, it is important to note that under normal circumstances you cannot make a wine that has more of an attribute than is present in the initial grape material. Figure 9.9 shows an example of the kind of relationship commonly found between parameters in the grape compared with the same parameter in the wine. On first glance, it can appear a fairly messy relationship, especially if a typical regression line is drawn through the points (solid line). However, a more useful interpretation is gained by drawing the relationship along the line represented by the maximum amount of the compound in the wine for any given value in the grape (dotted line). This signifies that in these situations there has been efficient and functionally complete extraction from the grape into the wine, whereas in the other cases there was incomplete extraction.

Phenolic compounds vary in their ability to be extracted in an aqueous solution, as you might find in a freshly crushed grape must and pre-fermentation maceration. Those that are glycosylated tend to be solubilized into the juice, while the aglycone and other hydrophobic compounds are less so. At the other end of fermentation, the presence of ethanol aids in the extraction of those hydrophobic compounds, which is one of the purposes of a post-fermentation maceration.

As just one, simplified, example of the extraction process, along with other winemaking procedures that change flavonoid and stilbene content, we can see that many factors can change how the materials are transferred to the final wine in the bottle.

### *Grape handling*

The method of harvest can influence extraction of grape compounds into the juice. Because of the localization of flavonoids in the berry (vacuoles of the

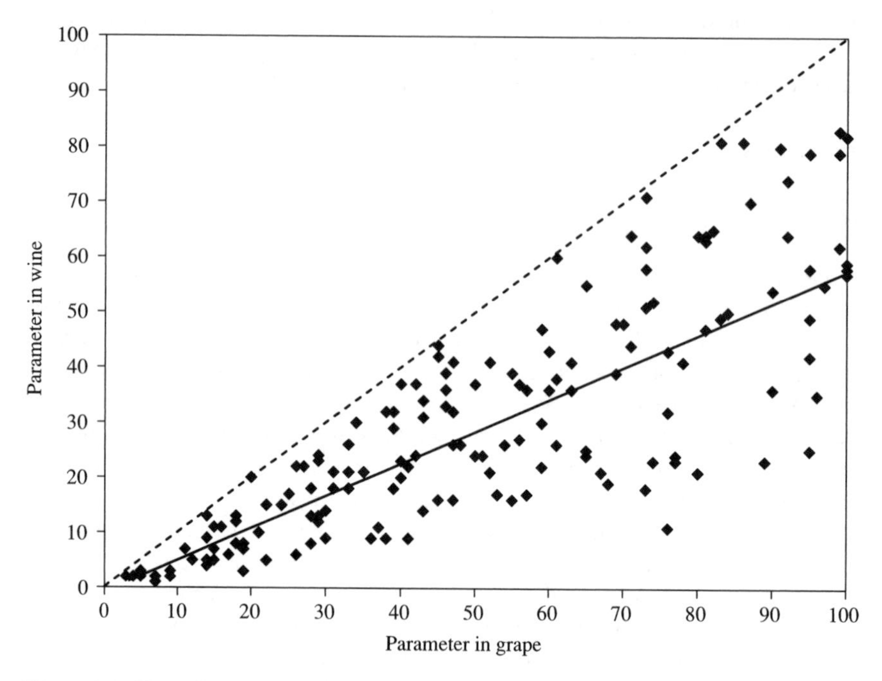


*Figure 9.9* Hypothetical relationship between a measured parameter in the grape versus that in the wine. Solid line indicates regression through all points, representing an average of the relationship. The dotted line represents approximately 100% extraction of material from the grape to the wine.

skin cells), extraction into the must or wine occurs only with skin contact. Hand-harvesting is recognized as the most gentle way to handle grapes, as each cluster remains largely intact, with a minimal loss of juice or breakage of berry skins until arrival at the winery.

Machine harvesting uses a principle of mass inertia to remove berries, setting up a back and forth motion (sometimes up and down) of the canopy to tear the berry away from the pedicel, or sometimes the pedicel will break off at the cluster rachis. The berries drop to a conveyor belt and are carried up and over to be emptied into a waiting trailer. The net effect is that most berries have broken skin and are partially crushed.

Once tissue has been broken, phenolics, such as flavonoids and stilbenes, are subject to oxidizing reactions, which can result in their net loss, as they polymerize and fall out of solution. This effect can be minimized by the addition of sulphites to the harvested berries and by carrying out harvest at the coolest times of the day or night. Mechanically harvested fruit can lead to wines that are more protein unstable and thus require more fining to ensure that a haze will not form during storage (Pocock and Waters 1998).

It is also worth noting that, compared with hand-harvested fruit, mechanically harvested fruit can have higher levels of material other than grapes

(MOG) in it, such as shoot pieces, leaves and parts of the cluster rachis. Green parts of the vine contain significant levels of flavans and flavonols, which, if extracted into the juice and wine, can result in wine stability problems later on (Somers and Ziemelis 1985). These effects can also take place if whole-bunch fermentations are used in the winery. Mechanical picking is also not discriminatory, meaning that if there is secondary crop that is not ripe in the canopy, it will be harvested along with the ripe fruit. This can result in lower flavonoid concentration in the harvested grapes as a whole.

This is not to say that there are no benefits to mechanical harvesting. The ability to harvest 24 hours a day and harvest large areas in a short amount of time (especially compared with managing hand-harvesting gangs) is invaluable in many growing areas that are short on labour or similar infrastructure, and the cost per tonne of harvesting is considerably lower. Additionally, in some cases the berries-only picking that a properly set-up mechanical harvester provides means that rachis material is left on the vine and does not end up in the fermentation vats. Some winemakers feel that this is a bonus due to the harsher and more bitter phenolics that tend to be extracted from the green rachis tissue (Durham 1997).

### *Pressing and fermentation*

Once at the winery, handling methods that are violent may shear grape skins and aid in the extraction of the berry contents. Because this kind of extraction can often lead to excessive bitterness and harshness in the wine, grapes are handled gently to minimize damage. Once through a destemmer-crusher, the grapes are transferred to the press or primary fermentation tanks.

For hand-harvested grapes destined for sparkling wine production, protecting the berries from damage is paramount. A minimum of skin contact is desired, as the presence of too many phenolics, particularly anthocyanins, but also catechins, is detrimental to the production of a quality sparkling wine made from red grapes. Whole bunches are usually loaded into a press (a hydraulic ram or a bladder press) and pressure is applied to split open the berry skins gently and let the juice run out. Skin contact is kept to an absolute minimum. Significant damage in red varieties results in anthocyanin extraction (detrimental to the production of a blanc de noir), and breakage in red or white varieties can result in flavan-3-ols being present in high concentrations, lending bitterness to the base wine.

Certain kinds of press design, such as the screw press, tend to cause more mechanical damage to the berries as they are compressed. Wines made with these types of presses rather than with the bladder or hydraulic ram type are more astringent and less pleasant, possibly due to the greater extraction of phenolics compared with gentler presses.

The presence of various maceration aids, such as enzyme preparations, can also cause enhanced extraction by breaking down the cell structures prior to pressing. As many of the flavonoid and stilbenic compounds are found in the cell vacuole, enzymatic degradation of the cell walls and the membranes surrounding the vacuole can result in more complete extraction from the grape solids. These can be used prior to fermentation (as part of the pre-fermentation maceration that is primarily an aqueous extraction); during fermentation; or towards the end of fermentation, when the ethanol level is higher and the aglycone and less-water soluble compounds come out into solution more easily. Parley *et al.* (2001) and Wightman *et al.* (1997) found that the use of enzymes during pre-fermentation maceration did not enhance extraction of anthocyanins from the grapes, but did result in increased colour in the finished wine, which was apparently due to an increase in polymeric pigment linked to enzyme use.

However, it should be noted that even with extraction that is deemed to be complete by winemaking standards, by no means have all the extractables been removed from the grape and into the wine. This is easily seen from the obvious colour left in red wine pressings after fermentation. Approximately 60–70% of the pigments remain in the berry solids (Van Balen 1984), along with other constituents. This has resulted in increasing interest from companies looking for natural sources of antioxidant compounds, as disposal of fermented grape solids is sometimes problematic.

The practice of separating some of the free-run juice from the grape solids before fermentation is an attempt to increase the skin to juice ratio, and thereby increase the final colour of the wine. However, because ordinarily such a low amount of the total anthocyanin is extracted from the grape solids, this may not result in an actual increase in wine colour as the extraction equilibrium is still met (Singleton 1972). However, this technique does result in higher wine concentrations of less abundant compounds such as resveratrol. Compounds in the skins (anthocyanins, quercetin and some flavans) are extracted quite quickly during the fermentation process, while those in the seeds require longer. The ethanol formed during fermentation also assists in the solubilization of the more hydrophobic compounds. Anthocyanins, for example, are more water soluble than quercetin. Price *et al.* (1996) found that Pinot noir anthocyanins (primarily malvidin) and quercetin are completely extracted by the fourth day of fermentation, while 18 days are required for catechin (this being eight days after the completion of fermentation). Resveratrol and the other stilbenes are also extracted quite readily, usually completely so when ethanol levels reach 2–3%.

### *Fining*

Since tannins, which result from the polymerization of flavans, are known for their ability to bind with proteins, the result of protein fining of wine should be apparent: additional protein reacts with the tannins present in the wine, causing them to precipitate. The net effect is a reduction in the amount of tannins (and, it is hoped, astringency) in the wine. While this is straightforward in a general sense, the complexity of the tannins found in wine and their ability to react with proteins and precipitate out makes it difficult to predict precisely how a wine will be affected by a protein addition. A variety of different proteins are used to fine wines, such as gelatine, isinglass, egg-white or milk; each has its own capacity to bind with tannins, and each is recognized as giving a particular character to the post-fining treatment. The profile of the wine being fined is another factor affecting the final outcome (Maury *et al.* 2001), and further complicates the relationship.

As the ability to separate and analyse tannin polymers is refined, so is our ability to determine what kind of tannins each treatment is affecting and what effect this will have on the astringency, bitterness, and other characteristics of the wine.

## Content in finished wine

Phenolics make up a relatively small proportion of total wine. About 0.025% of white wines and 0.15% of red wines are phenolic compounds, which corresponds to 250 mg/l and 1500 mg/l, respectively. Absolute amounts vary according to the numerous variables present between winegrowing areas, and even within them (e.g. different mesoclimates, varieties, vinification practices, etc.).

As mentioned previously, the flavonoids make up the bulk of the phenolics, though the non-flavonoids do contribute significantly in the form of bitterness, background aromas and flavours, and hydrolysable tannins, despite their relatively low concentrations.

### *Flavonoids*

In red wines, the flavonoids as a group are found at concentrations of around 1200 mg/l, which is many times the level of non-flavonoids (approximately 200 mg/l) (Singleton and Nobel 1976). As a result, the flavonoids have a much larger influence on wine quality, particularly in wines made from red grapes fermented on the skins. White and rosé wines have much lower concentrations of phenolics due to the less-complete extraction of skin and seed phenolics.

#### *Anthocyanins*

White wines have little to no anthocyanins present, almost by definition. In terms of the visual impact of a red wine they have a great presence, but make up only about 12% of the total flavonoid content (Singleton and Nobel 1976). In red wines, 300–500 mg/l is malvidin-3-glucoside. Rosé wines have 5–50 mg/l of anthocyanins, which is usually lower than that required for co-pigmentation to occur (Boulton 2001). Hence, rosé and red wines made from the same grapes will have both intensity and hue differences, the former due to the amount, and both the former and latter due to co-pigmentation. Pinot noir contains 150 mg/l of malvidin-3-glucoside, while a dark Merlot may have as much as 800 mg/l.

Young red wines contain 150 mg/l to greater than 800 mg/l of free anthocyanins (Singleton and Nobel 1976). As they age, this concentration falls to around 50 mg/l or less. Wines may still be decidedly red even when free anthocyanins fall to well below 50 mg/l, due to the contribution of complexes that bind anthocyanins and yet retain colour.

After three months to one year of ageing, about half of the anthocyanins are no longer in monomeric form, but are present as part of the polymeric compounds (Somers 1971). Older wines tend to have fewer free anthocyanins than younger wines, a relationship that carries on for wines of some age. Polymeric pigment content is very low in young wines, but tends to increase as they age; thus the polymeric pigments are quite important to the colour of aged wine.

### *Flavonols*

Flavonols are a relatively small portion of the total flavonoid content of a red wine, being found in concentrations of 20–50 mg/l concentration. However, since exposure of the fruit can have such a dramatic influence on production, fruit grown in non-shaded canopies can produce dramatically different amounts within a single cluster or even within a berry. In Oregon-grown Pinot noir, Price (1994) found that a single vineyard produced wines with a range in flavonol content of 5–30 mg/l depending on the exposure level of the fruit. Price also found Pinot noir wines with flavonol levels as high as 50 mg/l. A survey of Californian wines found that, distinct from other varietal wines, Pinot noir had the greatest variation in quercetin content (Waterhouse and Teissedre 1997). The other significant flavonols – kaempferol and myricetin – can be present in concentrations as high as 30 mg/l of wine (Boulton 2001), though more typically they are at levels below 15 mg/l.

The solubility of flavonols in particular seems to be influenced by other phenolics present in solution. Model wine solutions, which typically contain water, ethanol and acid, are poor carriers of flavones such as quercetin (Boulton 2001), while more complex solutions that have other phenolics, such as anthocyanins, can carry significantly more. There is a better relationship between the concentrations of quercetin aglycone and anthocyanins than between quercetin in the grapes or quercetin glycosides in the wines and anthocyanins in commercial wines (Price *et al.* 1996).

### *Flavans*

The content of catechin and related compounds in wines is, as already discussed, drastically altered by winemaking practice, which can lead to difficulties when trying to compare levels in wines from different regions due to wide ranges of concentrations. A global survey of catechin concentrations in commercial wines found a range of 22 mg/l (in Australian and South African Shirazes) to 208 mg/l (in Burgundy) (Goldberg *et al.* 1998). Curiously, Burgundian wines came out as having consistently higher levels of catechin and epicatechin levels in that particular study.

Pinot noir wines, regardless of where they come from, often have higher catechin levels than wines made from other grapes. Waterhouse and Teissedre (1997) found an average of 250 mg/l of catechin in Californian Pinot noir wines, while those of Cabernet Sauvignon and Cabernet franc tended to have levels nearer to 150 mg/l.

White wines, due to lack of skin and stem contact during fermentation compared with reds, have much lower levels of flavan-3-ols, with Chardonnay and Sauvignon blanc containing around 40 mg/l (Waterhouse and Teissedre 1997).

Catechin is the major flavan-3-ol found in grapes and wine. Epicatechin, the next most abundant compound, is found at approximately one-third the concentration of catechin, but it is noted that the relative concentrations between varieties is similar to that found for catechin (Waterhouse and Teissedre 1997).

### Stilbenes

Similarly to catechins, Pinot noir tends to be the wine with the highest levels of resveratrol – at least 2.5 times the levels found in some other varietal wines, and with higher variability (Lamuela-Raventós and Waterhouse 1993; Goldberg *et al.* 1995; Lamuela-Raventós *et al.* 1997; Waterhouse and Teissedre 1997; Pour Nikfardjam *et al.* 1999). Red wines typically have values under 5 mg/l. White wines, as might be expected, have much lower levels due to the nature of the white winemaking process, typically 10 times lower than those found in red wines.

As intimated earlier, aside from cultivar, climate and disease pressure can have an influence on grape, and thus wine, stilbene content. Surveys of wines produced in many countries have shown that those from regions with more humid and disease-prone growing seasons tend to have higher resveratrol content than those from areas with warmer and drier growing seasons (Goldberg *et al.* 1995). Once in the grapes at harvest, the stilbene content of wine is mostly the result of efficiency of extraction and other wine-processing manipulations.

However, harvested grapes, while not ripening once removed from the plant, can manufacture stilbenes in response to an elicitation event. By irradiating the fruit with UV, allowing a day or two for the stilbenes to accumulate, and then processing the fruit, you could end up with a wine that has significantly greater resveratrol and other stilbene content than wine made from grapes processed immediately. Although there seem to be health benefits from the consumption of resveratrol found in wine, it is not likely that this method of boosting its concentration in the product will find favour.

## References

Alessandro, M., Di Marco, S., Osti, F., Cesari, A. (2000) Bioassays on the activity of resveratrol, pterostilbene and phosphorous acid toward fungi associated with esca of grapevine. *Phytopathologia Mediterranea*: **39**, 357–65.

Almafitano, C., Evidente, A., Surico, G., Tegli, S., Bertelli, E., Mugnai, L. (2000) Phenols and stilbene polyphenols in the wood of esca-diseased grapevines. *Phytopathologia Mediterranea*: **39**, 178–83.

Arichi, H., Kimura,Y., Okuda, H., Baba, K., Kozawa, M., Arichi, S. (1982) Effects of stilbene components of the roots of *Polygonum cuspidatum* Sieb. et Zucc. on lipid metabolism. *Chemical and Pharmaceutical Bulletin*: **30**, 1766–70.

Beggs, C. J., Schnieder-Ziebert, U., Wellman, E. (1986) UV-B radiation and adaptive mechanisms in plants, in *Stratospheric Ozone Reduction, Solar Radiation and Plant Life* (eds R. C. Worrest and M. M. Caldwell). Springer-Verlag, Berlin, pp. 235–50.

Boulton, R. (2001) The co-pigmentation of anthocyanins and its role in the colour of red wine: A critical review. *American Journal of Enology and Viticulture*: **52**, 67–87.

Bullard, R. W., Garrison, M. V., Kilburn, S. R., York, J. O. (1980) Laboratory comparisons of polyphenols and their repellent characteristics in bird resistant sorghum grains. *Journal of Agriculture and Food Chemistry*: **28**, 1006–11.

Caldwell, M. M. (1981) Plant response to solar ultraviolet radiation, in *Physiological Plant Ecology. I. Responses to the Physical Environment* (eds O. L. Lange, P. S. Nobel, C. B. Osmond, H. Zeigler). Encyclopedia of Plant Physiology, New Series, Volume 12A. Springer-Verlag, New York, pp. 169–98.

Cheynier, V. and Rigaud J. (1986) HPLC separation and characterization of flavonols in the skins of *Vitis vinifera* var. Cinsault. *American Journal of Enology and Viticulture*: **37**, 248–52.

Creasy, G. L., Pool, R. M., Creasy, L. L. (1987) *Effects of ethephon and light exposure on coloration of table grape cultivars grown in the Finger Lakes region.* Report submitted to the New York Grape Production Research Fund Committee, New York.

Creasy, L. L. and Coffee, M. (1988) Phytoalexin production potential of grape berries. *Journal of the American Society for Horticultural Science*: **113**, 230–4.

Dercks, W. and Creasy, L. L. (1989) The significance of stilbene phytoalexins in the *Plasmopara viticola*-grapevine interaction. *Physiological and Molecular Plant Pathology*: **34**, 189–202.

Durham, J. (1997) Cape Mentelle Vineyards' approach to tannin management in Cabernet Sauvignon. *Proceedings of the ASVO Seminar on Phenolics and Extraction*, 9 October 1997, Adelaide, South Australia, pp. 48–9.

Gao, Y. and Cahoon, G. A. (1994) Cluster shading effects on fruit quality, fruit skin colour, and anthocyanin content and composition in Reliance (*Vitis* hybrid). *Vitis*: **33**, 205–9.

Geissman, T. A. (1962) *The Chemistry of Flavonoid Compounds*. Macmillan, New York.

Gholami, M. and Coombe, B. G. (1995) Occurrence of anthocyanin pigments in berries of the white cultivar Muscat Gordo Blanco (*Vitis vinifera* L.). *Australian Journal of Grape and Wine Research*: **1**, 67–70.

Glories, Y. (1984) La couleur des vins rouges. 2eme partie. Mesure origine et interprétation. *Connaissance des Vignes and des Vins*: **4**, 253–71.

Goldberg, D. M., Yan, J., Ng, E., Diamandis, E. P., Karumanchiri, A., Soleas, G., Waterhouse, A. L. (1995) A global survey of *trans*-resveratrol concentrations in commercial wines. *American Journal of Enology and Viticulture*: **46**, 159–65.

Goldberg, D. M., Karumanchiri, A., Tsang, E., Soleas, G. J. (1998) Catechin and epicatechin concentrations of red wines: Regional and cultivar-related differences. *American Journal of Enology and Viticulture*: 49, 23–34.

Hillis, W. E. and Ishikura, N. (1968) The chromatographic and spectral properties of stilbene derivatives. *Journal of Chromatography*: 32, 323–36.

Jeandet, P., Sbaghi, M., Bessis, R., Meunier, P. (1995) The potential relationship of stilbene (resveratrol) synthesis to anthocyanin content in grape berry skin. *Vitis*: 34, 91–4.

Keller, M., Steel, C. C., Creasy, G. L. (2000) Stilbene accumulation in grapevine tissues: Developmental and environmental effects. *Acta Horticulturae*: 514, 275–86.

Kimura, Y., Ohminami, H., Okuda, H., Baba, K., Kozawa, M., Arichi, S. (1983) Effects of stilbene components of roots of *Polygonum* ssp. on liver injury in peroxidized oil-fed rats. *Planta Medica*: 49, 55–64.

Kubo, M., Kimura, Y., Shin, H., Haneda, T., Tani, T., Namba, K. (1981) Studies on the antifungal substance of crude drug (II) on the roots of *Polygonum cuspidatum* Sieb. et Zucc. (Polygonaceae). *Shoyakugaku Zasshi*: 35, 58–61.

Kuć, J. (1994) Relevance of phytoalexins – a critical review. *Acta Horticulturae*: 381, 526–39.

Lamuela-Raventós, R. M. and Waterhouse, A. L. (1993) Occurrence of resveratrol in selected California wines by a new HPLC method. *Journal of Agricultural and Food Chemistry*: 41, 521–3.

Lamuela-Raventós, R. M., Romero-Pérez, A. I., Waterhouse, A. L., Lloret, M., de la Torre-Boronat, M. C. (1997) Resveratrol and piceid levels in wine production and in finished wines, in *Wine: Nutritional and Therapeutic Benefits* (ed. T. R. Watkins). ACS Symposium series 661. American Chemical Society, Washington D.C., pp. 56–68.

Langcake, P. and McCarthy, W. V. (1979) The relationship of resveratrol production to infection of grapevine leaves by *Botrytis cinerea. Vitis*: 18: 244–53.

Langcake, P. and Pryce, R. J. (1976) The production of resveratrol by *Vitis vinifera* and other members of the Vitaceae as a response to infection or injury. *Physiological Plant Pathology*: 9, 77–86.

Maury, C., Sarni-Manchado, P., Lefebvre, S., Cheynier, V., Moutounet, M. (2001) Influence of fining with different molecular weight gelatins on proanthocyanidin composition and perception of wines. *American Journal of Enology and Viticulture*: 52, 140–5.

Mirabel, M., Saucier, C., Guerra, C., Glories, Y. (1999) Co-pigmentation in model wine solutions: occurrence and relation to wine aging. *American Journal of Enology and Viticulture*: 50, 211–18.

Parley, A., Vanhanen, L., Heatherbell, D. (2001) Effects of pre-fermentation enzyme maceration on extraction and colour stability in Pinot Noir wine. *Australian Journal of Grape and Wine Research*: 7, 146–52.

Plumb, G. W., De Pascual-Teresa, S., Santos-Buelga, C., Cheynier, V., Williamson, G. (1998) Antioxidant properties of catechins and proanthocyanidins: Effect of polymerisation, galloylation and glycosylation. *Free Radical Research*: 29, 351–8.

Pocock, K. F. and Waters, E. J. (1998) The effect of mechanical harvesting and transport of grapes, and juice oxidation, on the protein stability of wines. *Australian Journal of Grape and Wine Research*: 4, 136–9.

Pool, R. M., Creasy, L. L., Frackelton, A. S. (1981) Resveratrol and the viniferins, their application to screening for disease resistance in grape breeding programs. *Vitis*: **20**, 136–45.

Pour Nikfardjam, M., Rechner, A., Patz, C. D., Dietrich, H. (1999) Trans-resveratrol content of German wines. *Viticultural and Enological Sciences*: **54**, 17–20.

Price, S. F. (1994) Sun exposure and flavonols in grapes. PhD Thesis, Oregon State University, USA.

Price, S. F. and Watson, B. T. (1995) Preliminary results from an Oregon Pinot noir clonal trial, in *Proceedings of the International Symposium on Clonal Selection* (ed. J. M. Rantz). American Society for Enology and Viticulture, 21–22 June 1995, Davis, California, pp. 40–4.

Price, S. F., Breen, P. J., Valladao, M., Watson, B. T. (1995) Cluster sun exposure and quercetin in Pinot noir grapes and wine. *American Journal of Enology and Viticulture*: **46**, 187–94.

Price, S. F., Watson, B. T., Valladao, M. (1996) Vineyard and winery effects on wine phenolics – flavonols in Oregon Pinot noir, in *Proceedings of the 9th Australian Wine Industry Technical Conference* (eds C. S. Stockley, A. N. Sas, R. S. Johnstone, T. H. Lee). 16–19 July, 1995, Adelaide, South Australia. Winetitles, Adelaide, pp. 93–7.

Robinson, G. M. (1937) Leucoanthocyanins: Part III. Formation of cyanidin chloride from a constituent of the gum of *Butea frondosa*. *Journal of the Chemical Society* (London): **1937**, 1157–60.

Robinson, G. M. and Robinson, R. (1933) A survey of anthocyanins. III. Notes on the distribution of leucoanthocyanins. *Biochemical Journal*: **27**, 206–12.

Runge, F. F. (1821) Neueste phytochemische Entdeckungen zur Begründung einer Wissenschaft. *Phytochemie*: **2**, 245. (Lieferungen, Berlin.)

Scheffeldt, P. and Hrazdina, G. (1978) Co-pigmentation of anthocyanins under physiological conditions. *Journal of Food Science*: **43**, 517–20.

Schröder, G., Brown, J. W. S., Schröder, J. (1988) Molecular analysis of resveratrol synthase: cDNA, genomic clones and relationship with chalcone synthase. *European Journal of Biochemistry*: **172**, 161–9.

Siemann, E. H. and Creasy, L. L. (1992) Concentration of the phytoalexin resveratrol in wine. *American Journal of Enology and Viticulture*: **43**, 49–52.

Singleton, V. L. (1972) Effects on red wine quality of removing juice before fermentation to simulate variation in berry size. *American Journal of Enology and Viticulture*: **23**, 106–13.

Singleton, V. L. and Nobel, A. C. (1976) Wine flavor and phenolic substances, in *Phenolic, Sulfur, and Nitrogen Compounds in Food Flavors* (eds G. Charalambous and I. Katz). American Chemical Society Symposium Series, **26**, 47–70.

Smart, R. S. and Sinclair, T. R. (1976) Solar heating of grape berries and other spherical fruits. *Agricultural Meteorology*: **17**, 241–59.

Smart, R. E., Smith S. M., Winchester, R. V. (1988) Light quality and quantity effects on fruit ripening for Cabernet Sauvignon. *American Journal of Enology and Viticulture*: **39**, 250–8.

Somers, T. C. (1971) The polymeric nature of wine pigments. *Phytochemistry*: **10**, 2175–86.

Somers, T. C. (1998) *The Wine Spectrum*. Winetitles, Adelaide.

Somers, T. C. and Ziemelis, G. (1985) Flavonol haze in white wine. *Vitis*: **24**, 43–50.

Stafford, H. A. (1990) *Flavonoid metabolism*. CRC Press, Boca Raton, Florida.

Šubíková, V. (1991) Resveratrol accumulation in grapevine infected with grapevine vein necrosis disease. *Biologia Plantarum*: **33**, 287–90.

Thorngate, III, J. H. and Singleton, V. L. (1994) Localization of procyanidins in grape seeds. *American Journal of Enology and Viticulture*: **45**, 259–62.

Van Balen, J. (1984) Recovery of anthocyanins and other phenols from converting grapes into wines. MS Thesis, University of California, Davis.

Waterhouse, A. L. and Teissedre, P.-L. (1997) Levels of phenolics in California varietal wines, in *Wine: Nutritional and Therapeutic Benefits* (ed. T. R. Watkins). ACS Symposium series 661. American Chemical Society, Washington D.C. 20036. pp. 12–23.

Watkins, N. G. (1999) Bird behaviour in vineyards. Thesis for M. Appl. Sc., Lincoln University, Canterbury, New Zealand.

Watson, B. T., Price, S. F., Lombard, P. B., Creasy, G., Yorgey, B. (1992) Anthocyanin content of Oregon Pinot noir fruit and wine: Effects of vintage, fruit maturity, and viticultural practices. *American Journal of Enology and Viticulture*: **43**, 400.

Wellman, E. (1974) Gewebespezifische Kontrolle von Enzymen des Flavonoidstoffwechsels durch Phytochrom in Kotyledonen des Senfkeimlings (*Sinapsis alba* L.). *Berichte der Deutschen Botanischen Gesellschaft*: **87**, 275–9.

Wightman, J. D., Price, S. F., Watson, B. T., Wrolstad, R. E. (1997) Some effects of processing enzymes on anthocyanins and phenolics in Pinot noir and Cabernet Sauvignon wines. *American Journal of Enology and Viticulture*: **48**, 39–48.

Willstätter, R. and Zollinger, E. H. (1915) Untersuchungen über die Anthocyane: VI. Über die Farbstoffe der Weintraube und der Heidelberre. *Liebigs Annalen*: **408**, 83–109.

Wilson, M. F. and Whelan, C. J. (1990) The evolution of fruit colour in fleshy-fruited plants. *American Naturalist*: **136**, 790–800.

# 10 Modern biotechnology of winemaking

*R. S. Jackson*

## Introduction

Biotechnology refers to the use of micro-organisms (or their components) in the transformation of material for human benefit. Although the term is relatively recent, the use of micro-organisms in food preservation is ancient. The conversion of grapes into wine by *Saccharomyces cerevisiae* is probably the oldest confirmed example of biotechnology (McGovern *et al.* 1996). Until little more than 100 years ago, the action of microbes in these conversions was unknown.

The current interest in biotechnology stems not so much from our understanding of the role played by micro-organisms, but from recent and dramatic advances in our knowledge of inheritance and our ability to conduct intergeneric gene transfer – genetic engineering. Although the potential of such techniques is staggering, public distrust of new technologies has retarded its widespread application. Part of the difficulty involves the speed of technological advancement relative to our comprehension of its long-term impact. This is especially so where food or food products are involved. Because wine is marketed and generally viewed as a natural beverage produced by time-honoured procedures, there is considerable industry reluctance to forsake this image. Thus, the eventual practical value of current research into the genetic engineering of grapes and wine microbes is in considerable doubt. The procedures of genetic engineering are outlined below, along with the more traditional techniques of genetic modification. In addition, recent developments in the use of microbes or their enzymes in wine production, and advances in our understanding of microbes in winemaking are covered. However, a brief description of the basic stages involved in the transformation for grapes into wine will set the stage for the rest of the chapter.

## Wine production

As befitting one of the oldest examples of biotechnology, there is no single method of winemaking. The production method used depends on the type and style of wine intended. Nevertheless, some stages are, to varying degrees,

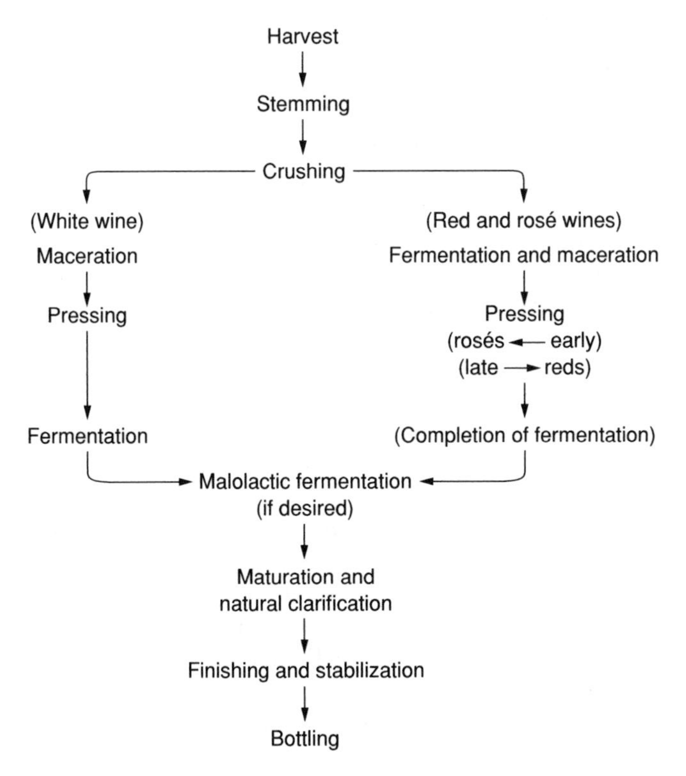


*Figure 10.1* Flow diagram of winemaking. (From Jackson 2000, reproduced with permission of Academic Press.)

involved in the production of all wines. Vinification is normally considered to begin when the grapes reach the winery. The principal steps in table wine production are presented in Figure 10.1.

The initial step typically involves stemming, during which the grapes are removed from the clusters. Grape crushing normally occurs simultaneously with stemming. Following crushing, the juice, skins and seeds (must) commence a process termed maceration (skin contact). Maceration expedites the extraction and generation of flavourants derived from the seeds and skins (pomace). Their solubilization initially occurs under the action of hydrolytic enzymes released from cells damaged during crushing. Enzymes also free nutrients required for fermentation. Finally, the pectic enzymes released assist in juice clarification.

Maceration is typically short (a few minutes to several hours) when producing white wines. The duration depends on the grape cultivar and the desired attributes of the wine. Juice that runs freely from the crushed grapes (free-run) is usually combined with that released by gentle pressing (press-run). Once maceration is complete, and the juice clarified, the juice is brought to the preferred fermentation temperature, and a yeast inoculum is added.

In the production of red wines, maceration is often prolonged and occurs in association with alcoholic fermentation. Although enzymatic action initiates maceration, its major role is taken over as alcohol begins to accumulate. This involves both ethanolic extraction and ethanolysis. Ethanol is particularly important in pigment (anthocyanin) and tannin solubilization. These phenolic compounds give red wines their colour and major flavour attributes. They also give red wines their ageing and mellowing characteristics. In addition, ethanol desorbs aromatic ingredients from the skins and grape pulp. After partial or complete yeast (alcoholic) fermentation, free-run wine is collected. Subsequently, the must is pressed to release additional wine (press fractions). These fractions are combined with the free-run in proportions determined by the type and style of wine desired.

Rosé wines are typically red wines made with a short maceration period. The duration depends primarily on the colour intensity wanted. Because of the short (< 24 h) maceration period, it typically occurs without significant ethanolic extraction.

Until comparatively recently, fermentation started spontaneously due to the action of indigenous yeasts. These either came from grape surfaces or were picked up from winery equipment. Today, fermentation usually involves the action of yeasts added to the juice or must. This assures a quick start to fermentation by a strain of known and selected characteristics, and minimizes the likelihood of microbial spoilage. Yeast strains are not only important to the efficiency of ethanol production, but also contribute significantly to the wine's fragrance.

After completing alcoholic fermentation, the wine may undergo a second, bacterial fermentation. Malolactic fermentation is particularly valuable in cool climatic regions, where excess acidity is a common problem. Malolactic fermentation is particularly beneficial for red wines if excess acidity augments the bitterness and astringency of tannins. For white wines, malolactic fermentation often has the secondary benefit of augmenting the wine's flavour. Nevertheless, the milder fragrance of white wines makes the presence of off-odours (occasionally produced during malolactic fermentation) more evident. In warmer climatic regions, where acid insufficiency tends to be a problem, malolactic fermentation is neither necessary nor desirable. To discourage its occurrence, the wine is clarified early, sulphited, and stored under cool conditions.

After fermentation, the young wine is protected from air exposure. This limits oxidation and microbial spoilage, while the wine slowly loses its yeasty odour, carbon dioxide in a supersaturated state escapes, and suspended material begins to precipitate. Certain aspects of the bouquet also start to develop. Although exposure to air is minimized, the slight uptake of oxygen during separation of the wine from accumulated sediment (racking) can be beneficial. Such exposure helps oxidize hydrogen sulphide (produced in yeast sediment), as well as improving colour stability in red wines.

Prior to bottling, the wine may be fined with the addition of one of several products to remove dissolved proteins and other colloidal materials.

Otherwise, these could generate haziness in the bottled wine. Fining may also be used to soften the wine's taste by removing excess tannins. Wines are commonly chilled and subsequently filtered to improve further clarification and stability.

A small dose of sulphur dioxide is generally given before bottling to limit oxidation and inhibit microbial spoilage. As an additional protection, sweet wines are usually sterile-filtered.

Newly bottled wines are stored for several months prior to sale and distribution to wholesalers. During this period, acetaldehyde (generated as a consequence of accidental oxygen uptake during bottling) is converted into non-aromatic compounds, and the wine's constituents continue to 'harmonize'.

## Genetic modification of microbial metabolism

Because of the importance of yeasts and bacteria to the sensory attributes of wine, most of the biotechnological interest in wine is associated with the control and modification of microbial activities. However, in contrast to most industrial fermentations, winemaking has thus far made little use of genetically engineered microbes. This is partially associated with the complex chemical origins of wine quality. It has made the definition of specific and detectable genetic quality improvements difficult. In addition, most wine fermentations are not pure, i.e. not under the control of a single organism. Even when fermentation occurs under the influence of inoculated yeasts, the initial phases involve the action of indigenous yeasts and bacteria that inhabit grapes or contaminate winery equipment. In addition, factors such as grape variety, fruit maturity and fermentation temperature are often considered of greater significance to quality than yeast strain. Finally, more is known about the chemical nature of the negative influences of spoilage microbes than about the positive sensory influences of beneficial yeasts and bacteria (Jackson 2000).

Genetically, properties controlled by one or a few linked genes are the most easily modified. For example, inactivating the gene that encodes sulphite reductase limits the production of $H_2S$ by yeasts. In contrast, enhancing the precipitation (flocculation) of yeasts at the end of fermentation is genetically more complex, and correspondingly difficult. Flocculation is regulated by several genes, epistatic (modifier) genes, and possibly cytoplasmic genetic factors (Teunissen and Steensma 1995). The major gene group (locus) encodes the synthesis of lectin-like cell-surface proteins that express late in colony growth. In addition, the method of flocculation may differ between yeast strains, as in the case of top- versus bottom-fermenting brewing yeasts (Dengis and Rouxhet 1997). Equally, important winemaking properties such as alcohol tolerance and fermentation ability at cool temperatures are under multigenic control. Because of this feature, they are ill-suited to modification by current genetic engineering procedures.

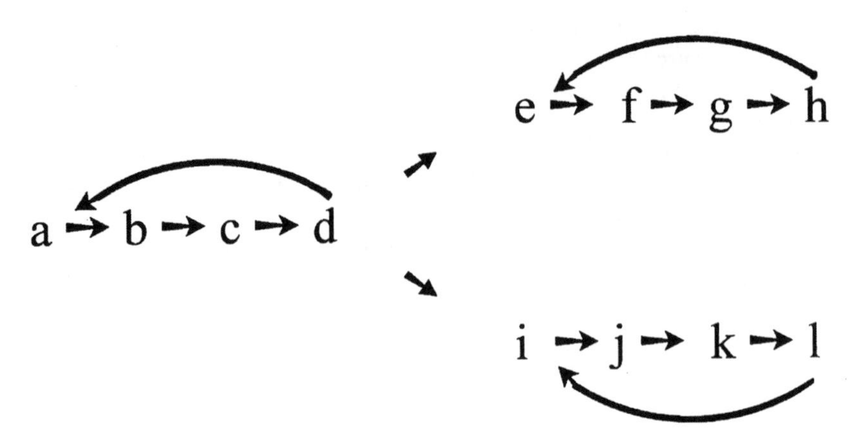


*Figure 10.2* Diagrammatic representation of a simple, two-branched, metabolic pathway, showing feedback inhibition by selected end products (back arrows).

Despite the apparent simplicity of modifying properties controlled by a single gene, adjusting the regulation of aspects of a metabolic pathway can have unanticipated effects (Guerzoni *et al.* 1985). Although we know much about the primary and secondary metabolism of *Saccharomyces cerevisiae*, our understanding of the intricacies of their control is still fragmentary. Disruptions are less likely if modification affects only the end-product of a metabolic pathway (Figure 10.2). In this situation, feedback control often affects only enzyme functions in the latter phases of the pathway. Consequently, it may not disrupt earlier stages in multiple-branched pathways. For example, incorporation of the gene expressing farnesyl diphosphate synthetase permits terpene synthesis by *S. cerevisiae* (Javelot *et al.* 1991). Because its insertion does not include any feedback mechanism, it can function without provoking accessory sensory effects. Terpene synthesis donates a muscat-like fragrance to the wine.

An additional complexity in breeding wine yeast entails the kaleidoscopic sequence of metabolic changes that occur throughout the successive stages in colony growth – lag, log, stationary and decline. Each phase involves continuous adjustment of enzyme and by-product concentrations to reflect the changing chemical composition of the fermenting juice (Jackson 2000; Pretorius 2000a). Compounds secreted in the early stages of fermentation may be reincorporated during later stages (Figure 10.3). This differs from most industrial fermentations in which growth conditions are adjusted to maintain the microbes at a particularly favourable point in their growth cycle (metabolic state). For wine, it is the relative and absolute concentrations of many aromatic compounds, produced during the various stages of colony growth, that is often of importance.

Various techniques may be used to modify the metabolic function of micro-organisms. The most elaborate is transformation – the incorporation

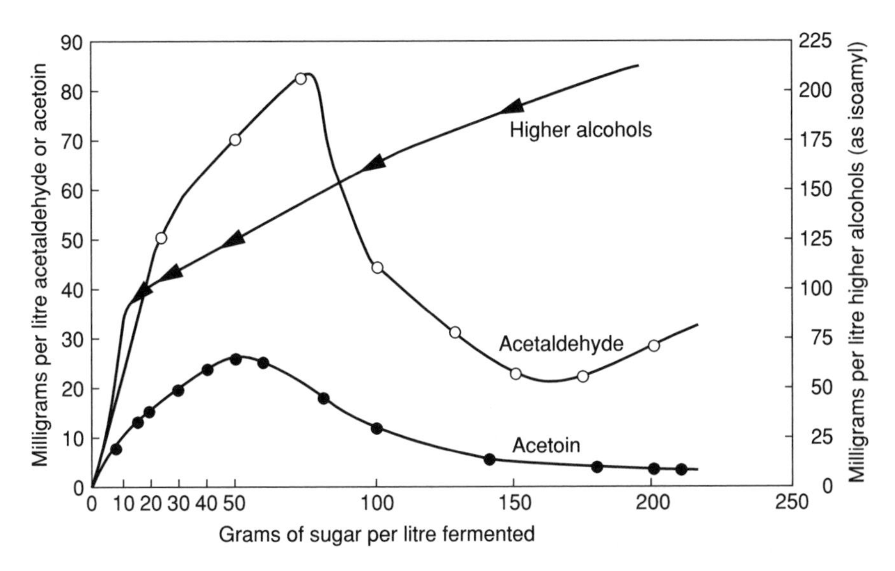


*Figure 10.3* Formation of acetaldehyde, acetoin, and higher alcohols during alcoholic fermentation. (From Amerine and Joslyn 1970, reproduced with permission of University of California Press.)

of a gene by genetic engineering. This can donate a property not already possessed by the organism. Nevertheless, useful genetic modulation often does not require this level of technical sophistication. As noted earlier, what may be needed is inactivation of one or several genes. In other situations, amplification of the gene (multiple copies) can produce the desired effect – by increasing synthesis of a product. Alternatively, disrupting a transport protein in the plasma membrane can lead to increased secretion. This can enhance production by deregulating synthesis. Any disruption in feedback control can lead to significant increases in synthesis of regulated compounds.

Researchers interested in improving the properties of wine yeasts and bacteria have several methods at their disposal (Pretorius 2000b). The simplest and most direct procedure involves selection of strains possessing desirable traits. Its success depends on the ease with which the presence of such traits can be identified and isolated. Isolation is greatly facilitated if a selective culture medium can be devised to permit growth of only those cells possessing the particular trait. In genetic engineering, the desired gene is typically combined with others, such as genes for antibiotic, herbicide, or heavy metal tolerance. Incorporation of an inhibitor into the growth medium suppresses (or kills) all cells except those possessing the resistance/tolerance gene (and the desired gene). Otherwise, individual cells must be physically isolated and separately investigated for presence of the attribute. When dealing with modification events (mutations or gene insertion) that occur once in $10^5$ to $10^9$ cells, this becomes an exceedingly arduous (and expensive) task.

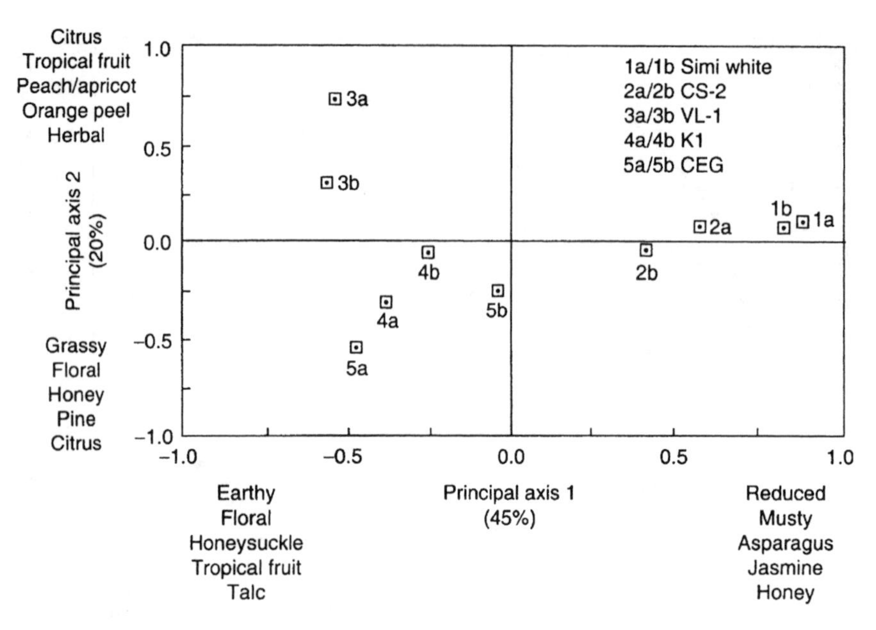


*Figure 10.4* Profile of aroma of a 'Riesling' wine (after 20 months) fermented with different yeast strains. (From Dumont and Dulau 1996, reproduced with permission.)

Selection tends to be most effective when dealing with genetically diverse populations. It has led to the isolation of strains with a diverse set of oenologically important properties, including those that significantly affect flavour development (Figure 10.4) (Cavazza *et al.* 1989; Grando *et al.* 1993).

Continued improvement, though, usually requires modifying the genetic make-up of the organism. For sexual organisms, such as *Saccharomyces cerevisiae*, this may involve procedures such as hybridization, backcrossing, and mutagenesis. When dealing with bacteria that do not possess standard sexual modes of genetic transfer, techniques such as transformation, somatic fusion, or genetic engineering are required.

Theoretically, breeding yeasts such as *Saccharomyces cerevisiae* should be comparatively simple, due to the extensive knowledge of its genetic make-up. The complete DNA sequence of *S. cerevisiae* has been known since 1996. However, *S. cerevisiae* normally exists in a diploid state; that is, each cell contains two copies of most genes. The diploid state can complicate improvement by masking the effects of desirable recessive genes. In haploid organisms, recessive genes are expressed directly because the cells contain only a single copy of each gene. Even more limiting is the rarity of meiosis – a precondition for sexual reproduction. In addition, when meiosis occurs, most of the haploid spores are non-viable – presumably due to the high frequency of aneuploidy in wine yeasts. Of the spores that germinate, most rapidly become diploid by mating with neighbouring cells. Typically, these are genetically identical because the mating-type gene oscillates between its two opposite allelic forms. This makes designed mating between selected strains particularly difficult. However, a technique developed by Bakalinsky and Snow (1990) may ease designed crossing by introducing a gene that prevents switching of the mating-type gene.

In situations where addition or amplification of a single property is desired, backcross breeding has traditionally been the preferred technique. It tends to induce the least disruption of existing combinations of desirable attributes. However, because of the difficulty of mating wine yeasts, genetic engineering is currently preferred if the intended gene can be isolated and cloned. For all practical purposes, mutation is effective only in inactivating genes.

Somatic fusion is one of the older procedures that can incorporate traits across species limits. It involves enzymatic cell-wall dissolution, maintenance of the protoplasts in an osmotically suitable medium, and mixing cells of both species in the presence of agents such as polyethylene glycol and calcium ions. These agents enhance cell fusion by promoting the physical union of cell membranes. On the reformation of cell walls, selection procedures are used to isolate the rare fused cells. Because the procedure incorporates many unnecessary or undesirable traits, and transformants are often genetically unstable, somatic fusion is rarely used today.

Potentially less disruptive is transformation. It incorporates only one or a few associated genes into the recipient. It has the advantage that donor and recipient organisms need not be closely related. With yeasts, protoplasts are immersed in a solution containing clones of the donor's gene, polyethylene glycol and calcium ions. Electropulses can increase DNA (gene) uptake. Conversely, intact cells may be transformed in the presence of lithium ions (see Pretorius 2000b). An alternative technique involves coating small particles with the gene(s) and propelling (shotgunning) the particles into cells.

Successful transformation requires insertion of the DNA segment into the host chromosome after uptake. Alternatively, the host gene may be spliced into a plasmid (self-replicating circular DNA) before transformation. This has been aided in *Saccharomyces cerevisiae* by the presence of an indigenous (2 μm) plasmid. Several modified *Escherichia coli* plasmids have been used as substitute gene vectors. The ability of plasmids to self-replicate, and make multiple copies of themselves, has eased the incorporation, expression, and maintenance of foreign genes in host cells.

Because cloned genes may not possess functional promoter or terminal signals (for transcription by RNA polymerases), these sequences must often be attached to the beginning and end of the gene sequence. Genes isolated from a taxonomically unrelated organism often possess promoter and terminator sequences poorly recognized by yeast RNA polymerases. Alternatively, promoter and terminator sequences may be missing if the gene has been cloned from a cDNA copy (derived from an mRNA transcript). The appropriate choice of promoter sequence can dramatically influence the efficiency of gene expression (Pretorius 2000b). Examples of efficient promoter and terminator signals are those from yeast PGK1 (phosphoglycerate kinase) and GAP (glyceraldehyde-3-phosphate dehydrogenase) genes. Secretion sequences may also need to be spliced into the gene cassette if the enzyme coded by the gene needs to be liberated from the cell to have its intended effect.

An example of successful gene transformation in wine yeasts is the insertion of the malolactic gene from *Lactococcus lactis*, along with the malate permease transport gene from *Schizosaccharomyces pombe* (Bony *et al.* 1997). Transformed cells can almost completely convert malic acid to lactic acid. Other incorporated genes include β-(1,4)-endoglucanase, urease and pectate lyase. If found safe and effective, these genetically modified strains could reduce the expense of occasionally adding enzyme preparations (see below). Van Rensburg and Pretorius (2000) provide a list of oenological genes incorporated into *Saccharomyces cerevisiae*. To date, the effectiveness of most of these transformants has not been tested under winemaking conditions.

When dealing with genetically complex features involving the interaction of multiple genes at several stages in colony growth, it is debatable whether genetic engineering can ever be effectively employed. For example, if deacidification were the only rationale for malolactic fermentation, incorporation of the malolactic gene along with malate permease could be useful. However, in most instances, malolactic fermentation is encouraged as much to provide the wine with a distinctive flavour as to deacidify the wine. In such cases, the simple conversion of malic acid to lactic acid would be inadequate. In addition, genetic engineering is still far more complex and expensive than traditional breeding techniques. Considerable costs are also associated with the need to establish safety for the US Food and Drug Administration or to obtain permitted status from the OIV (Organisation Internationale de la Vigne et du Vin).

Despite the obvious value of incorporating genes that enhance flavour development or donate the ability for malolactic conversion, other traits can be of equal or greater significance. For example, it is important that fermentation occurs rapidly and proceeds to completion (all fermentable sugars metabolized). This not only results in the generation of standard ethanol content, but also produces a wine that is dry and microbially stable. Many environmental factors can induce the premature termination of fermentation – a situation called stuck fermentation. Stuck fermentation may also result from yeasts being inactivated by the presence of contaminant yeasts that secrete a 'killer' protein. Under appropriate conditions, killer yeasts can replace inoculated strains at initial concentration as low as 0.1% (Jacobs and van Vuuren 1991). Genetic incorporation of killer genes (satellite dsRNAs) has successfully produced commercial wine yeast strains that are resistant to both the common K1 and K2 killer factors (Boone *et al.* 1990; Sulo *et al.* 1992). The possession of the killer factor protects the cell from the effects of the toxic protein it produces. Any feature that induces stuck fermentation opens the wine to potential spoilage.

Another potential use of genetic modification involves increased stress resistance. It provides the strain with enhanced tolerance to factors such as high sugar or alcohol content, or extreme temperature conditions. An understanding of the biochemical basis of these stress factors can pinpoint genes that need to be targeted for modification (see Bauer and Pretorius 2000). Without this information, improved tolerance is often pure happenstance.

The sugar concentration of most wine grapes at maturity ranges between 20% and 25%. This is sufficiently hypertonic to cause partial yeast-cell plasmolysis, and to delay the onset of fermentation (Nishino *et al.* 1985). High osmolarity can also reduce cell viability, limit cell division and increase alcohol toxicity. Strains of *S. cerevisiae* differ greatly in their sensitivity to sugar concentration. The genetic origins of this difference are unknown, but appear to be related to increased synthesis of, or reduced permeability of the cell membrane to, glycerol (see Brewster *et al.* 1993). Osmotolerance may also significantly affect the synthesis of aromatic compounds, such as acetic acid and fruit esters.

Although most strains of *S. cerevisiae* show considerable tolerance to ethanol, their improvement remains an important goal in wine research. As with osmotolerance, understanding the biochemical nature of ethanotolerance, and why it breaks down at high concentrations, could facilitate genetic improvement. It has been shown that enhanced glycerol and trehalose synthesis (Hallsworth 1998), substitution of ergosterol for lanosterol in the cell membrane, augmented membrane levels of phosphatidyl inositol (Arneborg *et al.* 1995), as well as ATPase activation and insertion of palmitic acid in the membrane, can increase ethanotolerance. For example, palmitic acid decreases membrane permeability (Mizoguchi 1998), which minimizes nutrient and co-factor loss, notably magnesium and calcium. Membrane function is also crucial for the retention of toxic substances stored in the vacuole (Kitamoto 1989). Ethanotoxicity, by reducing sugar uptake, can begin at contents as low as 2% (Dittrich 1977).

Although extreme temperature conditions can cause yeast death, growth and metabolic dysfunction occur well within toxic limits. High temperature tolerance appears to be partially dependent on the production of Hsp 104 (heat-shock protein 104). This limits or reverses the aggregation of essential cellular proteins (Parsel *et al.* 1994). Enhanced heat tolerance could reduce the need for expensive fermenter cooling during fermentation. In contrast, low temperature tolerance appears to be associated with an increase in the unsaturated fatty acid content of the plasma membrane (Rose 1989). Low temperate tolerance is especially desirable in white wine fermentations because it increases the synthesis and retention of fruit esters (notably isoamyl, isobutyl, and hexyl acetates). Greater ethanol and higher alcohol accumulation have also been associated with cooler fermentation temperatures. Because of the value of cold tolerance, the cryotolerant yeast *Saccharomyces uvarum* may be used in lieu of *S. cerevisiae*. Use of *S. uvarum* may also provide additional advantages, such as augmented glycerol content, enhanced synthesis of succinic acid, and reduced alcohol content (Massoutier *et al.* 1998).

## Ecology of yeast fermentation

Strain selection and genetic modification would be of little value were it not for independent progress in theoretical and applied science, which has permitted significant improvement in identification of both species and specific strains of yeasts and bacteria (usually by DNA fingerprinting). As a result, our knowledge of the ecology of yeasts in both vineyard and winery has increased dramatically. In addition, major advances in yeast propagation, storage, and reactivation (see pp. 240–1) have made the inoculation of grape juice and must with particular yeast strains commonplace. This has given the winemaker enhanced control over the characteristics shown by the wine. It has also markedly improved the average quality of wine.

One of the more intriguing recent discoveries about *Saccharomyces cerevisiae* is that it is not a common inhabitant of grape surfaces. In long-established vineyards, wine yeasts are isolated with rarity, and then only at the end of season (Mortimer and Polsinelli 1999). Even the original habitat of the wild progenitor of *S. cerevisiae* is uncertain – possibly the sap exudates of trees such as oak and pine (Phaff 1986). Most of the considerable yeast population found on grapes consists of members of the genera *Hanseniospora* (*Kloeckera*), *Candida*, *Pichia*, *Hansenula*, *Metschnikowia*, *Sporobolomyces*, *Cryptococcus*, *Rhodotorula*, and *Aureobasidium*. If active during fermentation, they usually grow only during its initial stages. The strains of *S. cerevisiae* that initiate spontaneous fermentation typically come from winery equipment (notably crushers, presses and sumps). Their surfaces are usually thinly coated with yeast from previous seasons.

Despite the low numbers of *Saccharomyces cerevisiae* on grapes, the acidic, hypertonic conditions of grape juice restrict the growth of most other yeasts. In addition, the rapid development of anaerobic conditions and the production of ethanol result in *S. cerevisiae* soon dominating and completing fermentation (Holloway *et al.* 1990). Nevertheless, some wild yeasts such as *Candida stellata* may persist throughout fermentation (Figure 10.5). This yeast can enhance the smooth mouth-feel of wine by contributing to glycerol production. Even in the initial stages of fermentation, wild yeasts may metabolize sufficiently to augment significantly the synthesis of acetic acid, glycerol and various esters (Ciani and Maccarelli 1998). These can either positively or negatively affect the wine's bouquet, depending on their concentration. For example, some strains of *Kloeckera apiculata* can spoil wine by generating up to 25 times the typical amount of acetic acid. Indigenous yeast activity is particularly important during the slow onset of white wine fermentations conducted at cool temperatures.

Recent investigations have shown that resident strains of *Saccharomyces cerevisiae* are often diverse, and their proportion often shifts during fermentation (Polsinelli *et al.* 1996). This also occurs if the must is inoculated with several strains of *S. cerevisiae* (Schütz and Gafner 1993).

One of the most significant changes in winemaking in recent years has been the increased use of specific yeast strains to initiate fermentation. Specific

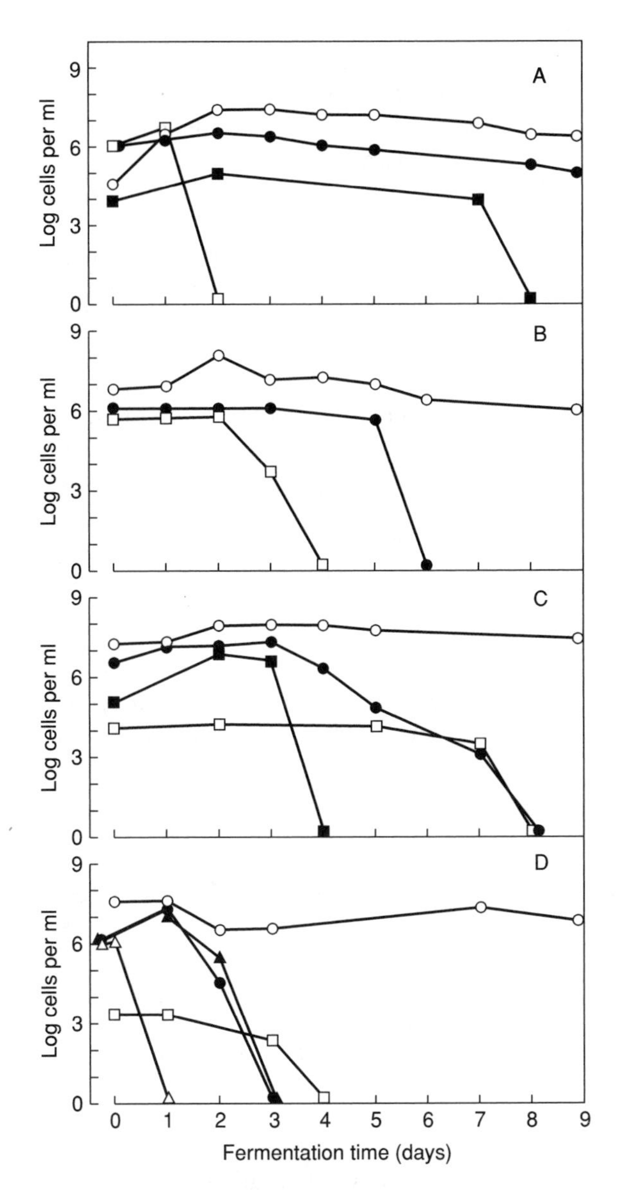


*Figure 10.5* Yeast numbers during fermentation of white (panels A and B) and red (panels C and D) wines. ○, *Saccharomyces cerevisiae*; ●, *Kloeckera apiculata*; □, *Candida stellata*; ■, *C. pulcherrima*; ▲, *C. colliculosa*; △, (*Hansenula anomala.* The initial population of *S. cerevisiae* comes predominantly from the inoculation conducted in the fermentations. (From Heard and Fleet 1985, reproduced with permission from the American Society for Microbiology.)

strains are available that can enhance varietal character, amplify fruit ester production, produce very low levels of acetic acid, hydrogen sulphide or urea, speed fermentation, possess enhanced alcohol tolerance, or efficiently reinitiate stuck fermentation. In addition, strains may be valuable in producing particular wine styles, notably carbonic maceration, late-harvest, or early-versus late-maturing reds. Finally, local strains may contribute to the character often thought to distinguish regional wines.

Some producers reject yeast inoculation, believing that the more complex interaction of yeast species in spontaneous fermentations partially provides the character their wines need to be distinctive (Henick-Kling *et al.* 1998). While this may be true, it also significantly increases the risk of microbial spoilage.

Recent investigation has shown not only the potential importance of indigenous (non-*Saccharomyces cerevisiae*) yeast, but also that related *Saccharomyces* spp. can play significant (and occasionally dominant) roles in winemaking. For example, *S. bayanus* is particularly tolerant of high sugar contents, and the low nitrogen, thiamine and sterol conditions that typify botrytized juice. Thus, it often conducts the fermentation of botrytized wines (see below). *S. bayanus* also frequently induces the second, in-bottle fermentation of sparkling wines. The cryotolerant nature of *S. uvarum* means it is well adapted to white wine fermentations. In addition, it has the supplemental benefits of increasing glycerol, succinic acid, 2-phenethyl alcohol, and isoamyl and isobutyl alcohol concentrations (Castellari *et al.* 1994; Massoutier *et al.* 1998).

Although much new information has been gathered in the past few years, aspects of the yeast ecology in specific wines still need clarification. A major example involves the population dynamics of yeasts growing in the pellicle (*flor*) that covers maturing *fino* sherries. In some instances, *Zygosaccharomyces fermentati* (*S. fermentati*) seems to dominate, while in other situations *S. bayanus* or strains of *S. cerevisiae* (i.e. *S. cerevisiae* f. *prostoserdovii*) predominate. Typically, the *flor* contains representatives of many genera, including *Pichia*, *Hansenula* and *Candida*. In addition, wines employing carbonic maceration or using either the recioto or governo processes require detailed succession studies.

## Active dry yeast production

Cultures of active dry wine yeast first became widely available in California in the early 1960s. By the late 1970s, several popular strains were available worldwide. This was a precondition for the adoption of yeast inoculation as a standard procedure in winemaking.

Wine yeast propagation employs techniques similar to those used for bakers' yeast (Reed and Nagodawithana 1991), with the exception of the presence of bisulphite. Bisulphite is incorporated to acclimatize cells to the sulphur dioxide typically added to crushed grapes to limit the activity of microbes derived from grapes and winery equipment.

Multiplication usually employs a fed-batch process in which the culture volume periodically increases as additional nutrient solution is added. The principal carbohydrate source is either molasses or corn syrup. Active aeration maximizes respiratory metabolism, favouring cell growth (biomass development), while minimizing alcohol production.

Once biomass accumulation is adequate, a compressed yeast cake is prepared using centrifugation or a vacuum rotary drum. The yeast cake is fragmented into small particles and dried to a moisture content of 5–7%. Packing under vacuum or inert gas ($N_2$ or $CO_2$) helps maintain yeast viability for at least one year.

Before use, cells must be rehydrated for about 20 minutes in warm (approximately 40° C) water or grape juice. This allows the cells to re-establish membrane function and cell activity suspended during drying.

## Yeast recycling and immobilization

Most winemaking involves separate lots of grape juice fermented to dryness. This is known as batch fermentation. In contrast, most industrial fermentations are continuous fermentations, where nutrients are added at a relatively constant rate once the desired metabolic state of the colony has been reached. The ferment volume remains stable by removing liquid at the same rate that nutrient solution is added. Such fermentations can remain operational for weeks or months. Unfortunately, the technique is ill-suited to wine production, where quality often depends on compounds produced throughout the various stages of colony growth and decline. Because of the cost efficiencies of continuous fermentation, though, several techniques are under investigation to reduce the costs associated with inoculating each separate batch of must with yeast.

One such system is called cell-recycle-batch fermentation (Rosini 1986). In the process, yeasts are isolated after each fermentation and are used to inoculate successive fermentations. Isolation may involve filtration, centrifugation, or spontaneous sedimentation. In addition to cost reduction, fermentation time is shortened, ethanol production slightly enhanced, and biosynthesis of sulphur dioxide reduced. However, acetic acid concentration tends to be increased. In addition, frequent monitoring for strain stability and microbial contamination is necessary. The latter requirements can be minimized if the yeasts are immobilized in calcium alginate (Suzzi *et al.* 1996).

The incorporation of microbes in a stable gel is gaining popularity in many industrial fermentations. One procedure injects a yeast–gel mixture through fine needles into a fixing agent, generating small beads of encapsulated yeasts (Fumi *et al.* 1988). Each bead may contain up to several hundred cells. Because of their mass, the beads readily settle following cessation of active $CO_2$ formation during fermentation. Efficient settling is of particular benefit in the second, in-bottle fermentation of sparking wine. Rapid collection of the yeast at the bottle neck can significantly reduce the expense of yeast removal. In addition, yeast entrapment appears to enhance cold tolerance and reduce the toxic effects of ethanol and acetic acid (Krisch and Szajáni 1997). Yeast encapsulation appears not to modify significantly the sensory attributes of sparkling wines (Hilge-Rotmann and Rehm 1990).

In a completely different application, immobilization has been studied to reduce the negative sensory effects of using *Schizosaccharomyces pombe* to conduct malolactic deacidification (Ciani 1995). It not only permits continuous deacidification (as wine passes through the chamber containing the immobilized cells), but also allows the winemaker easier regulation of the degree of deacidification.

## Malolactic fermentation

The role of malolactic fermentation in wine deacidification has long been known, as is its potential to spoil low-acid wines. The appreciation of its role in flavour development has, however, been slower to develop. Also of recent origin are significant advances in unravelling the complex ecology and physiology of the various lactic acid bacteria involved.

Despite these developments, there is still considerable controversy over the relative merits of malolactic fermentation. This arises partially because its major benefits occur in regions where the process happens infrequently (cool climates and acidic juice). Conversely, where it occurs spontaneously (warm regions where wines are often low in acidity), it has considerable potential for wine spoilage.

The most consistent consequence of malolactic fermentation is deacidification – the decarboxylation of a dicarboxylic acid (malic acid) to a monocarboxylic acid (lactic acid). By the elimination of one of the acidic groups, and changes in the dissociation constant, acidity falls and the pH rises. These may be associated with both indirect and direct sensory effects (Laurent *et al.* 1994). In wines of excessive acidity, these changes are usually beneficial. In wines of moderate to low acidity, the wine may taste 'flat', show increased tendencies to oxidize, lose colour, and develop off-odours.

Understanding the conditions that influence these opposing consequences has been complicated by the number of bacterial species involved and how wine chemistry affects their physiology. Species of at least three genera (*Oenococcus*, *Lactobacillus*, *Pediococcus*) may be active, and each differs in how it reacts to the low nutrient, acidic environment of wine. Only recently has it been shown definitively that malolactic fermentation can supply energy for growth (Henick-Kling 1995). Nevertheless, most of the metabolic energy comes from fermenting other compounds, notably sugars and amino acids. It is their fermentation by-products that produce the beneficial or detrimental olfactory consequences of malolactic fermentation. Figure 10.6 illustrates how different strains of *Oenococcus oeni* (the principal malolactic bacterium) affect the flavour attributes of wine. The buttery, nutty or toasty character

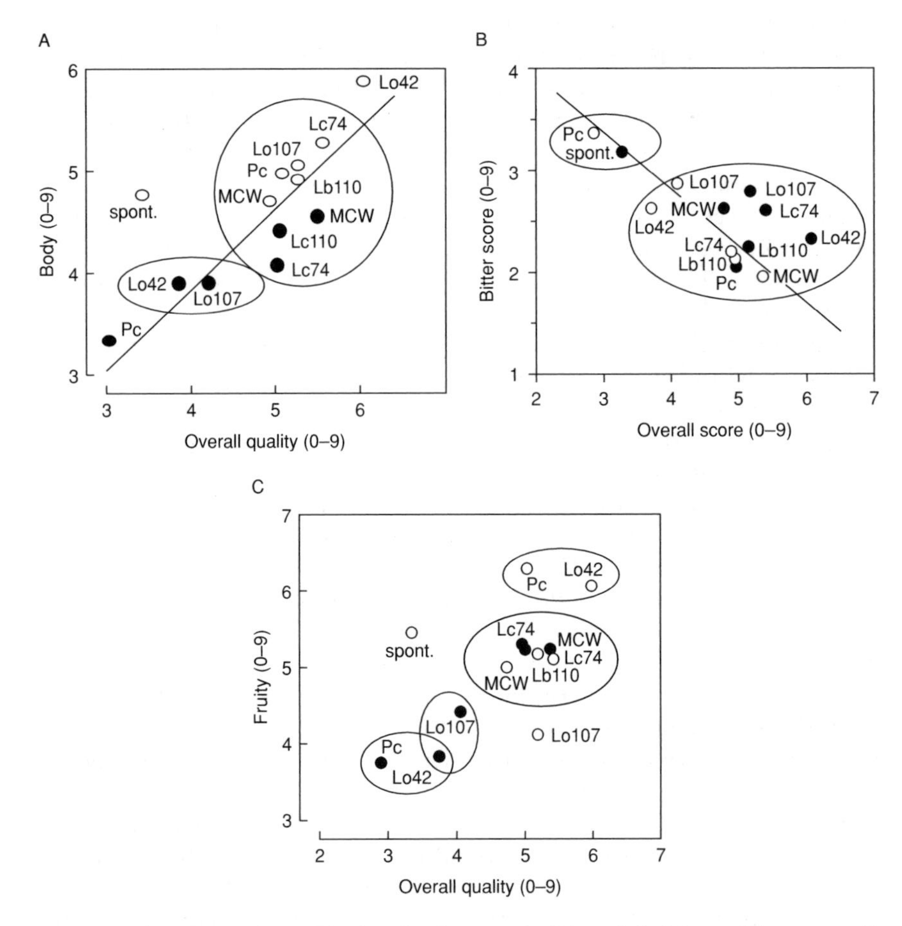


*Figure 10.6* Relationship of body (A), bitterness (B) and fruitiness (C) to overall quality of Cabernet Sauvignon wine fermented with various malolactic cultures, as evaluated by two taste panels composed of winemakers (●) and a wine research group (○). (From Henick-Kling *et al.* 1993, reproduced with permission.)

associated with diacetyl is a prominent example of the positive/negative sensory consequences of malolactic fermentation. Other flavourants occasionally produced at above sensory threshold values are acetaldehyde, acetic acid, acetoin, 2-butanol, diethyl succinate, ethyl acetate, ethyl lactate, and 1-hexanol. If too evident, they usually taint the wine.

For many years, it was thought that malolactic fermentation (in acidic wines) promoted microbial stability. An improved understanding of the physiology of lactic acid bacteria has shown that this view was in error. The microbial stability came from subsequent storage of the wine at cool temperatures, the addition of sulphur dioxide, and frequent racking (removal of sediment).

Ecological studies have shown that lactic acid bacteria are no more common on grape surfaces than is *Saccharomyces cerevisiae*. The natural habitat of the most important species, *Oenococcus oeni*, is unknown. Thus, as with *S. cerevisiae*, spontaneous induction of malolactic fermentation comes from inoculum derived from winery equipment. This is also the major source of contamination by undesirable lactic acid bacteria.

To encourage the activity of desirable strains, malolactic fermentation is increasingly being induced by inoculation with an active dry culture of *Oenococcus oeni*. These can be chosen to possess known flavour attributes. In addition, inoculation minimizes the activity of strains that may metabolize arginine to citrulline. Citrulline (as well as urea) can interact with ethanol to produce ethyl carbamate, a suspected carcinogen.

With the development of methods of storing and reactivating lactic acid bacteria, the selection and development of strains possessing particular sets of desirable traits has begun. To date, most of this activity has involved selection, with comparatively little direct genetic modification. Most of the genetic effort with lactic acid bacteria has been directed at transferring their malolactic genes to yeasts. The intention has been to produce a single organism that could conduct both alcoholic and malolactic fermentations. As noted, though, this does not address the olfactory modifications that are now often one of the principal reasons for promoting malolactic fermentation.

An alternative method of achieving malolactic fermentation involves passing wine slowly through a reactor lined with plates containing immobilized *Oenococcus oeni*. This provides faster malic acid decarboxylation, improved control over the degree of conversion (especially useful for wines above pH 3.5), and a reduction in the production of undesirable flavour compounds. Immobilization typically involves bacteria trapped in a gel, such as alginate, carrageenan, or polyacrylamide (see Diviès *et al.* 1994). Encapsulation limits cell division and infiltration into the wine. Multiplication is unnecessary as stationary phase cells actively decarboxylate malic acid. Major limitations to the application of encapsulation technology are its expense and short reactor-life.

In situations where only deacidification is desired, passing wine through a reactor containing immobilized malolactic enzyme and associated co-factors is a possibility. Because of the instability of the co-factor ($NAD^+$), the enzyme reaction mixture (maintained at a favourable pH in the reactor) must be separated from the wine. This can be achieved with the use of a differentially permeable membrane that divides the reactor into two chambers – one for the wine and the other for the enzyme solution (Formisyn *et al.* 1997).

## Special procedures involved in producing certain wines

### Sur lies *maturation*

*Sur lies* maturation is an old technique whose popularity and scientific investigation are recent. The procedure entails prolonged contact between the wine and autolysing (dead and dying) yeast cells. Typically, this occurs immediately following in-barrel fermentation. The large surface area to volume ratio of barrels favours the diffusion of yeast nutrients and flavourants into the wine.

During *sur lies* maturation, yeast membranes deteriorate, resulting in the release and activation of cellular hydrolytic enzymes. Among the compounds released are ethyl octanoate, ethyl decanoate, amino acids, and cell-wall mannoproteins. The latter bind polyphenolics dissolved by the wine from grape seeds and skins or wooden cooperage. This reduces bitterness and encourages early clarification and stabilization (Charpentier 2000; Dubourdieu 1995). Reducing the need for fining averts the associated loss of aromatics, such as the thiol varietal fragrances characteristic of 'Sauvignon blanc' wines. In addition, enzymes released during yeast autolysis hydrolyse grape glycosides, liberating bound aromatic compounds (Zoecklein *et al.* 1997).

*Sur lies* maturation is also an important stage in the maturation of sparkling wines. The prolonged yeast contact that follows the second (carbon dioxide-generating) fermentation is critical in producing the toasty bouquet that characterizes fine sparkling wines. In addition, cell wall mannoproteins (released during autolysis) help stabilize dissolved carbon dioxide, promoting the formation of long-lasting chains of bubbles. Mannoproteins also sustain (prolong) the release of aromatic compounds in wines (Lubbers *et al.* 1994).

### *Grape-cell fermentation (carbonic maceration)*

Carbonic maceration has been involved in the production of wine since time immemorial. Its role in most wine production has declined since the mid- to late 1800s due to the development of efficient crushers. In a few areas, though, its use has been retained, for example Beaujolais. Although an ancient process, the intricacies of its action were as unappreciated as they were unknown until the work of Flanzy and co-workers (1987).

Carbonic maceration has particular value in producing early-maturing, fruity red wines. Using standard techniques (Figure 10.1), it takes upwards of two years for most red wines to become readily drinkable. Early maturity (shortly after the completion of fermentation) often comes at a cost. Varietal distinctiveness is masked and the wine can have a short shelf-life. Beaujolais nouveau, is an example. It loses its distinctive raspberry-kirsch fragrance within about a year of production, and is often undrinkable after several years. This aspect is not consistently associated with carbonic maceration, though, as is evidenced by the long-ageing potential of traditionally produced red wine in Rioja. What distinguishes wines like Beaujolais nouveau is the application of carbonic maceration to all the grapes. Before the 1850s, essentially all wines would have involved partial carbonic maceration.

Carbonic maceration involves berry self-fermentation, before yeast and malolactic fermentation begins. For full carbonic maceration, the fruit is harvested with minimal breakage and intact grape clusters are placed in shallow fermenters. This minimizes the pressure exerted by the accumulated mass of grapes. Carbon dioxide gas may be flushed through the grapes to remove oxygen, favouring the rapid onset of autofermentation.

Although berry fermentation generates little alcohol (2–2.5%), the associated grape metabolism produces the distinctive flavour that typifies the process. Even after several decades of investigation, a chemical accounting of the carbonic maceration aroma is still incomplete. Nevertheless, certain aspects have been ascribed to ethyl cinnamate and benzaldehyde. Higher concentrations of ethyl decanoate, eugenol, methyl and ethyl vanillates, ethyl- and vinylguaiacols, and ethyl- and vinylphenols are typical (Ducruet 1984). Depending on the grape variety and fermentation temperature (Flanzy *et al.* 1987), upwards of 15–60% of the malic acid may be metabolized during carbonic maceration. This helps give the wine a smoother, less astringent taste.

### *Botrytization*

Because of the omnipresence of the grape pathogen *Botrytis cinerea*, wines made from infected grapes must have been made for centuries. Normally, these would have been unpalatable, at least to modern tastes. However, under special climatic conditions, infection leads to juice concentration and the potential for producing a sublimely sweet wine. The potential of this unique disease development (called 'noble rot') has been recognized for at least 350 years. It led to the fame of a special Hungarian white wine – Tokaji Essencia. The discovery of this transformation apparently occurred independently in Germany around 1750, and almost 100 years later in France. Although the association between these special white dessert wines and noble rot has been long known, an understanding of the climatology, biology and chemistry of their biotransformation is modern.

Climatological studies indicate that noble rot develops (in contrast to destructive bunch rot) when humid foggy nights alternate with dry, sunny days at harvest time. Cool humid conditions permit infection, but the daily drying limits fungal growth. As sporulation begins, the spore-bearing structures act as wicks, promoting dehydration. At maturity, water uptake from the vine is greatly restricted due to vascular disruption at the base of the fruit. What is unclear is whether the resulting juice concentration is what both retards *Botrytis* growth and limits the activity of secondary spoilage organisms, notably *Gluconobacter*, *Penicillium* and *Aspergillus*.

The importance of alternating humid/dry periods has been confirmed by its reproduction in the laboratory. These data have been used to produce botrytized wine under controlled conditions in commercial quantities (Nelson and Nightingale 1959). Although eminently successful, its expense has limited its adoption. Even juice inoculation followed by dehydration has been investigated as a means of commercializing botrytized wine production (Watanabe and Shimazu 1976).

During fruit dehydration, the concentration of compounds such as glucose, fructose, citric acid and malic acid increase. The amounts of other constituents, such as tartaric acid and ammonia, decrease due to fungal metabolism, despite the concentrating effect of water loss. The combined effects permit the production of a dessert wine without cloying sweetness. Part of the smooth mouth-feel of botrytized wines comes from the accumulation of glycerol during noble rotting.

One of the distinctive features of noble rotting is a loss in varietal aroma, especially that contributed by terpenes. *B. cinerea* actively metabolizes aromatic terpenes – such as linalool, geraniol and nerol – to less volatile compounds (Bock *et al.* 1986, 1988). Esterases produced by *Botrytis* may also degrade fruit esters, which give many young white wines their fruity character. Although seemingly an undesirable feature, the fungus compensates for these losses by generating its own set of flavourants. The most well-defined of these is sotolon. It possesses a sweet fragrance that, in combination with other aromatic compounds, can give the wine its distinctive honeyed, apricot-like fragrance (Masuda *et al.* 1984). Infected grapes also contain the mushroom alcohol, 1-octen-3-ol.

Surprisingly, *Botrytis cinerea* is implicated in the production of one of Italy's premium red wines (Recioto Amarone della Valpolicella) (Usseglio-Tomasset *et al.* 1980). A central aspect in the production of recioto (*appassimento*) wines involves the slow prolonged drying of grape clusters. Although the grapes appear healthy at harvest, nascent *Botrytis* infections slowly develop during storage. They generate the typical chemical changes associated with noble rotting (e.g. marked sugar, glycerol and gluconic acid accumulation). The glycerol content donates a smoothness that is rarely found in other full-bodied red wines.

*Recioto* wines possess a distinctive fragrance, resembling the oxidized phenol odour of tulip blossoms. This probably results from the action of laccase, a fungal polyphenol oxidase. Amazingly, laccase does not simultaneously degrade grape anthocyanins. Thus, the wine is red, rather than tawny brown, as would normally be expected.

## Flor *metabolism in sherry production*

Occasionally a yeast pellicle forms on the surface of wine. This is usually associated with wine spoilage. However, under a unique set of conditions, it permits the production of *fino* sherries. Without the development of the pellicle (*flor*), the solera-aged wine would develop into an *oloroso* sherry.

Until recently, the conditions that influenced whether a sherry developed into a *fino* or an *oloroso* were cloaked in mystery. However, the realities of modern commerce demanded that such a differentiation be controllable. Investigation has shown that the division depends largely on the degree to which the wine is fortified after fermentation. *Fino* (*flor*-influenced) sherries develop when the wine is fortified to between 15% and 15.5% alcohol, whereas *olorosos* develop when the wine is fortified to 18% alcohol. The higher alcohol content inhibits yeast action, and corresponding *flor* development. In contrast, the lower degree of fortification activates cell wall changes in yeasts. The result is that they float, forming a pellicle over the wine. Although not critical to pellicle formation, the frequent fractional blending employed in *fino* maturation provides the nutrients that maintain *flor* growth. In solera ageing (fractional blending), wine removed from one stage (*criadera*) is replaced by an equal volume from the next younger stage. Most *fino* soleras possess nine or more *criadera* stages (Jackson 2000).

As already noted, the population dynamics of the *flor* coating is complex, and is still in the early stages of being understood. Nevertheless, some if its effects on the wine's character have been clarified. Yeast respiration on the *flor* surface generates acetaldehyde, while limiting oxygen access to the underlying wine. The latter prevents undesirable wine oxidation. Associated with the frequent wine transfers, acetaldehyde reacts with ethanol, glycerol and other polyols, producing acetals. These probably contribute to the wine's complex oxidized character. In addition, *flor* yeasts synthesize terpenes such as linalool, *cis*- and *trans*-nerolidol, and *trans, trans*-farnesol (Fagan *et al.* 1981). Synthesis of $\gamma$-butyrolactones, such as sotolon, contribute to the characteristic nutty fragrance of *fino* sherries.

## Enzyme usage

Advances in industrial microbiology and chemical purification have permitted the isolation of enzymes in commercial quantities. Their use is now commonplace in many industries, including that of wine production. Filamentous fungi, notably *Aspergillus* and *Trichoderma* spp., are the primary sources for those enzymes authorized for wine use. Enzyme preparations may be added to facilitate wine clarification, decoloration, or dealcoholization, or to enhance flavour development and the release of anthocyanins from skins. However, because most commercial enzyme preparations are not fully purified, their effects often vary, depending on their source and conditions of use. A brief discussion of these applications is provided below. For additional details, see reviews by Canal-Llaubènes (1993) and Van Rensburg and Pretorius (2000).

### *Pectinase preparations*

Most wine grapes lose their pulpy texture and become juicy as they ripen. In some situations, however, juice extraction can be improved by the addition of pectinase immediately after crushing. In addition, pectinase can aid colour and flavour-release. Pectinase preparations may also be used after pressing to facilitate filtering and improve clarity. These effects result primarily from the degradation of grape pectins, the compounds that give fruit their pulpy character and juice a viscous attribute. Colloidal pectins can clog filters and retard the spontaneous settling of suspended particles. Pectinase preparations may also possess a macerating influence – causing cell death and tissue disintegration. This appears to result from cellular dysfunction induced by separation of the plasma membrane from the cell wall.

Although commercial pectinase preparations consist primarily of pectin lyase, they often contain additional enzymatic properties. Glucanases (e.g. hemicellulases and cellulases) often enhance juice or wine clarification (especially with grapes that possess significant amounts of viscous glucans resulting from *Botrytis* infection). In addition, preparations may possess cinnamoylesterase and glycosidase activities (Lao *et al.* 1997). Cinnamoylesterase can structurally modify certain phenolic acids, which can subsequently be metabolized to vinyl-phenols during fermentation. β-Glucosidase can release aromatic terpenes from fixed (non-volatile) states. Thus, pectinase preparations may modify wine flavour, occasionally with unanticipated effects. Glycosidase activity can occasionally also release anthocyanins from their glycosidic linkage, making them more susceptible to oxidation and permanent decoloration.

Because enzymatic additions can complicate protein stability, enzyme immobilization within an inert support is being investigated. Not only does it avoid protein addition, but it also provides superior control over enzyme action.

## Anthocyanase activity

Most white wines are produced from white grapes. Nevertheless, some ostensively white grapes, such as 'Gewürztraminer,' produce small amounts of anthocyanin. In addition, white wines may occasionally be produced from red-skinned cultivars. Decoloration, if necessary, is usually conducted after fermentation, because anthocyanin levels typically decline spontaneously during fermentation. However, if experience indicates that this is inadequate, anthocyanase may be added before fermentation. By removing the sugar moiety from anthocyanins, their solubility declines, promoting subsequent precipitation during fermentation. Loss of the sugar moiety also reduces coloration and makes the molecules more susceptible to oxidative decoloration.

Laccase, a fungal polyphenol oxidase, is being investigated as a potential decoloring agent. Laccase quickly oxidizes anthocyanins, turning them brown. These can be subsequently removed during fining. Because laccase is currently not a permitted food additive, immobilization can circumvent this limitation (Brenna and Bianchi 1994). In this form, laccase remains attached to a reactor, through which wine is passed during treatment.

### *Glucose oxidase/peroxidase activity*

Low-alcohol content wine is one of the rapidly growing segments of the wine industry. Dealcoholization may be achieved by many procedures (Jackson, 2000; Pickering 2001), most of which occur after alcoholic fermentation. A new enzymatic technique involves reducing the alcohol-producing capacity of the juice before fermentation (Villettaz 1987). The process entails the joint action of glucose oxidase and peroxidase. Glucose oxidase converts glucose to gluconic acid (leaving only fructose as the major fermentable substrate). Hydrogen peroxide, generated as a by-product of glucose oxidation, is inactivated by peroxidase. In the process, the alcohol-producing capacity of the juice is reduced to about one-half.

Because molecular oxygen is required for enzymatic dealcoholization, the juice becomes oxidized and turns brown. Although most of these oxidized colour compounds precipitate during fermentation, the wine retains a distinct golden colour. In the studies conducted by Pickering *et al.* (1999), taste is modified, but the fragrance and mouth-feel of the wines are not markedly affected.

### *β-Glycosidases preparations*

Grapes often possess an unexpressed flavour potential due to aromatic compounds bound in non-volatile glycosidic complexes. This applies especially to monoterpenes, but also affects norisoprenoids and volatile phenolic compounds. Thus, there is considerable interest in liberating these aromatics. Glycosidic linkages may be broken either by acidic or enzymic hydrolysis. Because of the slowness of acid-induced breakdown, and the flavour damage produced by heating to speed acidic hydrolysis, most of the attention has been directed towards enzymatic hydrolysis.

β-Glycosidases have garnered most of the attention, even though preparations containing α-arabinosidase, α-rhamnosidase, β-xylanosidase and β-apiosidase activities improve their effectiveness. This results from potential flavourants often being bound to other sugars as well as to glucose.

Enzyme additions are typically supplied at the end of fermentation, because sugars in the juice inhibit their catalytic action (Canal-Llaubènes 1993). Although enzyme activity can be stopped when desired, by the addition of agents like bentonite, immobilization provides even greater enzyme control (Caldini *et al.* 1994).

### *Urease*

Considerable concern was aroused when it was discovered that wines, especially those that were heated during processing, possessed a suspected carcinogen – ethyl carbamate (urethane). Ethyl carbamate forms as a by-product of a reaction between ethanol and urea. Urea can occur in wine as a result of arginine metabolism, being added as a nitrogen supplement to juice, or as a nitrogen fertilizer in vineyards. Although choice of yeast strain and elimination of urea use can reduce the accumulation of ethyl carbamate, addition of urease (Ough and Trioli 1988) can be used where past experience indicates that other control measures are insufficient.

## Microbial spoilage

### *Cork-derived taints*

Advances in the understanding and control of spoilage have been particularly important in relation to corky off-odours. It has been variously estimated that up to 1% of all wines are so affected. In most instances, it is due to trace contamination of cork with 2,4,6-trichloroanisole (TCA). Absorption of water vapour by the cork can release its TCA into the wine (Casey 1990). Additional sources of TCA tainting can come from oak barrels and the winery environment (Chatonnet *et al.* 1994).

Pentachlorophenol (PCP), used to control insects and wood rots in wineries and cork oak plantations, is a potential precursor of TCA. The pesticide can be metabolized by several common filamentous fungi to TCA (Maujean *et al.* 1985). In addition, some white-rot fungi (basidiomycetes) can directly incorporate chlorine into phenolic compounds. Subsequent metabolism can produce TCA even without treatment of the cork with hypochlorite. In most instances, though, the chlorophenols involved in TCA synthesis are probably generated as a consequence of treating corks with hypochlorite. This is suggested by the principal location of TCA in the outer surfaces of contaminated corks (Howland *et al.* 1997). Chloride ions can diffuse into the cork (notably through lenticels and fissures) if the stoppers are left in hypochlorite too long. Hypochlorite is frequently used as a surface sterilant and bleaching agent. Chlorophenols form when chloride ions react with cork phenolics. Subsequent microbial methylation can generate TCA.

Additional mouldy odours may originate from the growth of bacteria, filamentous fungi and, occasionally, from yeasts, found on cork (Jäeger *et al.* 1996). For example, *Penicillium roquefortii*, *P. citrinum*, and *Aspergillus versicolor* can produce several musty and mouldy smelling compounds (e.g. sesquiterpenes and octanol compounds). Contamination can occur before harvesting (bark on cork oaks), during seasoning of cork slabs, during storage of finished corks, or after insertion into the bottle. Additional taints may come from vanillin (derived from cork lignins) metabolized to guaiacol by *Streptomyces* spp. (Lefebvre *et al.* 1983). Guaiacol possesses a sweet, burnt odour.

Most of the fungi isolated from cork can be found on bark taken directly from the tree. In contrast, *P. roquefortii* appears to originate as a winery contaminant. It is seldom isolated from corks prior to delivery and storage in wine cellars. It is also one of the few organisms capable of growing through cork and directly tainting bottled wine (Moreau 1978). Other commonly isolated fungi, such as *Penicillium glabrum*, *P. spinulosum*, and *Aspergillus conicus* cannot grow in contact with wine. Thus, the lower two-thirds of the cork seldom yields viable fungi after a few months in-bottle (Moreau 1978).

Moisture retained by unperforated lead capsules favours fungal growth on the surface of corks. Although the growth seldom penetrates the cork, organic acids produced can speed capsule corrosion. Corrosion can dissolve lead, contaminating the neck and upper cork surfaces with lead salts.

Cork is seldom a source of spoilage yeasts and bacteria. If found, most isolates are bacteria in the endospore-forming genus *Bacillus*. Several *Bacillus* spp. have been linked to spoilage, notably *B. polymyxa*, in the conversion of glycerol to a bitter-tasting compound, acrolein (Vaughn 1955) and *B. megaterium*, associated with unsightly deposits in brandy (Murrell and Rankine 1979).

### *Yeast-induced spoilage*

Many yeast species have been implicated in wine spoilage, even *Saccharomyces cerevisiae* if it grows when and where it is not wanted (e.g. in bottles of semi-sweet wine). Epiphytic yeasts such as *Kloeckera apiculata* and *Metschnikowia pulcherrima* can spoil wine if they produce significant amounts of acetic acid, ethyl acetate, diacetyl, or 2-aminoacetophenone during the initial stages of fermentation. These organisms have also been implicated in the untypical ageing (UTA) off-odour of some German wines (Sponholz and Hühn 1996). The off-odour is characterized by a naphthalene-like odour.

Although off-odour production is typically associated with spoilage, haze production by itself is a fault. This has been a particular problem with *Brettanomyces* species. *Zygosaccharomyces bailii* can also produce flocculent and granular deposits. Because these yeasts can multiply in traces of wine left in bottling equipment, and are notoriously difficult to eliminate once established, strict hygiene is the only effective preventive measure.

Often more intractable are the off-odours produced by these organisms. *Z. bailii*, for example, can generate an intense buttery character due to marked diacetyl accumulation. *Brettanomyces* is the prime source of mousy off-odours (caused by 2-acetyltetrahydropyridines). Amazingly, some wine producers consider 'Brett' odours desirable, including those produced by ethylphenols and isobutyric, isovaleric and 2-methyl-butyric acids. Ethylphenols, for example, can donate smoky, phenolic, horse manure, spicy, medicinal, and woody taints.

### *Bacteria-induced spoilage*

Bacteria have been known to induce wine spoilage since the pioneering work of Pasteur (1866). The connection between particular species or genera and 'wine diseases' has also been known for years. Examples are *Lactobacillus brevis* with *tourne*, strains of *Lactobacillus brevis* and *L. buchneri* with *amertume*, some strains of *Oenococcus oeni* with mannitic fermentation, and other strains of *Oenococcus oeni* and *Pediococcus* with ropiness. High volatile acidity is a common feature of most forms of spoilage induced by lactic acid bacteria. Acetic acid can be produced as a by-product of the fermentation of citric, malic, tartaric, and gluconic acids, as well as hexoses, pentoses, and glycerol. What is recent is the discovery that acetic acid bacteria can both metabolize and grow under conditions that were formerly thought to inhibit their activity.

Acetic acid bacteria can oxidize ethanol to acetic acid. Although essential for commercial vinegar production, it makes wine undrinkable. Because acetic acid bacteria are obligate aerobes (cannot ferment), it was thought that they could not survive under anaerobic conditions (typical of wine). Recent studies have shown that they can use hydrogen acceptors other than molecular oxygen for respiration. This has forced researchers to reinvestigate the role of acetic acid bacteria in wine spoilage.

Because quinones can be used as substitutes for oxygen (Adlercreutz 1986), acetic acid bacteria have the potential to metabolize in barrelled or bottled wine. Of even greater importance is their growth using trace amounts of oxygen absorbed by wine during clarification and maturation (Joyeux *et al.* 1984; Millet *et al.* 1995). Significant growth under these situations could spoil the wine by the production of acetic acid and its ethyl ester, ethyl acetate.

Of the three species of acetic acid bacteria associated with grapes, only those in the genus *Acetobacter* multiply in wine. Typically *A. aceti* is the dominant species and that most associated with direct wine spoilage.

Spoilage by acetic acid bacteria during fermentation is rare, largely because most present-day winemaking practices restrict contact with air. Improved forms of pumping over and cooling have eliminated the major sources of must oxidation during fermentation. Also, a better understanding of stuck fermentation has limited its incidence, permitting the earlier application of techniques that reduce the likelihood of oxidation and microbial spoilage. However, the tendency to mature red wine in small oak cooperage, which has greater potential for oxygen uptake, has increased the potential for acetic acid bacterial spoilage. Used wooden cooperage is also a major source of bacterial contamination if improperly stored, cleansed and disinfected. Thus, it is not surprising that red wines have higher average levels of volatile acidity than white wines (Eglinton and Henschke 1999). Because typical dosages of sulphur dioxide are generally insufficient to inhibit the growth of acetic acid bacteria, additional techniques must be used for spoilage control. These include low pH, strict limitation of oxygen incorporation, and storing under cool conditions.

The popularity of *sur lies* maturation also poses the threat of increased acetic acid bacterial spoilage. To minimize the production of reduced sulphur off-odours in the lees, the wine is periodically stirred to incorporate oxygen. Although this promotes the oxidation of hydrogen sulphide, which can subsequently generate mercaptans, it can also supply oxygen for the growth of acetic acid bacteria.

Another aromatic compound sporadically associated with spoilage by acetic acid bacteria is acetaldehyde. Under most circumstances, acetaldehyde is rapidly respired to acetic acid. However, the enzyme that oxidizes it to acetic acid is sensitive to denaturation by ethanol. As a result, acetaldehyde may accumulate in highly alcoholic wines. Low oxygen tensions also favour the synthesis of acetaldehyde from lactic acid.

## Other microbial influences

Bacteria are often viewed, validly, as sources of serious spoilage. However, they also play beneficial roles in wine production. The most obvious example is malolactic fermentation. Nevertheless, bacteria play additional, if more indirect, roles in wine production. These include aspects such as the biological control of pests and disease, and possibly in the seasoning of oak.

Microbes are increasingly being used in biological pest control. For example, *Bacillus thuringiensis* has been used successfully to control several serious insect pests, notably the omnivorous leafroller (*Platynota stultana*) and the grape leaffolder (*Desmia funeralis*). In addition, *Bacillus megaterium* is a component of the biological control agent Greygold®, used to control *Botrytis cinerea*. This biofungicide also contains the filamentous fungus *Trichoderma hamatum* and the yeast *Rhodotorula glutinis*. The mycoparasite *Ampelomyces quisqualis* has also shown promise in the control of grapevine powdery mildew (Falk *et al.* 1995).

Early frost damage in the autumn often involves the action of bacteria, notably *Pseudomonas* and *Erwinia* spp. As temperatures decline in the autumn, the number of these bacteria increase on grape and leaf surfaces. Ice nuclei form particularly well around these bacteria, increasing the temperature at which frost damage begins, often by up to 2–4° C. This has led to the inoculation of grape surfaces with strains that are ice-nucleation deficient (Lindemann and Suslow 1987). A commercial preparation of ice-nucleation deficient *P. fluorescens* is marketed under the name BlightBan® A506.

In a separate sphere, the bacterium *Agrobacterium vitis* and related species have proven very successful as gene transporters in genetically engineering grapevines.

Finally, the seasoning of oak used in barrel-making involves beneficial microbial action. During weathering, bacteria and fungi grow within the wood, producing aromatic aldehydes and lactones from wood lignins. In addition, the wood-rotting fungus *Coriolus versicolor* produces polyphenol oxidases that degrade lignins, as well as inducing phenolic polymerization (Chen and Chang 1985). Although the outer few millimetres of the wood is contaminated by common saprophytic fungi and bacteria, the significance of their presence remains uncertain.

## References

Adlercreutz, P. (1986) Oxygen supply to immobilized cells. 5. Theoretical calculations and experimental data for the oxidation of glycerol by immobilized *Gluconobacter oxydans* cells with oxygen or *p*-benzoquinone as electron acceptor. *Biotechnology and Bioengineering*: **28**, 223–32.

Amerine, M. A. and Joslyn, M. A. (1970) *Table Wines, the Technology of their Production*, 2nd edn. University of California Press, Berkeley.

Arneborg, N., Høy, C.-E., Jørgensen, O. B. (1995) The effect of ethanol and specific growth rate on the lipid content and composition of *Saccharomyces cerevisiae* grown anaerobically in a chemostat. *Yeast*: **11**, 953–9.

Bakalinsky, A. L. and Snow, R. (1990) Conversion of wine strains of *Saccharomyces cerevisiae* to heterothallism. *Applied and Environmental Microbiology*: **56**, 849–857.

Bauer, F. F. and Pretorius, I. S. (2000) Yeast stress response and fermentation efficiency: How to survive the making of wine – A review. *South African Journal of Enology and Viticulture*: **21**, 27–51.

Bock, G., Benda, I., Schreier, P. (1986) Metabolism of linalool by *Botrytis cinerea*, in *Biogeneration of Aromas* (eds T. H. Parliament and R. Crouteau). ACS Symposium Series No. 317, American Chemical Society, Washington, D.C., pp. 243–53.

Bock, G., Benda, I., Schreier, P. (1988) Microbial transformation of geraniol and nerol by *Botrytis cinerea. Applied Microbiology and Biotechnology*: **27**, 351–7.

Bony, M., Bidart, F., Camarasa, C., Ansanay, L., Dulau, L., Barre, P., Dequin, S. (1997) Metabolic analysis of *Saccharomyces cerevisiae* strains engineered for malolactic fermentation. *FEBS Letters*: **410**, 452–6.

Boone, C., Sdicu, A.-M., Wagner, J., Degré, R., Sanchez, C., Bussey, H. (1990) Integration of the yeast $K_1$ killer toxin gene into the genome of marked wine yeasts and its effect on vinification. *American Journal of Enology and Viticulture*: **41**, 37–42.

Brenna, O. and Bianchi, E. (1994) Immobilised laccase for phenolic removal in must and wine. *Letters in Biotechnology*: **16**, 35–40.

Brewster, J. L., de Valoir, T., Dwyer, N. D., Winter, E., Gustin, M. C. (1993) An osmosensing signal transduction pathway in yeast. *Science*: **259**, 1760–3.

Caldini, C., Bonomi, F., Pifferi, P. G., Lanzarini, G., Galante, Y. M. (1994) Kinetic and immobilization studies on fungal glycosidases for aroma enhancement in wine. *Enzyme Microbial Technology*: **16**, 286–91.

Canal-Llaubènes, R.-M. (1993) Enzymes in winemaking, in *Wine Microbiology and Biotechnology* (ed. G. H. Fleet). Harwood Academic, Chur, Switzerland, pp. 447–506.

Casey, J. A. (1990) A simple test for cork taint. *Australian Grapegrower and Winemaker*: **324**, 40.

Castellari, L., Ferruzzi, M., Magrini, A., Giudici, P., Passarelli, P., Zambonelli, C. (1994) Unbalanced wine fermentation by cryotolerant *vs.* non-cryotolerant *Saccharomyces* strains. *Vitis*: **33**, 49–52.

Cavazza, A., Versini, G., DallaSerra, A., Romano, F. (1989) Characterization of six *Saccharomyces cerevisiae* strains on the basis of their volatile compounds production, as found in wines of different aroma profiles. *Yeast*: **5**, S163–7.

Charpentier, C. (2000) Yeast autolysis and yeast macromolecules? Their contribution to wine flavor and stability, in *Proceedings of the ASEV 50$^{th}$ Anniversary Annual Meeting, Seattle, Washington* (ed. J. M. Rantz). American Society for Enology and Viticulture, Davis, California, pp. 271–7.

Chatonnet, P., Guimberteau, G., Dubourdieu, D., Boidron, J. N. (1994) Nature et origine des odeurs de 'moisi' dans les caves, incidences sur la contamination des vins. *Journal International des Sciences de la Vigne et du Vin*: **28**, 131–51.

Chen, C.-L. and Chang, H. M. (1985) Chemistry of lignin biodegradation, in *Biosynthesis and Biodegradation of Wood Components*, (ed. T. Higuchi). Academic Press, New York, pp. 535–56.

Ciani, M. (1995) Continuous deacidification of wine by immobilized *Schizosaccharomyces pombe* cells: Evaluation of malic acid degradation rate and analytical profiles. *Journal of Applied Bacteriology*: **75**, 631–4.

Ciani, M. and Maccarelli, F. (1998) Oenological properties of non-*Saccharomyces* yeasts associated with wine-making. *World Journal of Microbiology and Biotechnology*: **14**, 199–203.

Dengis, P. B. and Rouxhet, P. G. (1997) Flocculation mechanisms of top and bottom fermenting brewing yeasts. *Journal of the Institute of Brewing*: **103**, 257–61.

Dittrich, H. H. (1977) *Mikrobiologie des Weines, Handbuch der Getränketechnologie*. Ulmer, Stuttgart.

Diviès, C., Cachon, R., Cavin, J.-F., Prévost, H. (1994) Immobilized cell technology in wine production. *Critical Reviews in Biotechnology*: **14**, 135–53.

Dubourdieu, D. (1995) Intérêts oenologiques et risques associés à l'élevage des vins blancs sur lies en barriques. *Revue Française d'Oenologie*: **155**, 30–5.

Ducruet, V. (1984) Comparison of the headspace volatiles of carbonic maceration and traditional wine. *Lebensmittelwissenschaft und -Technologie*: **17**, 217–21.

Dumont, A. and Dulau, L. (1996) The role of yeasts in the formation of wine flavors, in *Proceedings of the 4th International Symposium on Cool Climate Viticulture and Enology* (eds T. Henick-Kling *et al.*). New York State Agricultural Experimental Station, Geneva, New York, pp. VI-24–28.

Eglinton, J. M. and Henschke, P. A. (1999) The occurrence of volatile acidity in Australian wines. *Australian Grapegrower and Winemaker*: **426a**, 7–8, 10–12.

Fagan, G. L., Kepner, R. E., Webb, A. D. (1981) Production of linalool, *cis*- and *trans*-nerolidol, and *trans*, *trans*-farnesol by *Saccharomyces fermentati* growing as a film on simulated wine. *Vitis*: **20**, 36–42.

Falk, S. P., Gadoury, D. M., Pearson, R. C., Seem, R. C. (1995) Partial control of grape powdery mildew by the mycoparasite *Ampelomyces quisqualis*. *Plant Disease*: **79**, 483–90.

Flanzy, C., Flanzy, M., Bernard, P. (1987) *La Vinification par Macération Carbonique*. Institute National de la Recherche Agronomique, Paris.

Formisyn, P., Vaillant, H., Lantreibecq, F., Bourgois, J. (1997) Development of an enzymatic reactor for initiating malolactic fermentation in wine. *American Journal of Enology and Viticulture*: **48**, 345–51.

Fumi, M. D., Trioli, G., Colombi, M. G., Colagrande, O. (1988) Immobilization of *Saccharomyces cerevisiae* in calcium alginate gel and its application to bottle-fermented sparkling wine production. *American Journal of Enology and Viticulture*: **39**, 267–72.

Grando, M. S., Versini, G., Nicolini, G., Mattivi, F. (1993) Selective use of wine yeast strains having different volatile phenols production. *Vitis*: **32**, 43–50.

Guerzoni, M. E., Marchetti, R., Giudici, P. (1985) Modifications de composants aromatiques des vins obtenus par fermentation avec des mutants de *Saccharomyces cerevisiae*. *Bulletin de l'Office de la Vigne et du Vin*: **58**, 230–3.

Hallsworth, J. E. (1998) Ethanol-induced water stress in yeasts. *Journal of Fermententation and Bioengineering*: **85**, 125–37.

Heard, G. M. and Fleet, G. H. (1985) Growth of natural yeast flora during the fermentation of inoculated wines. *Applied and Environmental Microbiology*: **50**, 727–8.

Henick-Kling, T. (1995) Control of malo-lactic fermentation in wine: Energetics, flavour modification and methods of starter culture preparation. *Journal of Applied Bacteriology*: **79**, S29–S37.

Henick-Kling, T., Acree, T., Gavitt, B. K., Kreiger, S. A., Laurent, M. H. (1993) Sensory aspects of malolactic fermentation, in *Proceedings of the 8th Australian*

*Wine Industry Technical Conference* (eds C. S. Stockley, R. S. Johnstone, P. A. Leske, T. H. Lee). Winetitles, Adelaide, Australia, pp. 148–52.

Henick-Kling, T., Edinger, W., Daniel, P., Monk, P. (1998) Selective effects of sulfur dioxide and yeast starter culture addition on indigenous yeast populations and sensory characteristics of wine. *Journal of Applied Microbiology*: 84, 865–76.

Hilge-Rotmann, B. and Rehm, H.-J. (1990) Comparison of fermentation properties and specific enzyme activities of free and calcium-alginate entrapped *Saccharomyces cerevisiae*. *Applied Microbiology and Biotechnology*: 33, 54–8.

Holloway, P., Subden, R. E., Lachance, M. A. (1990) The yeasts in a Riesling must from the Niagara grape-growing region of Ontario. *Canadian Institute of Food Science Technology Journal*: 23, 212–16.

Howland, P. R., Pollnitz, A. P., Liacapoulos, D., McLean, H. J., Sefton, M. A. (1997) The location of 2,4,6-trichloroanisole in a batch of contaminated wine corks. *Australian Journal of Grape and Wine Research*: 3, 141–5.

Jackson, R. S. (2000) *Wine Science: Principles, Practice, Perception*, 2nd edn. Academic Press, San Diego.

Jacobs, C. J. and van Vuuren, H. J. J. (1991) Effects of different killer yeasts on wine fermentations. *American Journal of Enology and Viticulture*: 42, 295–300.

Jäeger, J., Diekmann, J., Lorenz, D., Jakob, L. (1996) Cork-borne bacteria and yeasts as potential producers of off-flavours in wine. *Australian Journal of Grape and Wine Research*: 2, 35–41.

Javelot, C., Girard, P., Colonna-Ceccaldi, B., Valdescu, B. (1991) Introduction of terpene-producing ability in a wine strain of *Saccharomyces cerevisiae*. *Journal of Biotechnology*: 21, 239–52.

Joyeux, A., Lafon-Lafourcade, S., Ribéreau-Gayon, P. (1984) Evolution of acetic acid bacteria during fermentation and storage of wine. *Applied and Environmental Microbiology*: 48, 153–6.

Kitamoto, K. (1989) Role of yeast vacuole in sake brewing (in Japanese). *Journal of the Brewing Society of Japan*: 84, 367–74.

Krisch, J. and Szajáni, B. (1997) Ethanol and acetic acid tolerance in free and immobilized cells of *Saccharomyces cerevisiae* and *Acetobacter aceti*. *Biotechnology Letters*: 19, 525–8.

Lao, C. L., López-Tamames, E., Lamuela-Raventós, R. M., Buxaderas, S., De la Torre-Boronat, M. C. (1997) Pectic enzyme treatment effects on quality of white grape musts and wines. *Journal of Food Science*: 62, 1142–4, 1149.

Laurent, M. H., Henick-Kling, T., Acree, T. E. (1994) Changes in the aroma and odor of Chardonnay due to malolactic fermentation. *Die Wein-Wissenschaft*: 49, 3–10.

Lefebvre, A., Riboulet, J. M., Boidron, J. N., Ribéreau-Gayon, P. (1983) Incidence des micro-organismes du liège sur les altérations olfactives du vin. *Science des Aliments*: 3, 265–78.

Lindemann, J. and Suslow, T. V. (1987) Competition between ice nucleation-active wild type and ice nucleation-deficient deletion mutant strains of *Pseudomonas syringae* and *P. fluorescens* Biovar 1 and biological control of frost injury on strawberry blossoms. *Phytopathology*: 77, 882–6.

Lubbers, S., Voilley, A., Feuillat, M. and Charpontier, C. (1994) Influence of mannoproteins from yeast on the aroma intensity of a model wine. *Lebensmittelwissenschaft und Technologie*: 27, 108–14.

Massoutier, C., Alexandre, H., Feuillat, M., Charpentier, C. (1998) Isolation and characterization of cryotolerant *Saccharomyces* strains. *Vitis*: **37**, 55–9.

Masuda, M., Okawa, E., Nishimura, K., Yunome, H. (1984) Identification of 4,5-dimethyl-3-hydroxy-2(5H)-furanone (Sotolon) and ethyl 9-hydroxynonanoate in botrytised wine and evaluation of the roles of compounds characteristic of it. *Agricultural and Biological Chemistry*: **48**, 2707–10.

Maujean, A., Millery, P., Lemaresquier, H. (1985) Explications biochimiques et métaboliques de la confusion entre goût de bouchon et goût de moisi. *Revue Française d'Oenologie*: **99**, 55–67.

McGovern, P. E., Glusker, D. L., Exner, L. J. and Voigt, M. M. (1996) Neolithic resinated wine. *Nature*: **381**, 480–1.

Millet, V., Vivas, N., Lonvaud-Funel, A. (1995) The development of the bacterial microflora in red wines during aging in barrels. *Journal des Sciences et Techniques de la Tonnellerie*: **1**, 137–50.

Mizoguchi, H. (1998) Permeability barrier of the yeast plasma membrane induced by ethanol. *Journal of Fermentation and Bioengineering*: **85**, 25–9.

Moreau, M. (1978) La mycoflore des bouchons de liège, son évolution au contact du vin: Conséquences possibles du métabolisme des moisissures. *Revue de Mycologie*: **42**, 155–89.

Mortimer, R. and Polsinelli, M. (1999) On the origins of wine yeast. *Research in Microbiology*: **150**, 199–204.

Murrell, W. G. and Rankine, B. C. (1979) Isolation and identification of a sporing *Bacillus* from bottled brandy. *American Journal of Enology and Viticulture*: **30**, 247–9.

Nelson, K. E. and Nightingale, M. S. (1959) Studies in the commercial production of natural sweet wines from botrytised grapes. *American Journal of Enology and Viticulture*: **9**, 123–5.

Nishino, H., Miyazakim S., Tohjo, K. (1985) Effect of osmotic pressure on the growth rate and fermentation activity of wine yeasts. *American Journal of Enology and Viticulture*: **36**, 170–4.

Ough, C. S. and Trioli, G. (1988) Urea removal from wine by an acid urease. *American Journal of Enology and Viticulture*: **39**, 303–6.

Parsel, D. A., Kowal, A. S., Singer, M. A., Lindquist, S. (1994) Protein disaggregation mediated by heat-shock protein HSP104. *Nature*: **372**, 475–8.

Pasteur, L. (1866) *Études sur le Vin*. Imprimerie Impériale, Paris.

Phaff, H. J. (1986) Ecology of yeasts with actual and potential value in biotechnology. *Microbial Ecology*: **12**, 31–42.

Pickering, G. J. (2001) Low- and reduced-alcohol wine: A review. *Journal of Wine Research*: **11**, 129–44.

Pickering, G. J., Heatherbell, D. A., Barnes, M. F. (1999) The production of reduced-alcohol wine using glucose oxidase-treated juice. Part III. Sensory. *American Journal of Enology and Viticulture*: **50**, 307–16.

Polsinelli, M., Romano, P., Suzzi, G., Mortimer, R. (1996) Multiple strains of *Saccharomyces cerevisiae* on a single grape vine. *Letters in Applied Microbiology*: **23**, 110–14.

Pretorius, I. S. (2000a) Tailoring wine yeast for the new millennium: novel approaches to the ancient art of wine making, in *Proceedings of the ASEV 50$^{th}$ Anniversary Annual Meeting Seattle, Washington* (ed. J. M. Rantz). American Society for Enology and Viticulture, Davis, California, pp. 261–70.

Pretorius, I. S. (2000b) Tailoring wine yeast for the new millennium: novel approaches to the ancient art of wine making. *Yeast*: **16**, 675–727.

Reed, G. and Nagodawithana, T. W. (1991) *Yeast Technology*, 2nd edn. Van Nostrand Reinhold, New York.

Rose, A. H. (1989) Influence of the environment on microbial lipid composition, in *Microbial Lipids* (eds C. Ratledge, S. G. Wilkinson). Volume 2, Academic Press, London, pp. 255–78.

Rosini, G. (1986) Wine-making by cell-recycle-batch fermentation process. *Applied Microbiology and Biotechnology*: 24, 140–3.

Schütz, M. and Gafner, J. (1993) Analysis of yeast diversity during spontaneous and induced alcoholic fermentations. *Journal of Applied Bacteriology*: 75, 551–8.

Sponholz, W. R. and Hühn, T. (1996) Aging of wine: 1,1,6-trimethyl-1,2-dihydronaphthalene (TDN) and 2-aminoacetophenone, in *Proceedings of the 4th International Symposium on Cool Climate Viticulture and Enology* (eds T. Henick-Kling *et al.*). New York State Agricultural Experimental Station, Geneva, New York, pp. VI-37–57.

Sulo, P., Michačáková, S., Reiser, V. (1992) Construction and properties of $K_1$ type killer wine yeast. *Biotechnology Letters*: 14, 55–60.

Suzzi, G., Romano, P., Vannini, L., Turbanti, L., Domizio, P. (1996) Cell-recycle batch fermentation using immobilized cells of flocculent *Saccharomyces cerevisiae* wine strains. *World Journal of Microbiology and Biotechnology*: 12, 25–7.

Teunissen, A. W. R. H. and Steensma, H. Y. (1995) The dominant flocculation genes of *Saccharomyces cerevisiae* constitute a new subtelomeric gene family. *Yeast*: 11, 1001–13.

Usseglio-Tomasset, L., Bosia, P. D., Delfini, C., Ciolfi, G. (1980) I vini Recioto e Amarone della Valpolicella. *Vini d'Italia*: 22, 85–97.

Van Rensburg, P. and Pretorius, I. S. (2000) Enzymes in winemaking: Harnessing natural catalysts for efficient biotransformations. *South African Journal of Enology and Viticulture*: 21, 52–73.

Vaughn, R. H. (1955) Bacterial spoilage of wines with special reference to California conditions. *Advances in Food Research*: 7, 67–109.

Villettaz, J. C. (1987) A new method for the production of low alcohol wines and better balanced wines, in *Proceedings of the 6th Australian Wine Industry Technical Conference* (ed. T. Lee). Australian Industrial Publishers, Adelaide, Australia, pp. 125–8.

Watanabe, M. and Shimazu, Y. (1976) Application of *Botrytis cinerea* for wine making. *Journal of Fermentation Technology*: 54, 471–8.

Zoecklein, B. W., Marcy, J. E., Jasinski, Y. (1997) Effect of fermentation, storage *sur lie* or post-fermentation thermal processing on White Riesling (*Vitis vinifera* L.) glycoconjugates. *American Journal of Enology and Viticulture*: 48, 397–402.

# 11 The identity and parentage of wine grapes

*R. M. Pinder and C. P. Meredith*

## Introduction

There are thousands of grapevine varieties. Traditional methods of identification include appearance, a technique known as ampelography that can easily be applied to some varieties such as Cabernet Sauvignon with its distinctive leaves. Successful ampelography depends upon an extensive knowledge and familiarity with the multitude of grape varieties, and upon the maintenance of stable distinguishing characteristics within the particular variety. Even expert ampelographers usually specialize in the varieties of one geographic region since no one person is able to identify all grape varieties. However, many varieties look alike, while the same variety may look quite different when grown in diverse locations, when diseased, or at various stages of the growing season. Fortunately, the world's great wines are produced from a relatively small number of classic European cultivars of *Vitis vinifera* L, but even this has not prevented a long history of speculation, debate and dispute about the identity and origin of varieties. This is particularly so in the New World, where virtually all grape varieties have been introduced from various parts of Europe, often acquiring new names in their new locations, names that have sometimes become confused and switched between varieties. Accurate identification is important not only to growers, vineyard owners and winemakers, but also to the marketing of wines since wine labelling laws and international trade regulations now demand that varietals be correctly identified.

The past decade has witnessed the addition of the science of DNA technology to the art of ampelography, helping to pin down the identification of varieties, their origin and their parentage (Meredith 2000; Sefc *et al.* 2001). This technology is similar to DNA profiling in humans, long used in forensic criminology and to establish genetic family relationships. Unlike its appearance, the DNA profile of a grapevine is stable and does not change when the vine is grown in different locations, when it is diseased or during the growing season. Not only is the identification of grapevines more objective, but DNA profiling can establish the original parents of a grape variety just as it can prove paternity in humans. DNA profiling is particularly suitable for grapevines because they are propagated not by seeds but vegetatively by cuttings or buds, resulting in a genetic identity for all individual vines within a particular variety. The original seed that began the variety was derived from a pollen parent and an egg parent. If those parents were cultivated and not wild varieties and if they are still in existence, DNA analysis can identify them. DNA profiling targets very specific and small regions of the DNA of grapevines, termed microsatellite markers, which exist in alternative forms in different varieties. A unique DNA profile for each grape variety can be produced using about six to eight markers, although for reliable parentage studies within closely related species at least 25 markers may be needed.

Eventually, it may even be possible to construct a sort of family tree, which will identify the parentage and origins of most of the major grape varieties used in making wine. DNA profiling has already produced some surprising findings (Meredith 2000):

- Cabernet Sauvignon is the offspring of Sauvignon Blanc and Cabernet Franc. It probably arose as a a chance cross-pollination in a vineyard in western France several centuries ago (Bowers and Meredith 1997).
- Virtually all of the varieties of north-eastern France, including such luminaries as Chardonnay, lesser lights like Gamay Noir and Aligoté, but also Melon de Bourgogne, Auxerrois and several others, were the offspring of two parents, Pinot, probably Pinot Noir for many varieties, and Gouais Blanc (Bowers *et al.* 1999a). Pinot is an ancient and noble variety, but Gouais Blanc, originating in Eastern Europe and once widespread in north-eastern France, was considered so mediocre in quality that several times it was banned.
- The Petite Sirah of California is almost entirely the French variety Durif, although some is an older and very similar French variety called Peloursin (Meredith *et al.* 1999). Durif originated as a seedling from a cross between Peloursin and Syrah, while Syrah itself is the chance offspring of two obscure grapes from south-eastern France, the red Dureza and Mondeuse Blanche. The Australian Shiraz is the same variety as the French Syrah.
- The Californian red variety Zinfandel is the same variety as the Primitivo of southern Italy but is different from its long suspected double the Plavac Mali from Dalmatia (Meredith 2000).
- Many old Pinot Blanc vines in California are actually the very similar French variety Melon de Bourgogne, which is used to make the wine Muscadet (Meredith 2000).
- Much of the Merlot grown in Chile is actually the old Bordeaux variety Carmenère (Meredith 2000).
- Many synonyms have been confirmed. For example, the Californian grapes Mataro, Valdepeñas and Black Malvoisie are indeed the European varieties Mourvèdre, Tempranillo and Cinsaut, respectively, while the Portuguese cultivar Moscatel de Setúbal is identical to the widely grown Muscat of Alexandria (Sefc *et al.* 2001).

## DNA profiling

The development of techniques for the molecular characterization of organisms has mushroomed in the past 20 years (Karp *et al.* 1998). Many of them have been applied to the differentiation of grapevine varieties, including isozyme analysis and several DNA-based methods (Sefc *et al.* 2001). None of the methodologies was ideal for various reasons, including some based upon DNA markers. Nevertheless, although the existence of simple sequence repeats (SSRs), also called microsatellite repeats, of nucleotides in plant nuclear DNA has been known for two decades (Delseny *et al.* 1983), and their significance as a major source of genetic variation was rapidly appreciated (Tautz *et al.* 1986) they did not become useful tools for grapevine genetics until the advent of polymerase chain reaction (PCR) technology, which allowed them to be targeted as single genetic loci. Microsatellites show the lowest degree of repetition of the three classes of repetitive DNA (Tautz 1993), and a typical microsatellite sequence consists of from five to about 100 tandem repeats of short and simple sequence motifs comprising one to six nucleotides, e.g. $(CA)_n$ or $(GATA)_n$. PCR-amplified microsatellite markers have the advantages of being locus-specific and highly polymorphic. Their high degree of reproducibility, coupled with profiles that are represented by allele sizes detected at the analysed loci and expressed in base pairs rather than as bands on an electrophoresis gel, means that genotype information can be expressed clearly and objectively. Results can therefore be exchanged between different laboratories as numbers rather than images. From a practical point of view, with appropriate controls to adjust for any possible small methodological differences between laboratories, microsatellite genotype information can be compared and grape varieties identified without having either actual plants or DNA samples of reference vines.

The pioneering investigations of the use of repetitive DNA sequences, including microsatellite markers, to identify grape varieties were performed in the laboratories of CSIRO in Australia by Thomas and colleagues (1993). They demonstrated an abundance of microsatellite sequences in grapevines that were suitable for identifying *Vitis vinifera* cultivars, while primer sequences for the polymorphisms were conserved across other *Vitis* species (Thomas and Scott 1993). Most importantly, microsatellite alleles were inherited in a co-dominant Mendelian fashion, confirming their suitability for genetic mapping and their potential use for identification of genetic relatedness (Thomas *et al.* 1994). The Australian group began with 26 *V. vinifera* cultivars (Thomas and Scott 1993), which they subsequently extended to 80 genotypes including rootstocks, wine grapes and table grapes (Thomas *et al.* 1994). The CSIRO currently maintains a database of DNA microsatellite profiles of about 200 genotypes.

This Australian beginning was quickly followed by the development of additional grapevine microsatellite markers around the world. The method was heavy on resources and time, since genomic libraries were constructed from a grapevine or rootstock cultivar, the library was screened with microsatellite probes, the microsatellite-containing clones were sequenced, PCR primers were designed from the sequences flanking the microsatellite, and finally the PCR conditions were optimized to characterize the particular microsatellite polymorphism. About 40 additional markers were developed in three different laboratories using the method (Thomas *et al.* 1993; Bowers *et al.* 1996, 1999b; Sefc *et al.* 1999), with a shortened procedure providing additional useful markers (Thomas and Scott 1994). In order to develop sufficient markers to map the *Vitis* genome more rapidly and make the technology useful to the viticulture community, the *Vitis* Microsatellite Consortium (VMC) was established in 1997, consisting of the private French company Agrogene and 21 research laboratories worldwide. To date, 333 new *Vitis* markers have been identified from a microsatellite-enriched genomic library, and the VMC has produced some 700 unique DNA sequences of the *Vitis vinifera* genome (Sefc *et al.* 2001). Additionally, all published primer sequences for grapevine SSR markers are available at the Greek *Vitis* database (http://www.biology.uoc.gr/gvd), while other databases are being developed (Sefc *et al.* 2001).

The first generation of grapevine SSR markers relied on the laborious screening of sequences selected from enriched genomic DNA libraries. More recently, expressed sequence tags (ESTs) have become available and have served as a rich source of microsatellite markers of any repeat type and motif. EST-derived markers are associated with the gene-rich regions of the genome and may prove very useful for mapping the *Vitis* genome. A large database of ESTs for grapevines has been reported (Ablett *et al.* 1998) and has been screened for microsatellite markers with the identification of more than 100 new microsatellites (Scott *et al.* 2000). Microsatellite markers, however they are derived, are highly polymorphic. Theoretically, five unlinked markers each with five equally frequent alleles could produce over 700 000 different genotypes (Bowers *et al.* 1996). In reality, however, in order to minimize the number of markers necessary for reliable differentiation and identification of cultivars, the most informative loci are selected (Sefc *et al.* 2001). A set of six specific markers has been agreed upon for international information exchange purposes, and is generally regarded as sufficient to differentiate most wine grape cultivars but not necessarily all members of *Vitis vinifera* such as seedless table grapes, which are highly interrelated (Crespan *et al.* 1999).

## Applications of DNA profiling

Grapevine genetics is now big business (Meredith 2001) and commercial certification of cultivars using DNA microsatellite profiles is a reality (Sefc *et al.* 2001). Genes that control important traits in grapes are being isolated and identified, while methods to introduce genes, either from other grapevines or from completely different plants and organisms, are well established. Targeted modification of existing cultivars has already taken place, focused particularly on improving resistance to viral diseases and fungal pathogens, but also on fruit quality. The wine attributes of classic cultivars have not so far been addressed. International and national controls on wine quality and composition will play important roles in the commercialization of genetically-modified grapevines. These are not topics that will be addressed here since other and more comprehensive sources are available (Meredith 2001; Roubelakis-Angelakis 2001). We will focus on the identification of wine grapes and their parentage.

### *Synonyms*

As grapevines have traveled around the world, with European varieties being introduced into the New World, their names have inevitably undergone local changes. Single varieties with different names are legion in the world of wine grapes (Sefc *et al.* 2001). Nevertheless, the particular *terroir* of a region – that unique combination of soil, climate, vineyard location and aspect – as well as the vineyard husbandry and winemaking technique will determine the nature of the wines even though they may contain the same cultivars. Many examples of synonymy can be found in California. In some cases a distinctly Californian name is used for a well-known European variety (Zinfandel and Primitivo, Black Malvoisie and Cinsaut), while in others a lesser-known European name is used instead (Mataro and Mourvèdre, Valdepeñas and Tempranillo). In the foregoing cases, the synonymy was already suspected or assumed on the basis of ampelographic observations, which were merely confirmed by microsatellite analysis. However, there have also been some cases of true confusion as for Californian Pinot Blanc and Chilean Merlot, much of which are actually Melon de Bourgogne and Carmenère, respectively (Meredith 2000).

Even within Europe many local names for grapevines have been shown by microsatellite analysis to be identical (Sefc *et al.* 2001). Confirmation of previous ampelographic suspicions was obtained for the Italian cultivars Refosco di Faedis and Refoscone (Cipriani *et al.* 1994) as well as for Favorita, Pigato and Vermentino (Botta *et al.* 1995). Similar confirmation was reported for the Austrian cultivar known as Morillon, which had long been suspected to be a synonym of Chardonnay (Sefc *et al.* 1998a). DNA profiling also confirmed the identity of the Croatian cultivars Plavina and Brajdica (Maletic *et al.* 1999). Many Portuguese cultivars previously suspected to be synonymous were verified in the same manner: in addition to the already mentioned identity of Moscatel de Setúbal with the widely grown Muscat of Alexandria – also shown to be the same as the Greek variety Moschato Alexandreias (Lefort *et al.* 2000) – previous assumptions regarding a number of other important cultivars were confirmed including several that are used in the making of some styles of Madeira, such as Boal Cachudo/Boal da Madeira/Malvasia Fina and Verdelho dos Açores/Verdelho roxo/Verdelho da Madeira together with others like Periquita/Castelão Francês/João de Santarém/

Trincadeira found in several of Portugal's top red wines (Lopes *et al.* 1999). In some cases, DNA profiling is at odds with supposed synonymy, as in the case of the Italian cultivar Croatina and the Croatian variety Hrvatica (both names translate as 'Croatian girl'), where microsatellite markers differed at several loci (Maletic *et al.* 1999).

Unexpected synonyms, where ampelographers did not suspect the identity of cultivars, are also common. DNA profiling has shown that the Croatian varieties Teran Bijeli and Muscat Ruza Porecki, grown in the province of Istria, are identical to the north Italian cultivars Prosecco and Rosenmuskateller respectively, while the Croatian variety Moslavac turned out to be identical to the Hungarian cultivar Furmint (Maletic *et al.* 1999). Subsequent ampelographic studies prompted by these results supported the DNA findings, which was also the case for the synonymy between the Greek cultivars Moschata Mazas and Moschato Kerkyras (Lefort *et al.* 2000). Within the context of a much larger Italian study of local biodiversity, Grando and colleagues (2000) demonstrated the identity of the Trentino grapevines Vernaccia Nera and Francesa Nera with the better known French varietals from Bordeaux, Merlot and Carmenère, respectively.

Finally, it appears that some cultivars reported as synonyms are in reality closely related plants. Seedlings from self-pollinated vines of Sangiovese, the Italian red grape varietal famous for its use in many classic Tuscan wines including Chianti and Vino Nobile di Montepulciano as well as most of the so-called super-Tuscan wines, are difficult to differentiate phenotypically from the mother plant but are genetically distinct according to microsatellite markers (Filippetti *et al.* 1999). Although only produced for research purposes, such seedlings suggest that it is possible that some grapevine cultivars may consist of more than one genotype as a result of mixed plantings in the past and the propagation of closely related plants that were phenotypically similar to each other.

## *Parentage and pedigree*

The grapevines of today have arisen in various ways over many centuries. Some of them are the result of the domestication of wild vines, firstly in the ancient sites of the Middle East and only later in the vine growing regions of Europe, while others are the results of spontaneous crosses between wild vines and cultivars or of crosses between two cultivars. The arrival of DNA profiling has served to satisfy much of the long-standing curiosity about the historical origins of modern grapevine varieties. Identifying the parentage of modern grapevines is possible if the parents were themselves cultivars and still exist in cultivation or collections. However, the ancient wild vines involved in the original crosses no longer exist and are therefore not available for molecular analysis. Recent pedigree studies have included the two great wine grapes of the world, Cabernet Sauvignon (Bowers and Meredith 1997) and Chardonnay (Bowers *et al.* 1999a), as well as other cultivars from

*Table 11.1* Microsatellite alleles in Zweigelt and its parents, St Laurent and Blaufränkisch (data from Sefc *et al.* 1997)*

| *Locus* | *St Laurent* | *Zweigelt* | *Blaufränkisch* |
|---|---|---|---|
| ssrVrZAG 7[†] | 157:157 | 155:157 | 155:155 |
| ssrVrZAG 15 | 175:177 | 165:175 | 165:165 |
| ssrVrZAG 21 | 200:206 | 202:206 | 202:206 |
| ssrVrZAG 25 | 225:236 | 225:236 | 225:225 |
| ssrVrZAG 30 | 149:151 | 147:151 | 147:149 |
| ssrVrZAG 47 | 163:167 | 157:163 | 157:172 |
| ssrVrZAG 64 | 139:163 | 139:159 | 139:159 |
| ssrVrZAG 67 | 126:152 | 126:139 | 139:149 |
| ssrVrZAG 79 | 238:246 | 236:238 | 236:250 |

* Numbers refer to the allele sizes in each base pair.
† ssr = simple sequence repeat.

France (Bowers *et al.* 1999a), Portugal (Lopes *et al.* 1999), and Central Europe (Sefc *et al.* 1998b).

Of all the molecular methods available for pedigree studies the method of choice has been microsatellite marker analysis. Microsatellite markers are transmitted in a co-dominant Mendelian manner, whereby each of the parents in a cross contributes one allele per locus to the offspring. A prime example in modern times is the Austrian red wine grape variety Zweigelt, which is a controlled cross between the older cultivars St Laurent and Blaufränkisch (Sefc *et al.* 1997). At each microsatellite locus, one allele in Zweigelt is inherited from each of the parents (Table 11.1). However, such prior knowledge of the parents is rare in grapevine genetics, and there is also usually little information on the chronological order of appearance of the different cultivars. In contrast to human parentage analysis, it is therefore necessary to include data from a large number of unlinked loci in order to be confident about identification. When microsatellite markers of the candidate offspring are compared within all possible sets of three cultivars to identify pairs that could have contributed to its alleles, even 10 markers may result in false positives. Analysis of at least 25 markers is recommended for reliable pedigree analysis in closely related cultivars (Sefc *et al.* 2001).

### *Cabernet Sauvignon*

The pioneering study of grapevine parentage also produced the most unexpected result (Bowers and Meredith 1997): Cabernet Sauvignon, long thought to be related because of its morphological similarities to the red wine grape of Bordeaux and the Loire valley, Cabernet Franc, had a surprising second parent in the shape of a white wine grape from precisely the same regions of France, the Sauvignon Blanc. Although the name Cabernet Sauvignon combines elements of both names and there is a more superficial morphological resemblance between Sauvignon Blanc and Cabernet Sauvignon than between the two Cabernet cultivars, it is most likely that the similarity in names comes from their resemblance to wild vines, Sauvignon being derived from the French word '*sauvage*', meaning 'wild'. All three cultivars are used to make classic Bordeaux; Cabernet Sauvignon mostly for Médoc and Graves, Cabernet Franc mainly for right bank wines such as St Emilion, and Sauvignon Blanc in both dry and sweet Bordeaux whites including Barsac and Sauternes. Both Cabernet Franc and Sauvignon Blanc have also long been associated with the Loire valley red and white wines such as Chinon, Sancerre and Pouilly-Fumé.

Bowers and Meredith (1997), at the University of California at Davis, applied the microsatellite marker method at 30 polymorphic loci to 51 European cultivars including most of the major wine grape varieties. No pair other than Cabernet Franc and Sauvignon Blanc could have been the parents, with the next nearest possibilities being excluded by at least three loci. The cumulative likelihood ratio of the probability of the observed Cabernet Sauvignon alleles if Cabernet Franc and Sauvignon Blanc were the true parents versus the probability of those same alleles if two random cultivars were the actual parents was greater than $10^{14}$. Furthermore, the likelihood of the Cabernet Sauvignon alleles was much higher if Cabernet Franc and Sauvignon Blanc were the true parents rather than their close relatives in the form of parents, full siblings or offspring. Cabernet Franc and Sauvignon Blanc are genetically dissimilar, sharing only 12 of 56 alleles at 28 loci. The microsatellite marker findings were confirmed by the use of isozyme data as well as information from other DNA methods like amplified restriction fragment polymorphism (AFLP) and restriction fragment length polymorphism (RFLP).

All three varieties, both parents and offspring, have been cultivated in Bordeaux since at least the seventeenth century, while Cabernet Franc appears to have been grown for considerably longer. The cross between Cabernet Franc and Sauvignon Blanc is unlikely to have been deliberate; there is no rationale, and it probably took place no later than the seventeenth century, which precedes the earliest reports of deliberate plant crossing. The most likely explanation is a spontaneous cross between vines in adjacent vineyards or even the same vineyard, since red and white cultivars were often grown next to each other. Cabernet Sauvignon seems to have developed from a single seedling, based upon identical microsatellite genotypes for clones (Bowers and Meredith 1997), while chloroplast SSR markers have indicated that Sauvignon Blanc was the chloroplast donor and therefore the female partner in the cross (Sefc *et al.* 2001).

### *Chardonnay and other grapes of north-eastern France*

Further surprises were in store when the pedigree of the great French white wine grape Chardonnay was explored. Chardonnay, and its noble red equivalent Pinot Noir, are used to make the classic wines of Burgundy and

*Table 11.2* The 16 progeny cultivars of north-eastern France, all being full siblings from the parents Gouais Blanc and Pinot (Bowers *et al.* 1999a)

| | |
|---|---|
| Aligoté | Gamay blanc Gloriod |
| Aubin vert | Gamay noir |
| Auxerrois | Knipperlé |
| Bachet noir | Melon de Bourgogne |
| Beaunoir | Peurion |
| Chardonnay | Romorantin |
| Dameron | Roublot |
| Franc noir de la Haute Saône | Sacy |

Champagne, and like the Cabernet Sauvignon have been exported to many of the world's wine regions. Other, less prestigious, grapes, such as Aligoté for white and Gamay Noir for red, are used in the Beaujolais and other parts of southern Burgundy, while Melon de Bourgogne, despite its name, is the basis for the western Loire valley white wine Muscadet.

Microsatellite analysis studies by an American group in Davis, with collaboration from a French group in Montpellier, performed on 322 grapevine varieties grown in France, revealed that 16 of them are full siblings, all the progeny of a single pair of parents, Pinot and Gouais Blanc (Bowers *et al.* 1999a). The progeny varieties encompass all of those grown in north-eastern France today and include such luminaries as Chardonnay, lesser lights like Gamay Noir and Aligoté, but also Melon de Bourgogne, Auxerrois and several other cultivars used in various regional appellations (Table 11.2). Many of these varieties were first thought to have originated outside France and to have been introduced from the Middle East, Dalmatia and Italy, while others were incorrectly suspected to be close relatives for morphological reasons. Both parental varieties were once widely grown in north-eastern France, and Pinot, in its various guises as blanc, gris, meunier and noir, remains a major cultivar in modern French viticulture. Pinot Noir, Pinot Gris and Pinot Blanc, while treated as discrete varieties by winemakers because of their different fruit colours, are simply colour mutants of the same variety and all share an identical microsatellite genotype, indicating that they originated from a single individual seedling. Meunier, the third grape variety used in Champagne with Chardonnay and Pinot Noir, is also considered a discrete variety by winemakers but is actually a chimeric mutant of Pinot.

It is therefore very plausible that the highly valued Pinot should be the parent of successful offspring cultivars. It has been in Burgundy since probably before the time of the Roman conquest of Gaul, and it has a long history in other parts of north-eastern France. However, the humble Gouais Blanc as a parent of quality wine varieties is entirely unexpected since it has always been regarded as mediocre and, although widespread in the Middle Ages, was banned at various times and in particular regions. It exists today only in cultivar collections, with the exception of small plantings in the Valais region of Switzerland, and is believed to be an eastern European variety also called Heunisch weiss, which may have been brought to France by the Roman emperor Probus as a gift to the Gauls from his homeland of Dalmatia.

Samples of 51 cultivars were taken from vineyards at the University of California at Davis, and the remaining 271 from the collection of the Institut National de la Recherche Agronomique at Domaine de Vascal near Montpellier (Bowers *et al.* 1999a). After an initial screening of all cultivars at 17 microsatellite loci, microsatellite alleles were compared within all possible sets of three in order to identify pairs of cultivars that could have contributed the alleles of the third variety. A subset of cultivars was further analysed at 15 additional microsatellite loci. On the basis of these 32 loci, 16 cultivars had microsatellite alleles consistent with their being the progeny of a single pair of parents, Pinot and Gouais Blanc, with likelihood ratios $10^{12}$ to $10^{15}$ times more probable than for two random parents and 447 to 28 000 times more probable than full relatives of either Pinot or Gouais Blanc. The parents are genetically dissimilar, sharing only 20 of 64 alleles at the 32 loci, consistent with an eastern European origin for Gouais Blanc. The large number of successful progeny may reflect the genetic distance of the parents, although this has not been the case with Cabernet Franc and Sauvignon Blanc, which also have a large genetic distance.

It is likely that separate crossings produced the progeny at different times and places. Both parents were widespread from the Loire valley to Champagne, Alsace and Burgundy, and the progeny are all historically associated with north-eastern France. However, both Pinot and Gouais Blanc were grown in other parts of Europe and crosses may have occurred in places other than France. Many of the crosses are very old – Chardonnay, Gamay Noir and Melon de Bourgogne date from the early Middle Ages – while the others have been mentioned in literature from 100 to 400 years ago. Nine of the progeny cultivars have light-coloured fruit, four are blue-black or violet-black, and two have intermediate pink fruit. Gouais Blanc has yellow-gold berries. The parent of the dark-berried cultivars – Bachet Noir, Beaunoir, Franc Noir de la Haute Saône and Gamay Noir – must have been Pinot Noir. However, Pinot Noir is heterozygous for berry colour, so it is impossible to be certain which of the three Pinots – Blanc, Gris or Noir – was the parent in the other cases. Pinot Noir has always been the most common form in France, so it is the more probable candidate.

### *Portuguese cultivar pedigrees*

Portuguese vineyards are rich in grapevine cultivars that do not occur elsewhere, although they do contain some that are identical to their Spanish counterparts, while modern popular varieties like Chardonnay and Cabernet Sauvignon are rapidly being introduced. The microsatellite profiles of a large Portuguese collection have been searched not only for synonyms (see pp. 264–5) but also for pedigree associations (Lopes *et al.* 1999). Boal Ratinho is a white grapevine from the Carcavelos region west of Lisbon and is one of several used in the classic fortified sweet wine of the region. It seems to be the progeny of a cross between Malvasia Fina, also known in other parts of Portugal as Boal Cachudo or Boal da Madeira, and Síria, which is also called Roupeiro or Crato Branco. Both Boal Ratinho and its parents have been grown in Portugal since ancient times, and the cross must have occurred spontaneously in vineyards where the Malvasia Fina and Síria were grown next to each other.

### *Cultivars from Central Europe*

The parentage of many central European cultivars is ancient and complicated, but recent genetic profiling studies have allowed the reconstruction of a four-generation pedigree detailing the close family relationships between nine varieties (Sefc *et al.* 1998b). Microsatellite profiles have rejected some historical associations and have also identified an improbable parent for one of the principal modern cultivars used in winemaking. Traminer, an ancient and noble grape first mentioned in the fourteenth century and still used in modern winemaking both in central Europe and elsewhere in the world, and Österreichisch Weiss, an ancient variety of no economic importance in today's wine world, are the parents of Silvaner. Now widely grown in Alsace, Germany and Austria, Silvaner was historically regarded as the product of a wild vine first selected from the banks of the river Danube. It now seems likely that it emerged as a result of an ancient spontaneous cross between Traminer and Österreichisch Weiss in a vineyard in the eastern part of Austria.

Traminer is also one parent of the Austrian cultivar Rotgipfler in a cross with Roter Veltliner. Two further crosses of Roter Veltliner with Silvaner led to the cultivars Neuburger and Frühroter Veltliner. Neuberger was originally thought to be the progeny of a natural cross between Pinot Blanc and Silvaner, but the parental role of Pinot Blanc was definitively ruled out by the microsatellite profiles. Surprisingly, the same parents, Silvaner and Roter Veltliner, also gave rise to a second cultivar, Frühroter Veltliner, which has an additional synonym of Malvasier.

On a final note, the ancient and mediocre variety Österreichisch Weiss, identified as a parent of Silvaner and a grandparent of Neuburger and Frühroter Veltliner, is also great-grandparent to a white wine grape variety Jubiläumsrebe which was selected in the 1920s for the production of dessert wine. Originally regarded as a cross between Blauer Portugieser and Blaufränkisch, Jubiläumsreber was shown by microsatellite analysis to be the offspring of Grauer Portugieser and Frühroter Veltliner. There are therefore some similarities between Österreichisch Weiss and Gouais Blanc, in that both are ancient and mediocre varieties that have contributed to the pedigree of many of Europe's prime wine-producing varietals: truly an odd couple.

*Petite Sirah*

The Californian red wine cultivar known as Petite Sirah has long suffered the criticism that its name is a gross misnomer as it is quite unrelated to the noble French Rhône varietal Syrah, although in the right hands excellent Rhône-style wines can be produced. Microsatellite studies at 25 loci demonstrated that the vast majority of vines grown under the name of Petite Sirah are identical to the southern French cultivar Durif, although some were identical to the related cultivar Peloursin (Meredith *et al.* 1999). Durif is morphologically very similar to Peloursin, and has been reported to be either a seedling or a selection of Peloursin produced in 1880 in France. Although neither Durif nor Peloursin are highly regarded in France, and have in fact fallen virtually into oblivion as far as winemaking is concerned, Durif is now identified as a result of the microsatellite analysis to be the progeny of the noble Syrah and the humble Peloursin. Syrah itself also has humble origins, being the result of a spontaneous cross between two cultivars of little renown, the red Dureza and the white Mondeuse Blanche from south-eastern France. However, it is indeed a parent of Petite Sirah.

## Concluding remarks

These are exciting times for those interested in the history of oenology and viticulture. The use of DNA profiling on the basis of microsatellite markers has revolutionized the field in less than a decade. Family pedigrees of grapevines can be established, synonyms identified and previous ampelographic or historical assignments of identity or parentage confirmed or denied. Authentication and certification of cultivars and wine musts is possible, while the expanding world of grapevine genetics opens up a plethora of possibilities, both good and bad depending upon one's views, on genetically modified organisms. Even the good – the introduction of genes from other grape varieties or plants in an effort to reduce disease loss and pesticide usage – may not be accepted by national and international bodies responsible for control of wine quality let alone by the general public. There is always the temptation to tinker with the wine attributes of a cultivar or even to make entirely new varieties by genetic engineering rather than traditional crossing.

Genetic linkage maps for the grapevine are being honed, and we can eventually expect to see a family tree that will identify the pedigree of the major wine grape varieties. While no other method or technology, including ampelography, has provided so much useful information about grapevine genetics in so short a period of time as DNA microsatellite markers, other markers have already been characterized and their potential is under scrutiny. Microsatellite markers are not always sufficiently sensitive, especially when identifying closely related cultivars or somatic mutants.

International collaborations will continue to play a major role in the future, especially in genetic mapping studies. Eventually, all of the grapevine genetic resources will have been analysed and most of the data on microsatellite alleles will be available through a network of web-based databases specific to national germplasm collections. Markers will be sought that are closely linked to useful genes to guide breeding studies and to permit isolation of those genes using map-based cloning strategies. Just a decade ago the grapevine was considered a poor genetic organism, but the door to the grapevine genome is now open.

## References

Ablett, E. M., Lee, L. S., Henry, R. J. (1998) Analysis of grape ESTs. *Acta Horticulturae*: **528**, 273–5.

Botta, R., Scott, N. S., Eynard, I., Thomas, M. R. (1995) Evaluation of microsatellite sequence-tagged site markers for characterizing *Vitis vinifera* cultivars. *Vitis*: **34**, 99–102.

Bowers, J. E. and Meredith, C. P. (1997) The parentage of a classic wine grape, Cabernet Sauvignon. *Nature Genetics*: **16**, 84–7.

Bowers J. E., Dangl, G. S., Vignani, R., Meredith, C. P. (1996) Isolation and characterization of new polymorphic simple sequence repeat loci in grape (*Vitis vinifera* L.). *Genome*: **39**, 628–33.

Bowers, J., Boursiquot, J.-M., This, P., Chu, K., Johansson, H., Meredith, C. (1999a) Historical genetics: The parentage of chardonnay, gamay, and other wine grapes of northeastern France. *Science*: **285**, 1562–5.

Bowers, J. E., Dangl, G. S., Meredith, C. P. (1999b) Development and characterization of additional microsatellite DNA markers for grape. *American Journal of Enology and Viticulture*: **50**, 243–6.

Cipriani, G., Frazza, G., Peterlunger, E., Testolin, R. (1994) Grapevine fingerprinting using microsatellite repeats. *Vitis*: **33**, 211–15.

Crespan, M., Botta, R., Milani, N. (1999) Molecular characterization of twenty seeded and seedless table grape cultivars (*Vitis vinifera* L.). *Vitis*: **38**, 87–92.

Delseny, M., Laroche, M., Penon, P. (1983) Detection of sequences with Z-DNA forming potential in higher plants. *Biochemical and Biophysical Research Communications*: **116**, 113–20.

Filippetti, I., Silvestroni, O., Thomas, M. R., Intrieri, C. (1999) Diversity assessment of seedlings from self-pollinated Sangiovese grapevines by ampelography and microsatellite DNA analysis. *Vitis*: **38**, 67–71.

Grando, M. S., Frisinghelli, C., Stefanini, M. (2000) Genotyping of local grapevine germplasm. *Acta Horticulturae*: **528**, 183–7.

Karp, A., Ingram, D. S., Isaac, P. (eds) (1998) *Molecular Tools for Screening Biodiversity*. Chapman and Hall, London.

Lefort, F., Anzidei, M., Roubelakis-Angelakis, K. A. and Vendramin, G. G. (2000) Microsatellite profiling of the Greek Muscat cultivars with nuclear and chloroplast SSR markers. *Quaderni della Scuola di Specializzazione in Scienze Viticole ed Enologiche*: **23**, 56–80.

Lopes, M. S., Sefc, K. M., Eiras Dias, E., Steinkellner, H., Laimer Da Câmara Machado, M., Da Câmara Machado, A. (1999) The use of microsatellites for germplasm management in a Portuguese grapevine collection. *Theoretical and Applied Genetics*: **99**, 733–9.

Maletic, E., Sefc, K. M., Steinkellner, H., Kontic, J. K., Pejic, I. (1999) Genetic characterization of Croatian grapevine cultivars and the detection of synonymous cultivars in neighboring regions. *Vitis*: **38**, 79–83.

Meredith, C. (2000) North American geneticists untangle the vine variety web, in *The Oxford Companion to the Wines of North America* (ed. B. Cass). Oxford University Press, Oxford, pp. 57–9.

Meredith, C. P. (2001) Grapevine genetics: Probing the past and facing the future. *Agriculturae Conspectus Scientificus*: **66**, 21–5.

Meredith, C. P., Bowers, J. E., Riaz, S., Handley, V., Bandman, E. B., Dangl, G. S. (1999) The identity and parentage of the variety known in California as Petite Sirah. *American Journal of Enology and Viticulture*: **50**, 236–42.

Roubelakis-Angelakis, K. A. (ed.) (2001) *Molecular Biology and Biotechnology of Grapevine*. Kluwer, Amsterdam.

Scott, K. D., Eggler, P., Seaton, G., Rosseto, M., Ablett, E. M., Lee, L. S., Henry, R. J. (2000) Analysis of SSRs derived from grape ESTs. *Theoretical and Applied Genetics*: **100**, 723–6.

Sefc, K. M., Steinkellner, H., Wagner, H. W., Glössl, J., Regner, F. (1997) Application of microsatellite markers to parentage studies in grapevines. *Vitis*: **36**, 179–83.

Sefc, K. M., Regner, F., Glössl, J., Steinkellner, H. (1998a) Genotyping of grapevine and rootstock cultivars using microsatellite markers. *Vitis*: **37**, 15–20.

Sefc, K. M., Steinkellner, H., Glössl, J., Kampfer, S., Regner, F. (1998b) Reconstruction of a grapevine pedigree by microsatellite analysis. *Theoretical and Applied Genetics*: **97**, 227–31.

Sefc, K. M., Regner, F., Turetschek, E., Glössl, J., Steinkellner, H. (1999) Identification of microsatellite sequences in *Vitis riparia* and their applicability for genotyping of different *Vitis* species. *Genome*: **42**, 367–73.

Sefc, K. M., Lefort, F., Grando, M. S., Scott, K. D., Steinkellner, H., Thomas, M. R. (2001) Microsatellite markers for grapevine: A state of the art, in *Molecular Biology and Biotechnology of Grapevine* (ed. K. A. Roubelakis-Angelakis). Kluwer, Amsterdam, pp. 433–64.

Tautz, D. (1993) Notes on the definition and nomenclature of tandemly repetitive DNA sequences, in *DNA Fingerprinting, State of the Art* (eds S. D. J. Pena, R. Chakraborty, J. T. Epplen, A. J. Jeffreys). Birkhauser Verlag, Basle, pp. 21–8.

Tautz, D., Trick, M., Dover, G. (1986) Cryptic simplicity in DNA is a major source of genetic variation. *Nature*: **322**, 652–3.

Thomas, M. R. and Scott, N. S. (1993) Microsatellite repeats in grapevine reveal DNA polymorphisms when analysed as sequence-tagged sites (STSs). *Theoretical and Applied Genetics*: **86**, 985–90.

Thomas, M. R. and Scott, N. S. (1994) Microsatellite sequence tagged site markers: simplified technique for rapidly obtaining flanking sequences. *Plant Molecular Biology*: 12, 58–64.

Thomas, M. R., Matsumoto, S., Cain, P., Scott, N. S. (1993) Repetitive DNA of grapevine: classes present and sequences suitable for cultivar identification. *Theoretical and Applied Genetics*: **86**, 173–80.

Thomas, M. R., Cain, P., Scott, N. S. (1994) DNA typing of grapevines: A universal methodology and database for describing cultivars and evaluating genetic relatedness. *Plant Molecular Biology*: **25**, 939–49.

# 12 Wine and migraine

*M. Sandler*

> One gentleman, a most intelligent member of our (medical) profession, who had suffered all his life from this complaint (megrim), but otherwise enjoyed excellent health, told me that for 30 years or more he could never take the smallest quantity of wine (and he mentioned the sacramental wine as an instance) . . . without infallibly producing a headache.
>
> Edward Liveing (1873)

> . . . though I cannot gainsay the observations of a patient who insists that his migraines come after eating ham or chocolate . . . I must regard the interpretation of this empirical fact as exceptionally tricky. I am not convinced that a migraine can ever be ascribed to a specific food-sensitivity, and I would suspect any association of the two to the establishment of a conditioned reflex.
>
> Oliver Sacks (1993)

An association between wine and migraine has been identified since time immemorial. Rose (1997) refers to Celsus (25BC–AD50) who wrote 'the pain . . . is contracted . . . by drinking wine' and lists other historical instances of apparent wine sensitivity. If such anecdotal accounts are valid, their mechanism is far from obvious. Migraine itself is an undoubted fact, but the existence of dietary migraine, as the phenomenon has come to be known, is controversial (Sandler *et al.* 1995): it is the purpose of this paper to inquire into its plausibility as a clinical entity.

## Role of 5-hydroxytryptamine

The cause of migraine is unknown. Despite some challenging recent hypotheses based on a succession of convincing experimental and biochemical clues, we still have no clear idea of the physicochemical progression of events leading to the well-known clinical presentations of the disease. Even so, there can now be little doubt that the monoamine, 5-hydroxytryptamine (5-HT, serotonin), plays an important role in migraine pathogenesis; indeed, the first pointers to its possible involvement appeared more than 40 years ago (Kimball *et al.* 1960; Sicuteri *et al.* 1961). Kimball and colleagues administered 5-HT directly to affected subjects during an attack and, despite not inconsiderable side-effects, noted significant attenuation of headache. The point was thrust home by the case of Hopf *et al.* (1992), who recorded one of nature's sophisticated experiments. They described a patient with migraine with aura, whose headaches subsided when a 5-HT-secreting carcinoid tumour developed and returned after its surgical removal.

Clinical findings such as these appeared against a counterpoint of intense laboratory investigation. Following an early observation by Gaddum and Picarelli (1957), Peroutka and Snyder (1979) laid the basis for our modern identification and classification of the multiple family of 5-HT receptors (Bradley *et al.* 1986). The scene was set for the work of Humphrey and colleagues (1990), who turned their attention to the development of a suitable 5-HT analogue and, from their efforts, the first of the 'rational' modern treatments of migraine, sumatriptan, emerged. Workers from other drug companies followed hotly in pursuit and a number of variations on the sumatriptan theme duly emerged (for brief review, see Sandler 1995).

## Role of other pharmacological agents

Despite such enormous progress in migraine therapy, albeit achieved empirically, the origin and mechanisms of an attack remain largely in the realms of speculation. To interpret all observed phenomena in terms of 5-HT and its receptors would obviously, given the present state of our knowledge, be a gross oversimplification. There is no question that the new drugs work, but do they do so by a pharmacological trick? And if they do act via one or more of the many 5-HT receptors, which? Umberto Eco (1989) in *Foucault's Pendulum*, said: 'For every complex problem, there's a simple solution, and it's wrong.' Thus Goadsby *et al.* (1988) have demonstrated increased concentrations of another neurotransmitter, calcitonin-gene related peptide (CGRP), in jugular venous blood throughout a migraine attack, decreasing in concentration after sumatriptan administration (Goadsby and Edvinsson 1993). The patient of Goltman (1935–36), and the clinically very similar patient of Lance (1995), both had a bony defect in the skull through which the brain protruded during a migraine attack, a possible manifestation of the oedema-producing action of CGRP. This effect is antagonized by sumatriptan (Humphrey *et al.* 1991).

A profusion of other chemical systems has also been implicated. Nakano *et al.* (1993), for example, detected increased amounts of substance P in migraine platelets. Sandler and colleagues (Sandler *et al.* 1970; Glover *et al.* 1977; Sandler 1978) monitored a deficit of platelet monoamine oxidase throughout a migraine attack. Three different groups (Littlewood *et al.* 1982; Launay *et al.* 1988; Jones *et al.* 1995) identified reduced platelet phenolsulphotransferase M activity in platelets of affected subjects. Peatfield *et al.* (2002) noted a decrease in circulating diamine oxidase in migrainous patients compared with controls. Following the suggestion of Edmeads (1991), several groups (e.g. Farkkila *et al.* 1992; Gallai *et al.* 1994) have demonstrated a substantial increase in circulating endothelin during a migraine episode. This work is particularly interesting in the light of the recent observations of Corder *et al.* (2001) who demonstrated that red wine contains specific components that inhibit endothelin production, at least *in vitro*, and might thus be expected to attenuate an episode if this peptide does play a role in attack generation. It is interesting to recollect that Maimonides, in the twelfth century, actually recommended wine as a treatment for migraine: 'People who suffer from a strong midline headache . . . are overtly benefited by drinking undiluted wine either after a meal or during the meal . . . Also feed them bread or toast [soaked] in undiluted wine . . .' (Rosner 1993). As discussed below, however, the popular notion is that red wine may *initiate* a headache in susceptible subjects.

### *Migraine triggers*

Unknown ionic mechanisms (Moskowitz and Macfarlane 1993) may initiate the whole migraine sequence and form a crucial part of a migraine episode. What seems indisputable is that some individuals with migraine are more sensitive than others to certain triggering events, such as flickering light, noise, strong perfume (Lance 1993) or chemical substances, e.g. reserpine (Kimball *et al.* 1985), fenfluramine (Del Bene *et al.* 1977), fluoxetine (Larson 1993), paroxetine (Currie *et al.* 1995), pravastatin (Ramsay and Snyder 1998), zimelidine (Lance 1996) and m-chlorophenylpiperazine (MCPP) (Brewerton *et al.* 1988), although a triggering effect of the latter could not be confirmed by Gordon *et al.* (1993).

A prominent addition to this list is nitric oxide. Olesen and colleagues (Olesen *et al.* 1993; Thomsen *et al.* 1993) administered nitroglycerine to migraineurs and controls, producing an apparent true migraine sequence only in previous migraine sufferers. Although they themselves considered that the nitric oxide so generated might be central to the disorder, the fact remains that headache did not supervene immediately but took something of the order of an hour to appear. Thus, nitric oxide must probably take its place as yet another migraine-triggering agent: as will be seen in the discussion of dietary migraine below, different dietary triggering agents may vary in the lag-period between administration and migraine onset.

## Migraine aura

More controversial, perhaps, is the status of the migraine aura which may now, in the view of many (e.g. Welch *et al.* 1993; Lauritzen 1994), be equated with Leão's spreading depression (Leão 1944). In the opinion of the writer, the migraine aura itself may well be emerging as merely another triggering event for a migrainous headache: the reason why some individuals are more sensitive than others to such triggering events is the mystery at the heart of migraine.

With regard to the migraine aura, some interesting byways have recently emerged. More than 30 years ago, Heyck (1969, 1970) claimed that multiple small transitory intracerebral arteriovenous aneurysms exist that may be of importance in the pathogenesis of migraine. This finding could not be replicated by later workers. In the meantime, however, the startling statistic emerged that a patent cardiac foramen ovale occurs in about a quarter of the population and is the commonest cause of a right-to-left shunt (Hagen *et al.* 1984); in retrospect, it is quite possible that Heyck's subjects were so affected. Three groups have now reported an increased incidence of right-to-left shunts in patients with migraine with aura (Del Sette *et al.* 1998; Anzola *et al.* 1999; Wilmshurst *et al.* 2000). Among divers with decompression illness in particular, those with large right-to-left shunts have a higher incidence of migraine with aura after diving and in everyday life than those with no shunt or a smaller shunt (Wilmshurst and Nightingale 2001). These episodes may be associated with paradoxical gas embolism, and Wilmshurst *et al.* (2000) postulate that the lungs may normally act as a filter for trigger substances in the venous circulation that can initiate an attack of migraine with aura if they reach the brain in sufficiently high concentrations. Conversely, the lungs themselves may release substances that trigger migraine attacks.

Thirty years ago, Sandler (1972) put forward the concept that migraine may, to some extent, be a pulmonary disease. This interpretation was based on the experimental observations of Alabaster and Bakhle (1970), who noted a release of vasoactive substances, including prostaglandins and 'slow-reacting substance A' (SRSA) (later identified as a leukotriene) into the efferent circulation of the rabbit lung after perfusion of the afferent circulation with 5-HT or tryptamine. The hypothesis, which has never been adequately tested in humans, is now of particular topical interest because of a new possibility: Sandler (1996) has suggested that nerve growth factor (NGF) is a powerful candidate for the pain-producing substance responsible for migrainous headache, and it now seems sensible to suggest that this protein be sought in the pulmonary efferent blood supply throughout a migraine attack.

The prima facie evidence for NGF involvement is circumstantial but strong. NGF administration to both neonatal and mature rats causes profound hyperalgesia (Lewin *et al.* 1993), and this phenomenon also occurs in mice (Lewin and Mendell 1993). In addition, transgenic mice that overexpress NGF in their skin manifest gross hyperalgesia (Lewin and Mendell 1993). In humans, intravenously administered recombinant human NGF results in muscle pain, while subcutaneous administration leads to hyperalgesia at the injection site, persisting for up to eight weeks (Petty *et al.* 1994). Intracerebroventricular NGF has been associated with severe abdominal pain in Alzheimer patients (Olson *et al.* 1992). Even so, it must be remembered that pain responses of the type produced by NGF are not entirely specific:

administration of certain cytokines, for instance, may also result in hyperalgesia (Watkins *et al.* 1994). Ciliary neurotrophic factor (CNTF), apparently quite toxic to humans (Barinaga 1994), appears to number ability to generate pain among its disadvantages (Lindsay 1995). Despite these provisos, it is the author's view that NGF might be a prime candidate for further investigation within this context.

It should be added that the 'pulmonary filtration' hypothesis of Wilmshurst *et al.* (2000) and the 'pulmonary release' hypothesis of Sandler (1972) may not be incompatible: whereas the former points to a mechanism for generation of a premigrainous triggering event – the aura – deriving from the initiation of spreading depression by minute gas bubbles reaching the cerebral cortex, the latter is more concerned with the release of headache-generating vasocactive substances from the pulmonary vascular bed into the efferent pulmonary circulation.

## Dietary migraine

As mentioned in the introduction to this chapter, anecdotal accounts of dietary initiation of migraine attacks in some individuals have punctuated the scientific literature since time immemorial. Even so, serious doubt exists in the minds of many neurologists as to whether the condition really exists as a clinical entity (e.g. Blau 1992; Sacks 1993) or whether the phenomenon is merely a manifestation of the placebo effect. Socrates said: 'Let no one persuade you to give the drug against headache to him who before has not opened his soul to your treatment' (Plato, *Charmides*). And even when an experiment appears to be carefully controlled, there may be pitfalls: Strong (2000), pondering on why some dietary migraine patients claim to get headaches from placebos, pinpointed the control gelatine capsule, composed of animal and vegetable protein, as a possible migraine trigger!

Earlier in this chapter, a number of chemical triggers of migraine were listed, including reserpine, fenfluramine, and MCPP, and, in general, their action is accepted without demur by workers in this field. Then there are single chemical substances employed in foodstuffs – aspartame (Lipton *et al.* 1989; Van Den Eeden *et al.* 1994) and vanillin (Saint Denis *et al.* 1996), for example – and few would have a problem in believing that they act as headache triggers in particular subgroups of the migraine population.

It is with the ingestion of complex chemical mixtures, foodstuffs of only partially known composition, that the main problems of plausibility arise. In her original and, indeed, groundbreaking paper, Hanington (1967) was persuaded that the tyramine content of cheese is responsible for the headache that follows its ingestion in some migraineurs. Several subsequent investigators were unable to replicate her finding of a migraine-triggering effect of tyramine capsule administration, and the tyramine story has duly languished. It is not quite dead, however, for Peatfield *et al.* (1983) did find that intravenous tyramine administration is followed by *slight* headache in a proportion of subjects, without a blood pressure rise sufficient to account for it. It appears, however, that tyramine *ingestion* is only associated with headache in patients treated with monoamine oxidase inhibitors, the so-called cheese effect. Even so, the popular press has enthusiastically adopted tyramine as the prime cause of dietary migraine, and tyramine-free diets now abound. Specialized cookery books for migraineurs are still being published, in which tyramine-free diets form an important plank in the author's platform (e.g. Marks and Marks 2000).

For believers in the concept of dietary migraine, the commonest initiators of an attack are alcoholic beverages (Peatfield 1995), and the largest and most notorious subgroup in this category has red wine as its trigger (Peatfield *et al.* 1984) – although even here there is conflict: on the continent of Europe, white wine is viewed as the major culprit (Relja *et al.* 1993; Henri 1996; Tournier-Lasserve 1996). Peatfield (1995) holds that cheese/chocolate and red wine sensitivity in particular have closely related mechanisms. It should be noted that there is very little tyramine in red wine (Hannah *et al.* 1988).

Against this background of uncertainty, it seemed important to try to obtain objective evidence for the existence of the condition and, accordingly, our group carried out two double-blind trials. In the first (Littlewood *et al.* 1998), we divided a subset of patients who thought their headaches were initiated by red wine into two groups, giving the first a heavily disguised red wine mixture and the second a similarly disguised vodka mixture of identical alcohol content. The subjects claimed to be unable to tell the difference but only those ingesting red wine (9 out of 11) developed headache with associated migraine symptoms. The second trial (Gibb *et al.* 1991) involved patients who claimed that chocolate set off their migraine episodes. They were again divided into two groups and were fed, respectively, real and mock chocolate of similar taste which they were unable to distinguish between. Compared with the wine experiment, a smaller but still significant proportion of those fed authentic chocolate, but not placebo, developed a headache attack, again with associated symptoms of migraine. What was particularly interesting, as mentioned above (Littlewood *et al.* 1988, Gibb *et al.* 1991), was the difference in lag period between ingestion and headache onset: about three hours in the red wine experiment and 22 hours after chocolate. Whatever the biochemical sequence separating the trigger from the event, it seems likely that the difference between migraineurs and normal subjects will turn out to be enhanced sensitivity to an initiating agent.

Again as mentioned above, a recent instalment in the red wine story (Corder *et al.* 2001) is the presence of specific polyphenol components identified in red wine that are able to inhibit endothelin-1 synthesis. These might be expected to attenuate migraine attacks if the raised level of circulating endothelin during these episodes is relevant to the sequence of clinical signs. It is tempting to speculate, as the authors do, that the inhibitory effect might be responsible for the French paradox, the lower death rate from coronary heart disease in France compared with the UK, despite a comparable dietary intake of saturated fats. It should be noted, however, that the observations of Corder *et al.* have only been performed *in vitro* on cultured bovine aortic cells. Before the data can be extrapolated to humans, we need to known considerably more about the *in vivo* effect of the polyphenols in question, their absorption and bioavailability. We should remember the case of resveratrol, a highly vaunted compound of high *in vitro* promise, which failed to achieve therapeutic fulfilment *in vivo* (Goldberg and Soleas 2002).

It may well be, of course, that alcohol itself, rather than red wine specifically, is responsible for the French paradox (Klatsky 2002). And then again, in a recent new departure, it seems likely that light to moderate alcohol ingestion is associated with a lower risk of dementia in individuals aged 55 years or over (Ruitenberg *et al.* 2002). It is gratifying that a recreational drug which, in moderation, is capable of providing substantial social benefit and support may also improve our actual physical health on a long-term basis.

## References

Alabaster, V. A. and Bakhle, Y. S. (1970) The release of biologically active substances from isolated lungs by 5-hydroxytryptamine and tryptamine. *British Journal of Pharmacology*: **40**, 582P–583P.

Anzola, G. P., Magoni, M., Guindani, M., Rozzini, L., Dalla Volta, G. (1999) Potential source of cerebral embolism in migraine with aura: a transcranial doppler study. *Neurology*: **52**, 1622–5.

Barinaga, M. (1994) Neurotrophic factors enter the clinic. *Science*: **264**, 772–4.

Blau, J. N. (1992) Migraine triggers: practice and theory. *Pathologie et Biologie*: **40**, 367–72.

Bradley, P. B., Engle, G., Feniuk, W., Fozard, J. R., Humphrey, P. P. A., Middlemiss, D. N. (1986) Proposals for the classification and nomenclature of functional receptors for 5-hydroxytryptamine. *Neuropharmacology*: **25**, 563–76.

Brewerton, T. D., Murphy, D. L., Mueller, E. A., Jimerson, D. C. (1988) Induction of migraine-like headaches by the serotonin agonist m-chlorophenylpiperazine. *Clinical Pharmacology and Therapeutics*: 43, 605–9.

Corder, R., Douthwaite, J. A., Lees, D. M., Khan, M. Q., Viseu dos Santos, A. C., Wood, E. G., Carrier, M. J. (2001) Endothelin-1 synthesis reduced by red wine. *Nature*: 414, 863–4.

Currie, A., Ryman, A., McAllister-Williams, R. H. (1995) Exacerbation of migraine with paroxetine: case report. *Human Psychopharmacology*: **10**, 349–50.

Del Bene, E., Anselmi, B., Del Bianco, P. L., Fanciullacci, M., Galli, P., Salmon, S., Sicuteri, F. (1977) Fenfluramine headache, in: *Headache: New Vistas* (ed. F. Sicuteri). Biomedical Press, Florence, pp. 101–9.

Del Sette, M., Angeli, S., Leandri, M., Ferriero, G., Bruzzone, G. L., Finocchi, C., Gandolfo, C. (1998) Migraine with aura and right-to-left shunt on transcranial doppler: a case-control study. *Cerebrovascular Diseases*: **8**, 327–30.

Eco, U. (1989) *Foucault's Pendulum* (English translation by W. Weaver). Harcourt Brace, Jovanovich, Orlando.

Edmeads, J. (1991) ET and EDRF: implication for migraine. *Headache*: **31**, 127.

Farkkila, M., Palo, J., Saijonmaa, O., Fyhrquist, F. (1992) Raised plasma endothelin during acute migraine attacks. *Cephalalgia*: 12, 383–4.

Gaddum, J. H. and Picarelli, Z. P. (1957) Two kinds of tryptamine receptor. *British Journal of Pharmacology*: 12, 323–8.

Gallai, V., Sarchielli, P., Firenze, C., Trequattrini, A., Paciaroni, M., Usai, F., Palumbo, R. (1994) Endothelin-1 in migraine and tension-type headache. *Acta Neurologica Scandinavica*: 89, 47–55.

Gibb, C. M., Davies, P. T. G., Glover, V., Steiner, T. J., Rose, F. C., Sandler, M. (1991) Chocolate is a migraine-provoking agent. *Cephalalgia*: 11, 93–5.

Glover, V., Sandler, M., Grant, E., Rose, F. C., Orton, D., Wilkinson, M., Stevens, D. (1977) Transitory decrease in platelet monoamine oxidase activity during migraine attacks. *Lancet*: i, 391–3.

Goadsby, P. J. and Edvinsson, L. (1993) The trigeminovascular system and migraine: studies characterizing cerebrovascular and neuropeptide changes seen in humans and cats. *Annals of Neurology*: 33, 48–56.

Goadsby, P. J., Edvinsson, L., Ekman, R. (1988) Release of vasoactive peptides in the extracerebral circulation of humans and the cat during activation of the trigeminovascular system. *Annals of Neurology*: 23, 193–6.

Goldberg, D. M. and Soleas, G. J. (2002) Resveratrol: biochemistry, cell biology and the potential role in disease prevention, in *Wine: A Scientific Exploration* (eds M. Sandler and R. Pinder). Taylor & Francis, London, pp. 160–98.

Goltman, A. M. (1935–36) The mechanism of migraine. *Journal of Allergy*: 7, 351–5.

Gordon, M. L., Lipton, R. B., Brown, S. L., Nakraseive, C., Russell, M., Pollack, S. Z., Korn, M. L., Merriam, A., Solomon, S., van Praag, H. M. (1993) Headache and cortisol responses to m-chlorophenylpiperazine are highly correlated. *Cephalalgia*: 13, 400–5.

Hagen, P. T., Scholz, D. G., Edwards, W. D. (1984) Incidence and size of patent foramen ovale during the first 10 decades of life: an autopsy study of 965 normal hearts. *Mayo Clinic Proceedings*: 59, 17–20.

Hanington, E. (1967) Preliminary report on tyramine headache. *British Medical Journal*: 2, 550–1.

Hannah, P., Glover, V., Sandler, M. (1988) Tyramine in wine and beer. *Lancet*: i, 879.

Henri, P. (1996) quoted by Lance, J. W. (1996) Discussion remark in *Migraine: Pharmacology and Genetics* (eds M. Sandler, M. Ferrari, S. Harnett). Chapman & Hall, London, pp. 134–5.

Heyck, H. (1969) Pathogenesis of migraine. *Research and Clinical Studies in Headache*: 2, 1–28.

Heyck, H. (1970) The importance of arterio-venous shunts in the pathogenesis of migraine, in *Background to Migraine* (ed. A. L. Cochrane). Third Migraine Symposium, Heinemann, London, pp. 19–25.

Hopf, H. C., Johnson, E. A., Gutmann, L. (1992) Protective effect of serotonin on migraine attacks. *Neurology*: 42, 1419.

Humphrey, P. P. A. and Feniuk, W. (1991) Mode of action of the antimigraine drug sumatriptan. *Trends in Pharmacological Sciences*: 12, 444–6.

Humphrey, P. P. A., Feniuk, W., Perren, M. J., Connor, H. E., Oxford, A. W., Coates, I. H., Butina, D. (1990) GR43175, a selective agonist for the 5-$HT_1$-like receptor in dog isolated saphenous vein. *British Journal of Pharmacology*: 94, 1123–32.

Jones, A. L., Roberts, R. C., Colvin, D. W., Rubin, G. L., Coughlin, M. W. H. (1995) Reduced platelet phenolsulphotransferase activity towards dopamine and 5-hydroxytryptamine in migraine. *European Journal of Clinical Pharmacology*: **49**, 109–14.

Kimball, R. W., Friedman, A. P., Vallejo, E. (1960) Effect of serotonin in migraine patients. *Neurology*: **10**, 107–11.

Klatsky, A. L. (2002) Wine, alcohol and cardiovascular diseases, in *Wine: A Scientific Exploration* (eds M. Sandler and R. Pinder). Taylor & Francis, London, pp. 108–139.

Lance, J. W. (1993) Current concepts of migraine pathogenesis. *Neurology*, **43**, S11–15.

Lance, J. W. (1995) Swelling at the site of a skull defect during migraine headache. *Journal of Neurology, Neurosurgery and Psychiatry*: **59**, 641.

Lance, J. W. (1996) Discussion remark, in *Migraine: Pharmacology and Genetics* (eds M. Sandler, M. Ferrari, S. Harnett). Chapman & Hall, London, pp. 134–5.

Larson, E. W. (1993) Migraine with typical aura associated with fluoxetine therapy: case report. *Journal of Clinical Psychiatry*: **54**, 235–6.

Launay, J. M., Soliman, H., Pradalier, A., Dry, J., Dreux, C. (1988) Activités PST plaquettaires: le 'trait' migraineux? *Thérapie*: **43**, 273–7.

Lauritzen, M. (1994) Pathophysiology of the migraine aura. The spreading depression theory. *Brain*: **117**, 199–210.

Leão, A. A. P. (1944) Spreading depression of activity in the cerebral cortex. *Journal of Neurophysiology*: **7**, 359–90.

Lewin, G. R. and Mendell, L. M. (1993) Nerve growth factor and nociception. *Trends in Neurological Science*: **16**, 353–9.

Lewin, G. R., Ritter, A. M., Mendell, L. M. (1993) Nerve growth factor-induced hyperalgesia in the neonatal and adult rat. *Journal of Neuroscience*: **13**, 2136–48.

Lindsay, R. M. (1995) Neuron saving schemes. *Nature*: **373**, 289–90.

Lipton, R. B., Newman, L. C., Cohen, J. S., Solomon, S. (1989) Aspartame as a dietary trigger of headache. *Headache*: **29**, 90–2.

Liveing, E. (1873) *On Megrim, Sick-headache and Some Allied Disorders: a Contribution to the Pathology of Nerve Storms*. Churchill, London, p. 45.

Littlewood, J., Glover, V., Sandler, M., Petty, R., Peatfield, R., Rose, F. C. (1982) Platelet phenolsulphotransferase deficiency in dietary migraine. *Lancet*: **i**, 983–6.

Littlewood, J. T., Gibb, C. Glover, V. Sandler, M., Davies, P. T. G., Rose, F. C. (1988) Red wine as a cause of migraine. *Lancet*: **i**, 558–9.

Marks, D. R. and Marks, L. (2000) *The Headache Prevention Cookbook*. Houghton Miffin, Boston.

Moskowitz, M. A. and Macfarlane, R. (1993) Neurovascular and molecular mechanisms in migraine headaches. *Cerebrovascular and Brain Metabolism Reviews*: **5**, 159–77.

Nakano, T., Shimomura, T., Kowa, H., Takahashi, K., Ikawa, S. (1993) Platelet substance P and serotonin in headache. *Proceedings of the VI Congress of the International Headache Society*, Paris, p. 85.

Olesen, J., Iversen, H. K., Thomsen, L. L. (1993) Nitric oxide supersensitivity: a possible molecular mechanism of migraine pain. *NeuroReport*: **4**, 1027–30.

Olson, L., Nordberg, A., von Holst, H., Bäckman, L., Ebendal, T., Alafuzoff, I., Hartvig, P., Herlitz, A., Lilja, A., Lundqvist, H., Langström, B., Meyerson, B.,

Persson, A., Viitanen, M., Winblad, B., Seiger, A. (1992) Nerve growth factor affects C-nicotine binding, blood flow, EEG, and verbal episodic memory in an Alzheimer patient (case report). *Journal of Neural Transmission*: 4, 79–95.

Peatfield, R. C. (1995) Relationships between food, wine and beer-precipitated migrainous headaches. *Headache*: 35, 355–7.

Peatfield, R., Littlewood, J. T., Glover, V., Sandler, M., Rose, F. C. (1983) Pressor sensitivity to tyramine in patients with headache: relationship to platelet monoamine oxidase and to dietary provocation. *Journal of Neurology, Neurosurgery and Psychiatry*: 46, 827–31.

Peatfield, R., Glover, V., Littlewood, J. T., Sandler, M., Rose, F. C. (1984) The prevalence of diet-induced migraine. *Cephalalgia*: 4, 179–83.

Peatfield, R. C., Fletcher, G., Rhodes, K., Gardiner, I. M., de Belleroche, J. (2002) Pharmacological analysis of red wine induced migrainous headaches. (In press.)

Peroutka, S. J. and Snyder, S. H. (1979) Multiple serotonin receptors: differential binding of [$^3$H]5-hydroxytryptamine, [$^3$H] lysergic acid diethylamide and [$^3$H] spiroperidol. *Molecular Pharmacology*: 16, 687–99.

Petty, B. G., Cornblath, D. R., Adornato, B. T., Chaudhry, V., Flexner, C., Wachsman, M., Sinicropi, D., Burton, L. E., Peroutka, S. J. (1994) The effect of systemically administered recombinant human nerve growth factor in healthy human subjects. *Annals of Neurology*: 36, 244–6.

Ramsey, C. S. and Snyder, Q. C. (1998) Altitude-induced migraine headache secondary to pravastatin: case report. *Aviation, Space and Environmental Medicine*: 69, 603–6.

Relja, G., Nider, G., Chiodo-Grandi, F., Kosica, N., Musco, G., Negro, C. (1993) Is red or white wine an important inducing factor in migraine attacks? *Proceedings of the VIth Congress of the International Headache Society, Paris*, p. 129.

Rose, F. C. (1997) Food and headache. *Headache Quarterly*: 8, 319–29.

Rosner, F. (1993) Headache in the writings of Moses Maimonides and other Hebrew sages. *Headache*: 33, 315–19.

Ruitenberg, A., van Swieten, J. C., Witteman, J. C. M., Mehta, K. M., van Duijn, C. M., Hofman, A., Breteler, M. M. B. (2002) Alcohol consumption and risk of dementia: the Rotterdam study. *Lancet*: 359, 281–6.

Sacks, O. (1993) *Migraine, Revised and Expanded*. Picador, London, p. 153.

Saint-Denis, M., Coughtrie, M. W., Guilland, J. C., Verges, B., Lemesle, M., Giroud, M. (1996) Migraine induite par la vanilline. *Presse Médicale*: 25, 2043.

Sandler, M. (1972) Migraine: a pulmonary disease? *Lancet*, i, 618–19.

Sandler, M. (1978) Implications of the platelet monoamine oxidase deficit during migraine attacks. *Research and Clinical Studies in Headache*: 6, 65–72.

Sandler, M. (1995) Migraine to the year 2000. *Cephalalgia*: 15, 259–64.

Sandler, M. (1996) The possible role of neurotrophins in migraine, in *Migraine: Pharmacology and Genetics* (eds M. Sandler, M. Ferrari, S. Harnett). Chapman & Hall, London, pp. 180–95.

Sandler, M. Youdim, M. B. H., Southgate, J., Hanington, E. (1970) The role of tyramine in migraine: some possible biochemical mechanisms, in *Background to Migraine* (ed. A. L. Cochrane). Third Migraine Symposium, Heinemann, London, pp. 103–15.

Sandler, M., Li, N.-Y., Jarrett, N., Glover, V. (1995) Dietary migraine: recent progress in the red (and white) wine story. *Cephalalgia*: 15, 101–3.

Sicuteri, F., Testi, A., Anselmi, B. (1961) Biochemical investigations in headache: increase in the hydroxyindoleacetic acid excretion during migraine attacks. *International Archives of Allergy*: **19**, 55–8.

Strong, F. C., III (2000) Why do some dietary migraine patients claim they get headaches from placebos? *Clinical and Experimental Allergy*: **30**, 739–43.

Thomsen, L. L., Iversen, H. K., Brinck, T. A., Olesen, J. (1993) Arterial supersensitivity to nitric oxide (nitroglycerin) in migraine sufferers. *Cephalalgia*: **13**, 395–9.

Tournier-Lasserve, E. (1996) Discussion remark in *Migraine: Pharmacology and Genetics* (eds M. Sandler, M, Ferrari, S. Harnett). Chapman & Hall, London, pp. 135.

Van Den Eeden, S. K., Koepsell, T. D., Longstreth, W. T., Jr, van Belle, G., Daling, J. R., McKnight, B. (1994) Aspartame ingestion and headaches: a randomised crossover trial. *Neurology*: **44**, 1787–93.

Watkins, L. R., Wiertelak, E. P., Goehler, L. E., Smith, K. P., Martin, D., Maier, S. F. (1994) Characterization of cytokine-induced hyperalgesia. *Brain Research*: **654**, 15–26.

Welch, K. M. A., Barkley, G. L., Tepley, N., Ramadan, N. M. (1993) Central neurogenic mechanism of migraine. *Neurology*: **43**, S521–5.

Wilmshurst, P. and Nightingale, S. (2001) Relationship between migraine and cardiac and pulmonary right-to-left shunts. *Clinical Science*: **100**, 215–20.

Wilmshurst, P. T., Nightingale, S., Walsh, K. P., Morrison, W. L. (2000) Effect on migraine of closure of cardiac right-to-left shunts to prevent recurrence of decompression illness or stroke or for haemodynamic reasons. *Lancet*: **356**, 1648–51.

# 13 Wine: protective in macular degeneration

*T. O. Obisesan*

## Introduction

Age-related macular degeneration (ARMD) is the leading cause of blindness in older adults (Leibowitz *et al.* 1980; Green and Enger 1993). Approximately 30% of people over the age of 65 years will suffer from this disorder. The presence of a high rate of ARMD in a fast-growing segment of any population will translate into higher demands for eye care services, and it is anticipated that associated health care costs will drastically increase. The definition of ARMD is broad and encompasses both mild degrees of atrophy as well as severe haemorrhagic disease. It is a bilateral progressive disease that deprives millions of older adults of central vision (Kashani 1990). The cardinal symptom is blurred or distorted vision, which may develop very gradually in the atrophic forms, or rather suddenly when subretinal neovascular vessels bleed or exude fluid.

Although the exact pathogenesis of ARMD is currently poorly defined, epidemiological, clinical and laboratory evidence suggests that many aetiological factors are involved. Hereditary influence, nutritional deficiency and solar radiation have been implicated (Stafford *et al.* 1984; Young 1988). Most of the factors implicated in the risk of cardiovascular disease (CVD) – age, sex, race, smoking, blood pressure, diabetes, body mass index, serum cholesterol, plasma concentrations of low-density lipoprotein (LDL), high-density lipoprotein (HDL) and apolipoprotein, regular vigorous exercise, regular aspirin use and social class – have all been found to influence the prospect of ARMD (de Lorimier 2000). Despite the observation of ARMD in non-human primates, attempts to find a good animal model have been unsuccessful (Stafford 1974; Stafford *et al.* 1984). There are currently no effective primary or secondary preventive strategies against this disorder. Tertiary prevention, using laser photocoagulation treatment, has only limited usefulness in those with the advanced form of ARMD (neovascular exudative form) and can only reduce, but not reverse, visual loss.

The paucity of scientific knowledge concerning the exact aetiology of ARMD, the association between ARMD and CVD risk factors, and the known attenuation of CVD risk factors by moderate wine consumption, formed the basis for the investigation of the influence of wine on ARMD. While a few studies reported an inverse association between moderate wine consumption and ARMD, others observed either minimal, but non-significant, effects, or lack of a relationship between wine consumption and ARMD. Interestingly, no study reported that moderate wine consumption has a deleterious action on the macula of the eye. This chapter will attempt to synthesize and present a brief synopsis of the limited current scientific knowledge on the association between wine consumption and ARMD.

## The influence of wine consumption on age-related macular degeneration

Alcohol has been part of human civilization for 6000 years, serving both dietary and socio-religious functions. Its production takes place on every continent, and its chemical composition is profoundly influenced by oenological techniques and climatic factors (Soleas *et al.* 1997a). As recently as the nineteenth century, alcohol was an essential daily staple. Up to that point, alcoholic beverages were relatively dilute and considered superior to water because of their lower risk of associated illnesses than water. They also provided important caloric supplementation and other nutrients, as well as serving as an alternative source of daily fluid intake. Alcohol, and most importantly wine, intake were relegated to social and recreational usage as a result of improved sanitation and water purification in more recent years.

Over the past several decades, attention has focused on the negative health consequences of alcohol. In addition to ocular anomalies among children with fetal alcohol syndrome, epidemiological studies indicate that chronic alcoholism is associated with a higher risk of cataract, keratitis and colour vision deficiencies. Not until the past few decades have we begun to explore the dichotomy of the health effect of alcohol. Many studies have attributed any beneficial effect of alcohol to red wine. Specifically, the last decade has witnessed a significant increase in our understanding of the health benefit of moderate wine consumption.

Despite renewed attention on the health consequences of alcohol, very few studies in the English-language literature have investigated the relationship between wine intake and the risk of developing ARMD (Ritter *et al.* 1995). Obisesan and colleagues (1998) were the first to show that moderate wine consumption is associated with reduced risk of developing ARMD, using a large population-based and representative sample (NHANES-1). The association of ARMD with hypertension, the deleterious effect of alcohol on hypertension (Ferris 1983), and its adverse effect on some forms of retinopathy (Hogan and Alvarado 1967; Young *et al.* 1984) formed the basis for the investigation of the association between wine consumption and ARMD. Surprisingly, the study showed alcohol to be associated with a reduced chance of developing ARMD. Additional analysis showed that wine, but not beer or spirits, had an inverse association with ARMD (Figure 13.1), indicating that

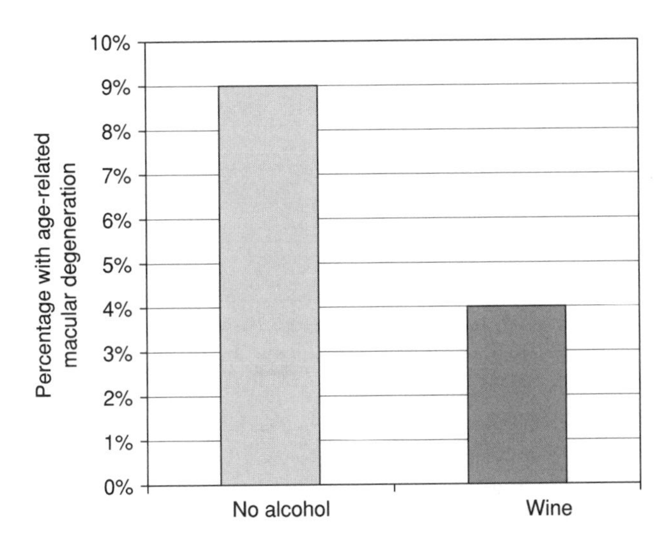


*Figure 13.1* Percentage of people with age-related macular degeneration by wine consumption in the First National Health and Nutrition Examination Survey (NHANES-1).

wine intake was responsible for the lower rate of ARMD observed with alcohol intake. The association between ARMD and wine consumption was maintained after adjustment for several confounding variables.

Although the evidence suggested that moderate amounts of alcohol in general may have a beneficial effect on ARMD, significant benefit was only noted among those who consume wine. A 34% reduction in the possibility of developing the disorder was observed with wine-drinkers, whereas, an 18% reduction was noted in spirit-drinkers. People consuming wine and beer, wine and spirits, and beer and spirits, had a 34%, 26% and 18% reduction in the chance of developing ARMD respectively, suggesting that the benefit of alcohol to the macula of the eye is mainly from wine and perhaps marginally from the consumption of spirits.

In a large prospective US male physician study, Ajani *et al.* (1999) reported a small possible effect (reduced or increased) in risk for low to moderate levels of alcohol intake. Consistent with trends reported in the literature, inclusion of alcohol information up to 84 months lowered the relative risk estimates for people reporting 2–4 drinks/week, but showed increased risk for those reporting higher intake. Unfortunately, the study concluded that there was no significant association, either overall or among those reporting various categories of alcohol intake at baseline, ranging from > 1/week to ≥ 1/day. Admittedly, according to Ajani and colleagues (1999), measures of alcohol consumption lack sufficient precision to evaluate the effect of dosage. Furthermore, the data were not stratified by alcohol type and therefore may not provide any additional insight into the relationship of ARMD with wine.

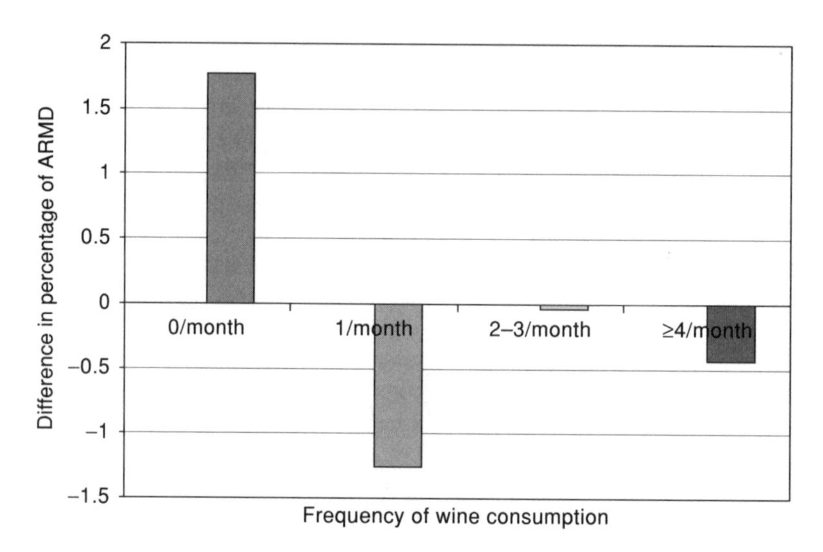


*Figure 13.2* Differences in the incidence of age-related macular degeneration between wine-drinkers and non-wine-drinkers in the Third National Health and Nutrition Examination Survey (NHANES-3, unpublished).

In another prospective study of female nurses and male health professionals aged 50 years and older, Cho *et al.* (2000) assessed history of alcohol intake at baseline and at follow-up using a validated semi-quantitative food-frequency questionnaire. In the study, separate analysis conducted for men and women to evaluate the relationship of alcohol (g/day) to the risk of ARMD led the authors to conclude that no specific type of alcoholic beverage provided protection against ARMD. Sub-analysis of the same data showed a trend towards reduced risk of ARMD among non-smokers, although not to a significant level. This suggests that part of the disagreement between studies may be attributed to confounding variables and methodological differences.

Recent analysis of data from the Third National Health and Nutrition Examination Survey (NHANES-3, unpublished) comparing percentage differences in the occurrence and absence of ARMD among non-drinkers and drinkers (categorized according to frequency of wine consumption) showed that a greater percentage of non-wine drinkers had ARMD than did wine-drinkers. Consistent with previous published work from the NHANES-1, the presence of ARMD was lowest among those consuming ≤ 1 drink per month (Figure 13.2). People who consumed 2–3 drinks per month and ≥ 4 drinks per month had lower rates of ARMD than non-drinkers, but higher rates than those consuming ≤ 1 drink per month. Collectively, the available evidence on moderate wine consumption points to an inverse association with ARMD.

### *Mechanism of the influence of wine on age-related macular degeneration*

Given the multiple aetiologies of ARMD as well as the many properties of alcohol and, more specifically, of wine, it is likely that more than one mechanism is involved in the protective effect of wine on the macula of the eye. Anti-inflammatory, antioxidant and antiplatelet properties of wine are potential candidates involved in these mechanisms. Although empirical evidence suggests that these mechanisms are interrelated, and may in fact augment each other, it is also possible that they are activated in a dose-dependent fashion.

#### *Platelet aggregability and fibrinolytic activity, and cardiovascular disease risk for age-related macular degeneration*

Although the exact mechanism by which wine protects against ARMD has not been clearly established, its association with CVD risk offers an important clue. Moderate alcohol intake is negatively associated with the risk of coronary artery disease and stroke (Gaziano *et al.* 1993). The literature suggests that ARMD may share the same pathophysiological pathway with cardiovascular diseases (Hirvelä *et al.* 1996). Recently, an association between body mass index (a known risk factor for cardiovascular disease) and occurrence of ARMD was reported ( Hirvelä *et al.* 1996). The positive association between non-use of postmenopausal oestrogen, cigarette smoking (Klein *et al.* 1993; Christen *et al.* 1996; Seddon *et al.* 1996), serum cholesterol level (Hirvelä *et al.* 1996) and ARMD, supports its common pathophysiological pathway with CVD.

While the pathophysiology of the association between cardiovascular disease risk and ARMD may not be entirely clear, one possible mechanism is the increased platelet aggregability (Christen *et al.* 1996) and serum fibrinogen (Sarks 1970, 1980) that occurs with some retinopathy (Sarks 1975, 1976, 1980). Additionally, high levels of serum cholesterol (Hirvelä *et al.* 1996) and dietary fat intake (Mares-Perlman *et al.* 1995) have been reported in people with neovascular exudative macular degeneration. Conversely, alcohol increases the prostacyclin/thromboxane ratio (Landolfi and Steiner 1984; Toivanen *et al.* 1984), decreases intraplatelet calcium and consequently reduces platelet aggregability (Toivanen *et al.* 1984). The effect of alcohol on fibrinolysis and platelet aggregability is consistent with its negative association with ischaemic stroke (Gill *et al.* 1986; Stampfer *et al.* 1988), whereas, the risk of haemorrhagic stroke is increased by moderate alcohol intake (Friedman and Kimball 1986; Stampfer *et al.* 1988).

Moreover, high-density lipoprotein cholesterol (HDL-C), which is protective against cardiovascular disease (Chumlea *et al.* 1992), increases with alcohol use (Kornzweig 1974; Chumlea *et al.* 1992). Growing evidence indicates that cigarette smoking is a risk factor for ARMD (Seddon *et al.* 1994). Because cigarette smoking increases platelet aggregability as well as ARMD, and because of the opposing effect of cigarette smoking and moderate wine consumption, the negative association between moderate wine consumption and ARMD appears logical. Interestingly, Abou-Agag and colleagues (2001a, b) recently showed that a low concentration of ethanol (0.1%) has different effects on platelet aggregability occurring in two phases – short (< 1 h) and sustained long-term (24 h) – by increasing surface-localized fibrinolytic activity and, consequently, up-regulation of t-PA, u-PA, and the candidate plasminogen receptor (PmgR). This supports our earlier finding that a relatively small dose of wine is required for its beneficial effect on the macula of the eye. The evidence of biological and clinical plausibility supports the observation that ARMD may yet be another phase of cardiovascular disease or its sequelae.

### *Antioxidation effect of wine and age-related macular degeneration*

First, oxidative damage has been reported to contribute to macular changes that result in ARMD (Seddon *et al.* 1994). Second, a predominant pro-oxidant effect of alcohol has been demonstrated in beer-drinkers and therefore supports the findings of Ritter *et al.* (1995) of a deleterious effect of beer on the macula of the eye. On the other hand, although antioxidant phenolic compounds are present in beer, wine (Singleton and Esau 1969), and spirits (Rosenblum *et al.* 1993), they are found in particularly high concentrations in red wine. This indicates that the effect of wine on the organ systems (e.g. the macula of the eye) may directly oppose that of beer, and thus impact negatively on the development of ARMD.

### *Anti-inflammatory effect of wine and age-related macular degeneration*

Recent studies suggest that inflammatory processes may be involved in the pathogenesis of coronary artery disease. This laid the groundwork for the investigation of levels of pro-inflammatory markers (C-reactive protein, $\alpha$1-globulins, $\alpha$2-globulins, transferrin and leucocyte count) and their association with alcohol consumption (Imhof *et al.* 2001; Sierksma *et al.* 2001). Among men in one study, alcohol showed a U-shaped association with C-reactive protein (CRP), with non-drinkers and heavy drinkers having higher levels. Concentrations of $\alpha$1-globulins and $\alpha$2-globulins were also inversely related to alcohol consumption. Because of the association between these markers of inflammation and coronary artery disease, it is believed that the anti-inflammatory effect of alcohol could contribute to the link between moderate alcohol consumption and lower CVD risk. Therefore, in view of the known common risk factor between CVD and ARMD, inflammatory processes may also play a role in the pathogenesis of ARMD. If this holds true, the anti-inflammatory properties of wine are likely to be involved in the reduction of risk associated with moderate wine consumption in some studies. It is entirely possible that the antioxidant and antiplatelet activities of wine occur at different doses and frequencies of wine consumption, and may provide some insight into the non-linear relationship between wine and ARMD as well as into the different outcomes observed in different studies. Future investigations must focus on establishing the exact amount and intake frequency of wine that is protective. Because the metabolism of wine is under genetic influence as well, such studies must investigate the interaction between genes, wine consumption and the influence of their combined effect on the development of ARMD.

### *Wine consumption, nutrition and age-related macular degeneration*

Higher intake of specific types of fat, including vegetable, monounsaturated and polyunsaturated fats and linoleic acid, rather than total fat intake, is associated with a greater risk of advanced ARMD. Conversely, diets high in omega-3 fatty acids and fish are inversely associated with the risk of ARMD when intake of linoleic acid is low (Seddon *et al.* 2001). Altogether, this suggests that dietary factors play a major role in the pathogenesis of ARMD. Evidence is available in support of the fact that the protective effect of wine on the macula is attributable to the nutritional content of red wine. For example, wine (especially red wine) contains a range of polyphenols with desirable biological properties. These include phenolic acids (p-coumaric, cinnamic, caffeic, gentisic, ferulic and vanillic acids), trihydroxystilbenes (resveratrol and polydatin), and flavonoids (catechin, epicatechin and quercetin). They are synthesized by a common pathway from phenylalanine involving polypeptide condensation reactions (Soleas *et al.* 1997a). Resveratrol (3,5,4′-trihydroxystilbene) is a naturally occurring antifungicide that confers disease resistance on plants (vines, peanuts and pines) and is also found in wine (Soleas *et al.* 1997b). Experiments *in vivo* and with animals have shown that this compound has multiple anti-atherosclerotic actions such as antioxidant activity, modulation of hepatic apolipoprotein/lipid synthesis, and inhibition of platelet aggregation. Although evidence supporting the absorption of resveratrol in the gastrointestinal tract is presently unavailable, red wine is known to be a major source of this molecule in the human diet. The most direct alcohol–nutrition–ARMD link to date suggests that ethanol exposure causes a decrease in docosahexaenoic acid levels (the most abundant fatty acids in the human retina) in feline retina (Pawlosky and Salem 1995). The cumulative evidence supports the contribution of the nutritional component of wine to ARMD risk reduction.

Alternatively, the deleterious effect of alcohol on several organ systems is well documented. There is reason to believe that the macula of the eye is not spared the devastating effect of excessive alcohol consumption. Chronic alcoholics tend to have nutritional problems and insufficiency of several vitamins. Alcohol also influences antioxidant vitamin E levels and therefore increases the risk of ARMD. Furthermore, evidence suggests that alcohol consumption is inversely associated with serum or tissue antioxidant nutrients such as β-carotene and lutein/zeaxanthin, even after controlling for dietary antioxidant intake (Stryker *et al.* 1988; Forman *et al.* 1995). In fact, Cho and colleagues (2000) reported adverse effects of white wine on early and dry forms of ARMD at high intake levels (> 2 drinks/day). The negative findings with high intake of white wine is not surprising and, in fact, correlates well with the established deleterious effect of alcohol at high doses. Undoubtedly, excessive intake of any alcoholic beverage will have negative effects on the macula of the eye as well.

## How much alcohol is protective?

Significant variation exists between studies of the association between wine consumption and ARMD. The variance appears to be related to the paucity of information on the exact amount and frequency of wine ingestion that is beneficial in reducing the risk of ARMD. This is a somewhat difficult question to answer because of the multidirectional fate of wine after ingestion. Several factors are known to affect the metabolism of wine, as well as of alcohol in general, after ingestion: dose; concentration; by-products; body and environmental temperature; consumption conditions, prandial state, drug–drug(s) interaction; and temporal patterns of wine intake, menstrual cycle and circadian rhythm. Individual characteristics such as gender, body weight, body water, increased tolerance from chronic alcohol intake and ethnicity all affect wine metabolism. Similarly, the volume of distribution and elimination time vary significantly from person to person, and therefore affect the health consequences of alcohol. For example, women develop higher blood alcohol levels than do men from a similar amount of alcohol intake.

Although an association has been reported between a few servings of alcohol and reduced cardiovascular disease risk, no such characterization currently exists for the relationship between wine consumption and reduced ARMD risk. In fact, very few studies have specifically examined the association between wine consumption and ARMD. Most of the prospective studies on this association have not focused on the question of how much of wine is protective, but have simply explored the possible effect of the qualitative association. It is therefore clear that the amount of wine that is beneficial is unknown. For example, two servings of beer per day were shown to be beneficial in reducing CVD risk, whereas about one serving of wine per month was shown to be most beneficial in reducing the risk of ARMD, based on an observation from NHANES-1 data, and supported by data from NHANES-3. In our earlier published work from NHANES-1, people consuming ≤ 12 drinks of alcohol per year were half as likely to develop ARMD as those in the control group (4% versus 9%) (Figure 11.1) (Obisesan *et al.* 1998).

## Other disagreements between studies

At present, there are disagreements between studies on whether wine consumption has an inverse relationship with ARMD. While a few studies reported a deleterious effect of alcohol, others reported a beneficial action or no association. For example, Smith and Mitchell (1996) did not find an adverse effect of wine on early or dry ARMD. In the Beaver Dam Eye Study, results were inconclusive because of the limited number of wine-drinkers in the study sample (Ritter *et al.* 1995). Similarly, the result of the Blue Mountain Eye Study was inconclusive since it did not account for other possible confounding variables (Attebo *et al.* 1996). Additionally, none of these investigations examined white and red wine separately, suggesting that several of the studies available in the literature were not designed to answer the question of whether wine consumption is protective against ARMD.

There are several additional explanations for the lack of consistency in findings on wine consumption and ARMD. First, the study design, sample size and analytical strategies varied widely. Arguably, for a single cause and effect relationship this may not necessarily present a concern. Unfortunately, this is not the case in ARMD, which is a disease with multiple risk factors. Furthermore, ARMD develops over several years, making it subject to the time effects of multiple risk factors. Second, very few of the studies had a sample size sufficient to make it possible to examine the effect on ARMD of the different alcohol types. As stated earlier in this chapter, there is a lack of uniformity in the characterization of the amount of alcohol and frequency of wine consumed. In fact, the exact amount of wine that is beneficial or otherwise remains poorly defined. Nonetheless, one thing is certain: studies that reported no association did not examine, or, indeed, did not have, the sample size required to examine the frequency of ingestion of sufficient wine to be protective (Obisesan *et al.* 1998).

Third, there is also lack of uniformity in the data stratification between the different studies, the classification of mild, moderate and high wine consumption. While some defined alcohol intake in terms of grams per day, others examined frequency of wine intake. Adjustments for confounding variables varied just as much. Because ARMD is likely to be a disease of multiple aetiology, and given the fact that alcohol consumption co-varies with other lifestyle risk factors, variations in study outcome are to be expected in this early phase of our understanding of the relationship. Furthermore, high intake of wine can be a surrogate marker for other lifestyle risk factors. For example, people who consume large quantities of alcohol (wine) are more likely to be smokers and to consume beer and spirits as well, all of which increase the risk of deleterious consequences of alcohol on ARMD, and, perhaps through direct effect, nutritional deficiency or other, as yet unknown, mechanisms. Unfortunately, all these factors are further complicated by heavy drinkers who are known to under-report alcohol intake to a substantial extent, resulting in underestimation of any true protective effect of moderate wine intake.

To test the effect of confounding factors, the authors of the most rigorous study on the association between wine consumption and ARMD to date conducted a sub-analysis on never-smokers and showed that, while the relative risk of developing ARMD did not change for total alcohol intake, a decrease in the development of ARMD in wine-drinkers was evident (Cho *et al.* 2000). Additionally, geographic differences in the occurrence of ARMD are known (Cruickshanks *et al.* 1997) and may interact with wine consumption, leading to additional variability in study outcome. Appropriate classification of this factor in the USA is made difficult by migration and, perhaps, interaction of alcohol metabolism with yet unknown factors. Furthermore, the demonstration of reduction in CVD risks among low–moderate users of wine that provided the template and validity for categorization of alcohol in many studies may be fatally flawed. This is because studies that found a beneficial effect of wine on ARMD did so with much lower frequency of wine intake. It is therefore imperative that future investigations depart from this dogma.

The rate of elimination of alcohol (including its vehicle, wine) from the body and the time required to reach peak concentration after an oral intake is subject to genetic variation (De la Paz *et al.* 1997). The lack of adjustment for heritability adds more variability to study outcomes. Alcohol elimination in general is controlled by alcohol dehydrogenase, an enzyme with known genetic polymorphism. Alcohol is oxidized to acetaldehyde by alcohol dehydrogenase (ADH) and cytochrome P-4502E1 (CYP2E1), and then to acetate by aldehyde dehydrogenase (ALDH) (Wan *et al.* 1998; Lee *et al.* 2001). Multiple forms and gene loci of human alcohol dehydrogenase (ADH, EC: 1.2.1.3) and aldehyde dehydrogenase (ALDH, EC: 1.2.1.3) in the major pathway of alcohol metabolism have been found and characterized in the past two decades (Harada 2001). Similarly, genetic polymorphism at the ADH2 locus has been shown to result in the inheritance of isoforms with different metabolic properties, resulting in inter-individual differences in alcohol disposition. Consequently, the influence of genes on the rate of elimination of wine and its combined effect on health consequences cannot be overemphasized. Studies are needed to examine the interaction of genetic variations of ADH and ALDH2 and wine consumption. Clearly, pointed methodologically thoughtful study design is needed to answer the question about the beneficial effect of wine and ARMD in diverse populations.

## Summary

The evidence suggests that moderate wine consumption is associated with reduced rates of ARMD. Taken together, the combination of the antioxidant effect of wine, its effect on platelet aggregability, intracellular calcium and its anti-inflammatory property may all be involved. Nonetheless, the exact amount of wine that is beneficial is unknown at the moment. Misclassification of ARMD, bias in alcohol consumption recall, geographic differences and associated environmental influence, and heritability offer additional explanations for the differential effect of wine on ARMD between different studies. It should also be emphasized that a large proportion of the studies on wine and ARMD to date either do not have the sample size or the information on wine consumption required to detect an association. Although age correlates highly with the occurrence of ARMD, it is unclear whether the influence of age is a reflection of cumulative environmental insults associated with reactive oxygen molecules and/or health behaviours, and therefore whether it is an additional confounder. The negative association with wine and the intake frequency that is involved in this association warrants further investigation.

With continued research, a strategy will become available for the prevention of ARMD among people for whom alcohol may be contraindicated. Ultimately, recommendations stemming from evidence-based medicine must be balanced against individual patients' clinical circumstances.

## Ackowledgement

This chapter is dedicated to my wife and children for their unwavering support during the earlier part of my academic career.

## References

Abou-Agag, L. H., Aikens, M. L., Tabengwa, E. M., Benza, R. L., Shows, S. R., Grenett, H. E., Booyse, F. M. (2001a) Polyphyenolics increase t-PA and u-PA gene transcription in cultured human endothelial cells. *Alcoholism, Clinical and Experimental Research*: 25, 155–62.

Abou-Agag, L. H., Tabengwa, E. M., Tresnak, J. A., Wheeler, C. G., Taylor, K. B., Booyse, F. M. (2001b) Ethanol-induced increased surface-localized fibrinolytic activity in cultured human endothelial cells: kinetic analysis. *Alcoholism, Clinical and Experimental Research*: 25, 351–61.

Ajani, U. A., Christen, W. G., Manson, J. E., Glynn, R. J., Schaumberg, D., Buring, J. E., Hennekens, C. H. (1999) A prospective study of alcohol consumption and the risk of age-related macular degeneration. *Annals of Epidemiology*: 9(3), 172–7.

Attebo, K., Mitchell, P., Smith, W. (1996) Visual acuity and the causes of visual loss in Australia. The Blue Mountains Eye Study. *Ophthalmology*: 103(3), 357–64.

Cho, E., Hankinson, S. E., Willett, W. C., Stampfer, M. J., Spiegelman, D., Speizer, F. E., Rimm, E. B., Seddon, J. M. (2000) Prospective study of alcohol consumption and the risk of age-related macular degeneration. *Archives of Ophthalmology*: 118(5), 681–8.

Christen, W. G., Glynn, R. J., Manson, J. E., Ajani, U. A., Buring, J. E. (1996) A prospective study of cigarette smoking and risk of age-related macular degeneration in men. *Journal of the American Medical Association*: 276(14), 1147–51.

Chumlea, W. C., Baumgartner, R. N., Garry, P. J., Rhyne, R. L., Nicholson, C., Wayne, S. (1992) Fat distribution and blood lipids in a sample of healthy elderly people. *International Journal of Obesity and Related Metabolic Disorders*: 16(2), 125–33.

Cruickshanks, K. J., Hamman, R. F., Klein, R., Nondahl, D. M., Shetterly, S. M. (1997) The prevalence of age-related maculopathy by geographic region and ethnicity. The Colorado-Wisconsin Study of Age-Related Maculopathy. *Archives of Ophthalmology*: 115(2), 242–50.

De la Paz, M. A., Pericak-Vance, M. A., Haines, J. L., Seddon, J. M. (1997) Phenotypic heterogeneity in families with age-related macular degeneration. *American Journal of Ophthalmology*: 124, 331–43.

de Lorimier, A. A. (2000) Alcohol, wine, and health. *American Journal of Surgery*: 180(5), 357–61.

Ferris, F. L., III (1983) Senile macular degeneration: review of epidemiologic features. *American Journal of Epidemiology*: 118(2), 132–51.

Forman, M. R., Beecher, G. R., Lanza, E., Reichman, M. E., Graubard, B. I., Campbell, W. S., Marr, T., Yong, L. C., Judd, J. T., Taylor, P. R. (1995) Effect of alcohol consumption on plasma carotenoid concentrations in premenopausal women: a controlled dietary study. *American Journal of Clinical Nutrition*: 62, 131–5.

Friedman, L. A. and Kimball, A. W. (1986) Coronary heart disease mortality and alcohol consumption in Framingham. *American Journal of Epidemiology*: 124(3), 481–9.

Gaziano, J. M., Buring, J. E., Breslow, J. L., Goldhaber, S. Z., Rosner, B., VanDenburgh, M., Willett, W., Hennekens, C. H. (1993) Moderate alcohol intake, increased levels of high-density lipoprotein and its subfractions, and decreased risk of myocardial infarction. *New England Journal of Medicine*: 329(25), 1829–34.

Gill, J. S., Zezulka, A. V., Shipley, M. J., Gill, S. K., Beevers, D. G. (1986) Stroke and alcohol consumption. *New England Journal of Medicine*: 315(17), 1041–6.

Green, W. R. and Enger, C. (1993) Age-related macular degeneration: histopathologic studies. *Ophthalmology*: 100(10), 1519–35.

Harada, S. (2001) Classification of alcohol metabolizing enzymes and polymorphisms – specificity in Japanese. *Nihon Arukoru Yakubutsu Igakkai Zasshi*: 36(2), 85–106.

Hirvelä, H., Luukinen, H., Läärä, E., Laatikainen, L. (1996) Risk factors of age-related maculopathy in a population 70 years of age or older. *Ophthalmology*: 103(6), 871–7.

Hogan, M. J. and Alvarado, J. (1967) Studies on the human macula. IV. Aging changes in Bruch's membrane. *Archives of Ophthalmology*: 77(3), 410–20.

Imhof, A., Froehlich, M., Brenner, H., Boeing, H., Pepys, M. B., Koenig, W. (2001) Effect of alcohol consumption on systemic markers of inflammation. *Lancet*: 357(9258), 763–7.

Kashani, A. A. (1990) Pathogenesis of age-related macular degeneration: embryologic concept. *Annals of Ophthalmology*: 22(7), 246–8.

Klein, R., Klein, B. E., Linton, K. L., DeMets, D. L. (1993) The Beaver Dam Eye Study: the relation of age-related maculopathy to smoking. *American Journal of Epidemiology*: 137(2), 190–200.

Kornzweig, A. L. (1974) Modern concepts of senile macular degeneration. *Journal of the American Geriatrics Society*: 22(6), 246–53.

Landolfi, R. and Steiner, M. (1984) Ethanol raises prostacyclin in vivo and in vitro. *Blood*: 64, 679–82.

Lee, H. C., Lee, H. S., Jung, S. H., Yi, S. Y., Jung, H. K., Yoon, J. H., Kim, C. Y. (2001) Association between polymorphisms of ethanol-metabolizing enzymes and susceptibility to alcoholic cirrhosis in a Korean male population. *Journal of Korean Medical Science*: 16, 745–50.

Leibowitz, H. M., Krueger, D. E., Maunder, L. R., Milton, R. C., Kini, M. M., Kahn, H. A., Nickerson, R. J., Pool, J., Colton, T. L., Ganley, J. P., Loewenstein, J. I., Dawber, T. R. (1980) The Framingham Eye Study monograph: An ophthalmological and epidemiological study of cataract, glaucoma, diabetic retinopathy, macular degeneration, and visual acuity in a general population of 2631 adults, 1973–1975. *Surveys in Ophthalmology*: 24 (supplement), 335–610.

Mares-Perlman, J. A., Brady, W. E., Klein, R., Van den Langenberg, G. M., Klein, B. E., Palta, M. (1995) Dietary fat and age-related maculopathy. *Archives of Ophthalmology*: 113(6), 743–8.

NHANES-1 (1971–1975) *National Health and Nutrition Examination Survey, 1971–1975: Ophthalmology, Ages 1–74 years*. US Department of Health and Human Services, Center for Disease Control, National Center for Health Statistics, Hyattsville, Maryland.

Obisesan, T. O., Hirsch, R., Kosoko, O., Carlson, L., Parrott, M. (1998) Moderate wine consumption is associated with decreased odds of developing age-related macular degeneration in NHANES-1. *Journal of the American Geriatrics Society*: 46(1), 1–7.

Pawlosky, R. J. and Salem, N., Jr (1995) Ethanol exposure causes a decrease in docosahexaenoic acid and an increase in docosapentaenoic acid in feline brains and retinas. *American Journal of Clinical Nutrition*: 61(6), 1284–9.

Ritter, L. L., Klein, R., Klein, B. E., Mares-Perlman, J. A., Jensen, S. C. (1995) Alcohol use and age-related maculopathy in the Beaver Dam Eye Study. *American Journal of Ophthalmology*: 120(2), 190–6.

Rosenblum, E. R., Stauber, R. E., Van Thiel, D. H., Campbell, I. M., Gavaler, J. S. (1993) Assessment of the estrogenic activity of phytoestrogens isolated from bourbon and beer. *Alcoholism, Clinical and Experimental Research*: 17(6), 1207–9.

Sarks, S. H. (1970) The fellow eye in senile disciform degeneration. *Transactions of the Australian College of Ophthalmologists*: 2, 77–82.

Sarks, S. H. (1975) The aging eye. *Medical Journal of Australia*: 2(15), 602–4.

Sarks, S. H. (1976) Ageing and degeneration in the macular region: a clinico-pathological study. *British Journal of Ophthalmology*: 60(5), 324–41.

Sarks, S. H. (1980) Drusen and their relationship to senile macular degeneration. *Australian Journal of Ophthalmology*: 8(2), 117–30.

Seddon, J. M., Ajani, U. A., Sperduto, R. D., Hiller, R., Blair, N., Burton, T. C., Farber, M. D., Gragoudas, E. S., Haller, J., Miller, D. T., Yannuzzi, L. A., Willett, W. (1994) Dietary carotenoids, vitamins A, C and E, and advanced age-related macular degeneration. *Journal of the American Medical Association*: 272, 1413–20.

Seddon, J. M., Willett, W. C., Speizer, F. E., Hankinson, S. E. (1996) A prospective study of cigarette smoking and age-related macular degeneration in women. *Journal of the American Medical Association*: 276(14), 1141–6.

Seddon, J. M., Rosner, B., Sperduto, R. D., Yannuzzi, L., Haller, J. A., Blair, N. P., Willett, W. (2001) Dietary fat and risk for advanced age-related macular degeneration. *Archives of Ophthalmology*: 119(8), 1191–9.

Sierksma, A., van der Gaag, M. S., Kluft, C., Hendriks, H. F. (2001) Effect of moderate alcohol consumption on fibrinogen levels in healthy volunteers is discordant with effects on C-reactive protein. *Annals of the New York Academy of Sciences*: **936**, 630–3.

Singleton, V. L. and Esau, P. (1969) Phenolic substances in grapes and wine, and their significance. *Advances in Food Research*: 1 (supplement), 1–261.

Smith, W. and Mitchell, P. (1996) Alcohol intake and age-related maculopathy. *American Journal of Ophthalmology*: **122**(5), 743–5.

Soleas, G. J., Diamandis, E. P., Goldberg, D. M. (1997a) Wine as a biological fluid: history, production, and role in disease prevention. *Journal of Clinical and Laboratory Analysis*: **11**(5), 287–313.

Soleas, G. J., Diamandis, E. P., Goldberg, D. M. (1997b) Resveratrol: a molecule whose time has come? And gone? *Clinical Biochemistry*: **30**(2), 91–113.

Stafford, T. J. (1974) Maculopathy in an elderly sub-human primate. *Modern Problems in Ophthalmology*: **12**, 214–19.

Stafford, T. J., Anness, S. H., Fine, B. S. (1984) Spontaneous degenerative maculopathy in the monkey. *Ophthalmology*: **91**(5), 513–21.

Stampfer, M. J., Colditz, G. A., Willett, W. C., Speizer, F. E., Hennekens, C. H. (1988) A prospective study of moderate alcohol consumption and the risk of coronary disease and stroke in women. *New England Journal of Medicine*: **319**(5), 267–73.

Stryker, W. S., Kaplan, L. A., Stein, E. A., Stampfer, M. J., Sober, A., Willett, W. C. (1988) The relation of diet, cigarette smoking, and alcohol consumption to plasma beta-carotene and alpha-tocopherol levels. *American Journal of Epidemiology*: **127**, 283–96.

Toivanen, J., Ylikorkala, O., Viinikka, L. (1984) Ethanol inhibits platelet thromboxane A2 production but has no effect on lung prostacyclin synthesis in humans. *Thrombosis Research*: **33**, 1–8.

Wan, Y. J., Poland, R. E., Lin, K. M. (1998) Genetic polymorphism of CYP2E1, ADH2, and ALDH2 in Mexican-Americans. *Genetic Testing*: **2**(1), 79–83.

Young, R. W. (1988) Solar radiation and age-related macular degeneration. *Surveys in Ophthalmology*: **32**(4), 252–69.

Young, R. J., McCulloch, D. K., Prescott, R. J., Clarke, B. F. (1984) Alcohol: another risk factor for diabetic retinopathy? *British Medical Journal*: **288**(6423), 1035–7.

# 14 Antimicrobial effects of wine

*M. E. Weisse and R. S. Moore*

## Historical perspective

Wine has been part of human culture for 6000 years, both as a mainstay of the diet and as part of social and religious functions (Soleas *et al.* 1997). It was not only the staple drink of the aristocracy, but diluted (three parts water to one part wine) it was consumed by the poor and even by children (Durant 1939; Quennell and Quennell 1954). The subject of much folklore and legend, its practical usefulness also became apparent. Take, for example, the tale of 'Four Thieves' Vinegar'. In Marseilles, in 1721, four condemned criminals were recruited to bury the dead during a terrible plague. The gravediggers proved to be immune to the disease. Their secret was a concoction they drank consisting of macerated garlic in wine. This immediately became famous as *vinaigre des quatre voleurs*, and is still available in France today (Block 1985).

The history of wine has been documented as far back as 2000 BC. In Babylon, salves were mixed with wine to treat various skin conditions. The Ancient Egyptians are also recorded to have used it for the treatment of other ailments, including asthma, constipation, epilepsy, indigestion, jaundice and depression. Perhaps as a deterrent to self-medication, various bizarre and somewhat unlikely ingredients were added, including pigs' eyes, bats' blood, dogs' urine and crocodile dung (Pickleman 1990).

Wine production is ubiquitous, its composition varying according to the grape cultivar and climate. It is therefore easy to understand how its production and usage could develop independently across many different cultures. Medicine in ancient India antedated that of Greece and the use of wine in the Hindu culture developed independently. Although many of the practices of these various groups are similar, it is very unlikely that the various practitioners knew of each other's teachings. This is one of the strongest arguments for regarding wine as beneficial to health. It seems improbable that so many different cultures would have adopted these attitudes and praised the health-giving properties of wine if it had little observable effect.

For many centuries, until the time of the early Greeks, healing the sick was the prerogative of priests and magicians. Wine and medicine were intertwined with religion and magic until the time of Hippocrates (460–380 BC) and the beginning of the birth of medicine as we know it. Hippocrates dissociated science from magic, and medicine became an independent profession. Treatment was based on rational observations of the response of illness to treatment, with wine being incorporated into regimens for both acute and chronic disease (Pickleman 1990). This is in direct conflict with other cultures of the time. In India, the practice of medicine, religion, culture and tradition were intricately woven together: nowhere else did religion mingle so much with private and public behaviour (Majno 1975).

In addition to Hippocrates, there were many others in early Greece and later in Greco-Roman times who extolled the virtues of wine. Homer (approximately 850 BC), in the *Iliad* and the *Odyssey*, discussed the use of wine in treating war wounds. Wine was also recommended for non-traumatic ailments, including depression. Socrates, Aristotle and particularly Plato all wrote of the importance of wine. Plato thought that it should be avoided by judges and those about to procreate, but especially recommended it for old men 'to lighten the sourness of old age'. Asclepiades (124–40 BC) included wine in his regimen of physical activity and attention to diet. Celsus (25 BC–AD 37) documented the many uses of wine and went so far as to differentiate various types of wine for specific ailments. Dioscorides (approximately AD 80), an army surgeon, used wine to anaesthetize his patients as well as cauterize their wounds (Pickleman 1990).

This use is also mentioned in the story of the 'Good Samaritan', who tended to a man left for dead by some thieves. Luke 10: 34 says: 'He dressed his wounds, pouring in oil and wine . . .'. Galen (AD 131–201), while caring for gladiators, covered their wounds with cloth soaked with wine to prevent pus from forming (Soleas *et al.* 1997). In addition, in cases of evisceration, he would bathe the bowels in wine before placing them back into the abdominal cavity to prevent peritonitis (Pickleman 1990). This is not unlike the practice of giving champagne for peritonitis after Caesarean section, mentioned in a 100-year study of obstetrics in Germany (Frobenius *et al.* 1996).

During the Dark Ages, western Europe saw the church take a more prominent part in health and medicine. This role of 'faith to heal' no longer utilized the medicines and procedures that had existed until that time. Pain and the ability to tolerate it were a major component of this philosophy, and writings no longer discussed possible ways to improve health. Many 'healers' appeared and began taking advantage of the situation, not always with good results. Happily, in due course, schools taught by clerical and lay teachers replaced the church hospitals, and medical treatises began to appear again. Trotula (eleventh century), for example, wrote about hygiene and obstetrics. He, too, discussed the diverse role wine could play in conditions ranging from uterine prolapse to croup (Pickleman 1990).

Later, Ambroise Paré (1510–1590) came to fame as a French battle surgeon. He experimented with various ways of treating the many wounds he encountered, while avoiding some of the more extreme measures common at the time, such as red hot pokers and boiling oil. Following Hippocratic theories, he believed pus was not a necessary part of the healing process and concocted many ointments containing mixtures of rose oil, egg yolk and other items mixed with wine. As a result, there was a marked decrease in the number of deaths due to battle injuries (Pickleman 1990).

For the next 200 years (the seventeenth to the nineteenth centuries), wine was once again utilized for any number of conditions. Pharmacopoeias were published in London (1618), the USA (1820) and France (1840), all of which included wine paired with ingredients ranging from ginseng to tobacco. In 1920, during the Prohibition era, all wine mixtures were dropped from the US publication, never to return. As in previous dark times, people turned to the church once again. There was a huge rise in the number of people interested in becoming religious since wine could be used only for worship or medicinal purposes. This trend disappeared with the lifting of Prohibition.

Today, numerous drugs and products are available, all requiring greater 'scientific proof' of efficacy. Many such trials are, perhaps, aiming to establish what has been known for thousands of years and discussed in the writings of the Talmud: 'Wine is the foremost of all medicines; wherever wine is lacking, medicines become necessary'.

## Enteric bacteria

*Salmonella*, *E. coli*, and *Shigella* are three of the most common bacterial causes of diarrhoea. In addition to the millions of cases of bacterial enteritis in the USA and western Europe annually, they cause more than half of all cases of diarrhoea in travellers to developing countries. These bacteria are not new to the human race – they have caused intestinal diseases for millennia.

Most cases of diarrhoea due to these bacteria stem from contamination of food or water. Although the germ theory of infection was not recognized until the seventeenth century, it has been long accepted that contaminated food and water could lead to disease. Whether by serendipity or design, it was recognized that drinking diluted wine with dinner was associated with health.

It is interesting that disease attributable to the shared communion cup has rarely been reported. The most effective barrier to transmission of germs during communion may be wiping the brim with a cloth rather than any intrinsic antibacterial activity of the wine itself, or any supernatural power (Managan *et al.* 1998).

In Ancient Greece, one part wine was mixed with three parts water (Quennell and Quennell 1954) and was the drink served with the main meal of the day. In Achaean Greece (1300–1100 BC), 'the staple drink, even among the poor and among children, is diluted wine' (Durant 1939). In Sparta (500–450 BC), 'the Spartans thought the madness of Cleomenes was caused by his having learned to drink wine without water from Scythians

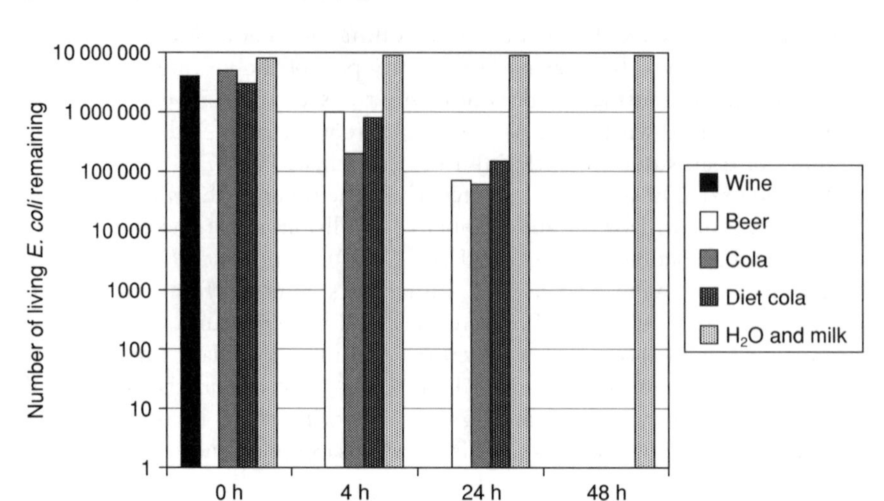


*Figure 14.1* Survival of *E. coli* bacteria after exposure to wine, beer, cola, diet cola, water and milk. Note that by 4 h, wine left no surviving bacteria, while beer and the two soft drinks required 24 h to achieve the same result. As expected, water and milk supported growth of the bacteria throughout the experiment. Similar results were seen for other diarrhoea-causing bacteria, *Salmonella* and *Shigella*.

who came to Sparta. The Greeks always diluted their wine with water' (Quennell and Quennell 1954). Over time, the practice had spread to North Africa, so that an author in Judea wrote in 150 BC: '. . . whereas mixing wine with water makes a more pleasant drink that increases delight, so a skilfully composed story delights the ears of those that read the work' (2 Maccabees 15: 39). It may have been because of wine's antibacterial effect that St Paul wrote in his second letter to Timothy: 'Stop drinking water only. Take a little wine for the good of your stomach, and because of your frequent illnesses' (2 Timothy 5: 23).

Wine's antibacterial effect on enteric pathogens has been studied by a variety of investigators in several countries. In 1988, Sheth *et al.* tested strains of *Salmonella*, *E. coli*, and *Shigella* in carbonated beverages (cola and diet cola), two alcoholic beverages (beer and wine), skimmed milk and non-chlorinated well water. The drinks were inoculated with bacteria, and colony counts of surviving bacteria were measured over two days. While milk and water promoted the growth and survival of the bacteria, the carbonated beverages decreased the colony counts of bacteria from greater than a million to fewer than 10 within 48 h. Wine eliminated the bacteria within 4 h (Figure 14.1). Sheth *et al.* (1988) ascribed part of the antibacterial activity to the low pH of the beverages, as water and milk have near-neutral pH and the others are quite acidic (Table 14.1). A further discussion of the antibacterial properties of wine follows at the end of this chapter.

*Table 14.1* Survival of *Salmonella*, *Shigella* and *E. coli* in various beverages, and relation to pH of the beverages (adapted from data of Sheth *et al.* 1988)

| | *pH* | *Bacteria surviving at 4 h* | *Bacteria surviving at 48 h* |
|---|---|---|---|
| Cola | 2.4 | 100 000 to 1 000 000 | <10 |
| Beer | 3.8 | 1 000 000 | <10 |
| Wine | 3.0 | 0 | 0 |
| Sour mix | 3.1 | 1 000 000 | 0 |
| Diet cola | 3.2 | 100 000 to 1 000 000 | 0 |
| Skimmed milk | 6.8 | >1 000 000 | >1 000 000 |
| Water | 6.2 | >1 000 000 | >1 000 000 |

Another laboratory investigation of wine's effect against intestinal bacteria was performed in Hawaii (Weisse *et al.* 1995). Suspensions of *Salmonella*, *E. coli* and *Shigella* were tested against red and white wine, tequila (diluted to 10% ethanol), 10% ethanol, tap water, and bismuth salicylate. For each of the bacterial preparations, wine proved superior to the other test solutions, decreasing the bacteria count from nearly a million to zero (no growth detected) within 20 min (Figure 14.2). Interestingly, 10% ethanol, an alcohol concentration similar to that of wine, had virtually no effect on the bacteria. Wine was also superior to bismuth salicylate, a medication proven to be effective in preventing traveller's diarrhoea. In another part of the same study, red and white wine were mixed with water contaminated with approximately one million *E. coli* bacteria per millilitre. The wine was mixed with the contaminated water in dilutions of 1:1, 1:2 and 1:4. In this study, white wine was marginally superior to red wine, and both were significantly better than bismuth salicylate. These data give a scientific basis to the practice of mixing water with wine that was common in ancient cultures. In addition, because even dilute wine decreased the bacterial count within 60 min and was superior to similar dilutions of bismuth salicylate, we surmised that wine taken with meals may decrease the possibility of contracting traveller's diarrhoea.

Another group of investigators took this theory to the laboratory and tested it in mice. Four groups of mice were fed $5 \times 10^7$ (50 000 000) colony-forming units (cfu) of *Salmonella enteritidis* at the same time as 10 ml/kg of red wine, white wine, 14% ethanol or water. The mice were followed for up to 12 days, and no difference was seen in survival in any of the groups. All the animals died of infection by day 12 (Sugita-Kunishi *et al.* 2001). The authors of this study concluded that while wine has activity against enteric bacteria in the test tube, it has no protective effect in animals. The problem with their conclusion is extrapolating results in mice to humans. We know that ingestion of approximately $10^5$ (100 000) cfu of *Salmonella enteritidis* is required to make human adults ill, while fewer colonies are required to infect children. As mice weigh 1/1000 as much as a child (20 g adult mouse

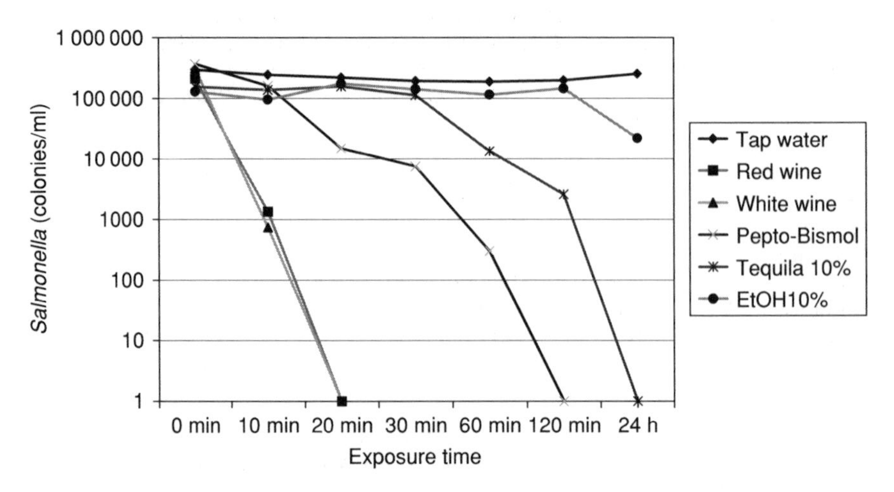


*Figure 14.2* Survival of *Salmonella* over time when exposed to water, red wine, white wine, Pepto-Bismol®, tequila diluted to 10% ethanol, and ethanol 10%. Both red and white wine eradicated the bacteria within 20 min, Pepto-Bismol® by 2 h, and the dilute tequila by 24 h. The 10% ethanol was only marginally better than water. It is interesting that the red and white wine, dilute tequila and 10% ethanol had strikingly different effects despite having very similar alcohol content.

versus 20 kg human child), using such a high level of bacteria in the mouse is not a true test. Even if the wine resulted in a considerable decrease in *Salmonella* count in the mouse stomachs, the high initial bacterial level was enough to cause a fatal infection. It may be that the same test done in humans could result in enough bacteria being killed in the stomach for the number of bacteria to fall below that needed for infection.

A description of an outbreak of food poisoning due to *Salmonella* in a home for the aged in Spain (Yanez Ortega *et al.* 2001) hoped to shed light on the protective powers of wine. An epidemiological investigation of the members of a nursing home was undertaken after an outbreak of food poisoning had been documented. Of the 128 inhabitants, 42 became sick and were found to have *Salmonella*. As the exposure appeared to have occurred at a party, each member of the home was asked what they had eaten there, whether they had drunk alcohol, and specifically whether they had drunk wine. The offending food was determined to have been a milk-based dessert, as nearly everyone who became ill had sampled the food. No one who had not eaten that food became ill. Wine-drinkers were 32% less likely to have become ill, although the sample size in the outbreak was so small that this protective effect could not be substantiated statistically. The study of this outbreak was limited with regard to the protective effect of wine, however, as the quantity of wine imbibed was not determined, nor was it established whether there was a temporal relationship between eating the offending food and drinking wine. It seems logical that larger volumes of wine imbibed in close temporal relation to the offending food may have a greater protective effect (Yanez Ortega *et al.* 2001).

Another outbreak of *Salmonella* food poisoning in Castellon, Spain (Bellido Blasco *et al.* 1996), showed a protective effect of alcohol, though not specifically of wine. The outbreak took place after a meal attended by 116 people. The investigation revealed that a sandwich of tuna, boiled eggs and vegetables was implicated. Among adults, a protective effect of alcohol was demonstrated, adjusted for age, gender and consumption of the implicated food. In this study, the beneficial effect of alcohol in preventing illness was highly statistically significant ($p = 0.007$). We cannot give wine all the credit, however, as no break-down of the types of alcohol imbibed was given (Bellido Blasco *et al.* 1996).

## *Staphylococcus aureus* and wound disinfection

Wine has been used for thousands of years as an antiseptic and constituent of poultices for wounds. Modern medical studies of wine as an antiseptic in clinical practice are unavailable, but we can look at laboratory evidence to judge wine's beneficial effect on wound infections.

The most common bacterium causing wound infection is *Staphylococcus aureus*. Although this micro-organism is a normal inhabitant of healthy skin, it accounts for the vast majority of all skin infections. It is the leading cause of wound infections, whether the wound was intentionally made (such as during surgery) or caused by accident or trauma.

The most important procedure in preventing infection is to irrigate the wound to remove foreign bodies such as dirt and other contaminating substances. After cleaning and irrigation, the topical application of antibacterial substances (antibiotics or antiseptics) shortens wound-healing time. A bandaged wound heals more quickly than an open one, and it seems that a small amount of moisture is also helpful. The presence of bacteria inhibits wound-healing, so application of an antibacterial agent and bandaging provide the best conditions for rapid healing and repair.

We conducted an experiment to test the ability of red wine, wine vinegar and ethanol to eradicate *Staphylococcus aureus*. Approximately five million bacteria were mixed with 3.8 ml each of the following test solutions: one-year-old Portuguese red table wine (9% ethanol), red wine vinegar, 10% ethanol, 40% ethanol, and sterile tap water as a control. The suspensions were mixed, and plated on nutrient agar at 0, 10, 20, 30 and 60 min, and 2 and 24 h. They were incubated under standard conditions and colony counts were made after 24 h. The 40% ethanol eliminated the bacteria immediately, wine vinegar rapidly eradicated them within 20 min, red wine decreased the colony counts from 200 000 to 1000 within 2 h, and 10% ethanol showed no effect at 2 h.

The 40% ethanol and wine vinegar were far superior to wine, so why not use them as a poultice constituent on wounds? The answer is simply because of the pain and burning associated with their use. Even the best antiseptic is of little benefit if no one can bear to use it! Wine, on the other hand, has sufficient antibacterial effect against *Staphylococcus aureus* to prevent it from causing infection, while also being acceptable to the patient.

## Cholera

Cholera is a severe form of contagious diarrhoea caused by the bacterium *Vibrio cholerae.* It has been a scourge for hundreds of years, occurring in epidemic waves that rapidly affected large areas of population. The first massive outbreak occurred in 1817 when the disease spread throughout India and into Russia. The second pandemic began in Russia in 1829. It spread across the Atlantic to New York and Montreal, and eventually invaded Latin America. The current seventh cholera pandemic began in Asia in 1961 and had spread to Africa by the mid-1970s. In 1991, cholera appeared in epidemic proportions in Peru and quickly spread throughout South and Central America. Within two years, more than 450 000 cases of cholera, resulting in more than 4000 deaths, were reported in the Americas. While this illustrates the magnitude of the epidemic disease, cholera continues to exist as a constant smouldering threat, especially in Asia and Africa.

The cholera bacillus does not invade the body as such, but causes its disease by producing a protein toxin. This toxin causes a profuse watery diarrhoea to such an extent that a patient can become severely dehydrated and die within 24 h. Fortunately, only about 1% of patients who are infected with the cholera bacillus develop severe disease, with most experiencing mild to asymptomatic infection, and approximately 10% having moderate diarrhoea without severe dehydration.

The history of wine's ability to protect against cholera goes back almost as far as the science of microbiology itself. In the epidemic that struck Paris in the late nineteenth century, it was noted that the street denizens, who subsisted largely on wine, seemed to be spared from the illness. Dr Alois Pick was intrigued by this observation, and set up experiments to examine it further. Cholera bacilli were added to wineskins containing red or white wine, water, or a wine–water mixture. The wine and wine–water mixtures eradicated the bacilli within 15 min, while the wineskins filled with water failed to have any effect (Majno 1975).

In a more modern approach to wine's protective effect against cholera, experiments were performed to assess its activity against cholera toxin. As mentioned above, severe illness from cholera is the result of massive fluid loss from the small intestine. The cholera toxin is made up of two parts, or subunits. When both penetrate cells of the small intestine, they induce an intense loss of sodium, chloride and water. This toxin-induced diarrhoea can be so severe that a patient can lose 10 litres of fluid per day through stool losses, making it almost impossible to stay hydrated. A preventive strategy to counteract this effect would be important in preventing or modifying the disease.

*Table 14.2* Relative amounts of intestinal fluid secretion after injection of test substance into isolated intestinal segments (Roberts *et al.* 2000).

| *Test substance* | *Amount of secretion (mg/cm)* |
|---|---|
| Normal saline (control) | 5–10* |
| Cholera toxin | 200–250 |
| Red wine + toxin | 10–20* |
| Boiled red wine + toxin | 10–15* |
| Grape juice + toxin | 120–150** |
| Apple juice + toxin | 180–230 |
| Ethanol 12% + toxin | 120–150** |

* Statistically significantly reduced secretion when compared with cholera toxin, grape juice + toxin, apple juice + toxin and ethanol 12% + toxin.
** Statistically significantly reduced secretion when compared with cholera toxin and apple juice + toxin.

Scientists in the USA designed experiments to look at wine's effect on cholera toxin (Roberts *et al.* 2000). Laboratory rats were anaesthetized and portions of their small intestine ligated, to produce six 10 cm loops, each separated by 1 cm. The blood supply to the bowel loops remained intact. Bowel segments were randomly injected with 1 ml of several different test solutions. These included: cholera toxin, red wine with and without cholera toxin, alcohol-free red wine with and without cholera toxin, grape juice with and without cholera toxin, ethanol 12% with and without cholera toxin, and apple juice with and without cholera toxin. These were all compared to a physiological control of normal saline (NaCl 0.9%). The results are summarized in Table 14.2. The saline controls as well as all the other test solutions were well absorbed in the absence of cholera toxin. Cholera toxin by itself caused a 33-fold increase in the amount of fluid secreted. Red wine and boiled red wine completely inhibited cholera toxin-induced secretion, grape juice and ethanol had smaller inhibitory effects, and apple juice had no effect (Roberts *et al.* 2000).

From these two experiments, performed 100 years apart, it appears that wine can protect against the severe effects of cholera through two different mechanisms.

## Helicobacter pylori

The micro-organism *Helicobacter pylori* is a major cause of gastritis, upper abdominal pain, and gastric and duodenal ulceration. In patients with *H. Pylori*-associated gastritis, treatment of the infection will ease the symptoms and help heal the inflamed stomach. Chronic gastritis or gastric ulceration puts patients at increased risk of developing cancer of the stomach.

Infection with *H. pylori* can occur during childhood where it can remit spontaneously or require treatment. Due to various lifestyle factors, specifically smoking, alcohol and coffee consumption, infection is more commonly acquired and eradicated throughout adulthood (Brenner *et al.* 1997). Because of the chronic inflammation caused by *H. pylori*, and the relationship between chronic gastritis and cancer of the stomach, *H. pylori* is considered carcinogenic. A Japanese study (Uemura *et al.* 2001) looked at more than 1500 patients who had: ulcers of the duodenum or stomach; or polyps or hypertrophy of the stomach; or dyspepsia (heartburn) without an ulcer. Approximately 80% had evidence of *H. pylori* infection, and 20% did not. Gastric cancers developed in 36 (2.9%) of the infected patients and in none of those uninfected. When looked at by specific categories, 21 (4.7%) patients with non-ulcer dyspepsia developed cancer, as well as 10 (3.4%) patients with previous gastric ulcers, and 5 (2.2%) with gastric hypertrophy or polyps. No gastric cancers were found in patients with duodenal ulcers alone (Uemura *et al.* 2001).

A number of studies has investigated the protective role of alcohol against infection by *H. pylori*. In a study performed in southern Germany, the relationship between alcohol consumption and active infection with *H. pylori* was examined. The amount and type of alcohol consumed was determined by a standardized questionnaire; infection with *H. pylori* was measured by a standard technique used for diagnosis. The study showed an inverse relationship between alcohol consumption and infection with *H. pylori*. This relationship was stronger for wine than for beer. The magnitude of this effect is considerable, with alcohol-drinkers only one-third as likely to be infected as non-drinkers (Brenner *et al.* 1999).

One study of Danish adult lifestyles (Rosenstock *et al.* 2000), among others, has shown similar results and supports the hypothesis that wine is effective in preventing *H. pylori* infection. As wine has been shown to prevent *H. pylori* infection, moderate wine-drinking may prevent stomach cancer. Drinkers of wine have been noted to have a 40% reduced risk of developing oesophageal cancers as well (de Lorimier 2000).

The mechanism by which wine protects against *H. pylori* disease is not completely clear. *H. pylori* infection causes ulcers by producing an enzyme called urease, which breaks urea down to ammonia. The ammonia causes a local increase in pH, which directly irritates the stomach cells (Czinn 1993). Wine-drinking results in an increase in the secretion of gastrin, an enzyme that increases gastric acid production. The increased gastric acid then lowers the pH of the stomach (Singer and Leffmann 1988). This may affect the local environment for the *H. pylori*.

In addition, wines appear to have a direct antibacterial activity against *H. pylori*. The antibacterial activity of 16 Chilean red wines (Cabernet Sauvignon, Cabernet Merlot, Cabernet Organic and Pinot Noir), and the active extracts of two randomly selected wines were assayed for their antibacterial activity on six strains of *H. pylori* isolated from gastric biopsies. All the red wines studied showed some antibacterial activity on each of the six strains. The active fraction of the two wines selected also showed good activity against the strains tested. The main active compound was identified as resveratrol (Daroch *et al.* 2001).

## Antibacterial mechanisms

We have given ample evidence that wine is active against a variety of bacteria. One of the questions that remains is: what are the substances in wine that give it its antibacterial properties?

### *Alcohol*

The most common assumption is that it is the alcohol in wine that kills bacteria. After all, alcohol has long been used as an antiseptic and is an effective disinfectant for wounds or healthy skin. The most common alcohol used as an antiseptic is isopropyl alcohol, or isopropanol (chemical structure: $CH_3$–$CH_2$–$CH_2OH$). Isopropanol replaced ethanol as an antiseptic because it has similar properties but is much cheaper to produce. High concentrations of alcohol are required to kill bacteria, and 70% isopropanol has been found to be the most active concentration against *Staphylococci* and other bacteria (Harrington and Walker 1903). Ethanol, or ethyl alcohol ($CH_3$–$CH_2OH$), is the alcohol found in wine, beer and spirits. Like isopropanol, it is strongly antibacterial at high concentrations, but its efficacy wanes as the percentage decreases (Block 2001). We found that ethanol 40% was almost instantaneously lethal against *Staphylococci* and enteric bacteria, but ethanol 10% had almost no effect on bacterial counts after 2 h of exposure. In the same experiment, red wine (9% ethanol) eradicated the enteric bacteria within 20 min and the *Staphylococci* within 2 h.

### *pH*

Another candidate for active antibacterial action is the low pH of wine. It was mentioned above that liquids with a low pH have a striking antibacterial effect when compared with neutral substances such as water or milk. A number of researchers have established that the low pH of stomach acid acts as a barrier to infection; indeed, travellers with decreased stomach acid production have an increased rate of traveller's diarrhoea (Evans *et al.* 1997; Khosla *et al.* 1993; Cook 1985). But the low pH cannot be the only answer. If so, cola (pH 2.4) would be superior to wine (pH 3.0). As seen in Table 14.1, wine eradicated bacteria within 4 h, while there were still viable bacteria in the cola after 48 h of exposure. So could it be the low pH in combination with the alcohol? Again, probably not. Investigators in Spain compared the bactericidal activity of a red wine (12.5% ethanol concentration, pH 3.5) with a solution of absolute ethanol diluted in a hydrochloric acid solution to a final ethanol concentration of 12.5% and pH 3.5. Red wine had a bactericidal effect against *Salmonella* that was much greater than that of the same ethanol concentration at the same pH. The wine decreased the *Salmonella* from $10^5$–$10^6$ colonies to zero within 5 min, while the test solution decreased the colony count to $10^2$ in 60 min. Dilutions of the test solutions again demonstrated wine's superior activity (Marimon and Bujanda 1998).

### *Sulphites*

Sulphites are added to wine to inhibit the oxidation of the ethanol to vinegar. Sulphites also have antibacterial properties at high levels, and are used in packaged foods to inhibit the growth of enteric bacteria, so perhaps it is the sulphite in wine rather than the wine itself that produces the antibacterial effect.

The survival of *Salmonella* in bone-meal (rendered animal by-product) was studied in varying concentrations of sulphites. No viable *Salmonella* was detected after 4 days of exposure to 5000 parts per million (ppm) of sodium metabisulphite. The minimum killing concentration was found to be 4000 ppm (Abalaka and Deibel 1990).

Metabisulphites have also been used in sausages, and a study in Ireland found metabisulphite at a concentration of 450 ppm to be an effective antibacterial (Scannell *et al.* 2000). The concentrations in wine are at the low level needed to serve as an antioxidant – an average of 12.1 ppm for white wine and 3.1 ppm for red wine (Sullivan *et al.* 1990). These levels are much too low to have effective antibacterial properties.

### *Polyphenols and flavonoids*

Researchers from Bordeaux in the 1950s pioneered the study of the many organic compounds found in wine (Masquelier and Jensen 1953). Phenols have potent antibacterial activity. Indeed, phenol was the first antiseptic discovered by Joseph Lister, the father of antisepsis. Oenoside (Figure 14.3), a polyphenol abundant in wine, has been demonstrated to have a powerful antibacterial effect. Oenoside is one of the principal pigments found in red grapes. In the grape, it is combined with sugars and is therefore inactive.

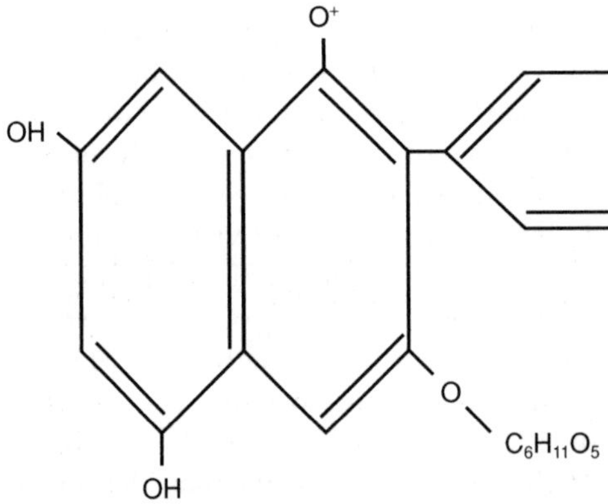

*Figure 14.3* A polyphenol, oenidol. This complex sugar is found in the grape, and is liberated by fermentation. It is antibacterial at low pH, such as is found in wine (Masquelier and Jensen 1953).

*Table 14.3* Antibacterial activity of Médoc wine according to its age*

| *Age of wine* | *Bactericidal index* |
|---|---|
| Fermented must | 0 |
| Wine of the same year | 11 |
| 3 year-old wine | 14 |
| 6 year-old wine | 16 |
| 9 year-old wine | 19 |
| 14 year-old wine | 17 |
| 23 year-old wine | 16 |
| 46 year-old wine | 12 |
| 56 year-old wine | 9 |
| 82 year-old wine | 6 |

* The bactericidal index refers to the dilution of wine in water. An index of 1 indicates weak activity in undiluted wine; an index of 20 indicates intense activity and corresponds to a dilution of 5% wine in purified water.

During fermentation, the sugar is cleaved and the active phenolic compound is liberated. The phenols and flavonoids are most active in acidic environments, and the low pH of wine promotes the antibacterial activity and is responsible for maintaining the pigment in solution. Experiments were performed with wine that had been passed over charcoal to remove the pigment. The antibacterial effect of the wine was completely eliminated by the removal of the pigment (Riberau-Gayon and Peynaud 1966). As would be expected, the antibacterial strength of the wine increases with age, up to a point (Table 14.3).

In addition to phenolic compounds, stilbenes and flavonoids are found to be plentiful in wine. These are all synthesized by a common pathway from the amino acid phenylalanine (Soleas *et al.* 1997) and are the substances that appear to lend wine most of its beneficial health properties – from antioxidant to antibacterial.

## References

Abalaka, J. A. and Deibel, R. H. (1990) Viability of *Salmonella* in bone meal. *Microbios*: 62, 155–64.

Bellido Blasco, J. B., Gonzalez Moran, F., Arnedo Pena, A., Galiano Arlandis, J. V., Safont Adsuara, L., Herrero Carot, C., Criado Juarez, J., Mesanza del Notarion, I. (1996) Outbreak of *Salmonella enteritidis* food poisoning. Potential protective effect of alcoholic beverages. *Medicina Clinica (Barcelona)*: 107, 655–6.

Block, E. (1985) The chemistry of garlic and onions. *Scientific American*: 252, 114–19.

Block, S. S. (2001) *Disinfection, Sterilization, and Preservation.* Lippincott, Williams & Wilkins, Philadelphia, USA. pp. 3–17.

Brenner, H., Rothenbacher, D., Bode, G., Adler, G. (1997) Relation of smoking and alcohol and coffee consumption to active *Helicobacter pylori* infection: cross-sectional study. *British Medical Journal*: 315, 1489–92.

Brenner, H., Rothenbacher, D., Bode, G., Adler, G. (1999) Inverse graded relation between alcohol consumption and active infection with *Helicobacter pylori*. *American Journal of Epidemiology*: **149**, 571–6.

Cook, G. C. (1985) Infective gastroenteritis and its relationship to reduced gastric acidity. *Scandanavian Journal of Gastroenterology – supplement*: **111**, 17–23.

Czinn, S. J. (1993) *Pediatric Gastrointestinal Disease: Pathophysiology, Diagnosis, Management*. W.B. Saunders, Philadelphia.

Daroch, F., Hoeneisen, M., Gonzalez, C. L., Kawaguchi, F., Salgado, F., Solar, H., Garcia, A. (2001) In vitro antibacterial activity of Chilean red wines against *Helicobacter pylori*. *Microbios*: **104**, 79–85.

de Lorimier, A. (2000) Alcohol, wine and health. *American Journal of Surgery*: **180**, 357–61.

Durant, W. (1939) *The Life of Greece*. Simon & Schuster, New York, pp. 45, 101.

Evans, C. A., Gilman, R. H., Rabbani, G. H., Salazar, G., Ali, A. (1997) Gastric acid secretion and enteric infection in Bangladesh. *Transactions of the Royal Society of Tropical Medicine and Hygiene*: **91**, 681–5.

Frobenius, W., von Maillot, K., Sauerbrei, W., Lang, N. (1996) Champagne administered by spoon for peritonitis after caesarean section. From the history of obstetrics in Erlangen – data from about 60 000 deliveries in 100 years. *Gynäkologisch-Geburtschilfliche Rundschau*: **36**, 212–20.

Harrington, C., Walker, H. (1903) The germicidal action of alcohol. *Boston Medical and Surgical Journal*: **148**, 548–52.

Khosla, S. N., Jain, N., Khosla, A. (1993) Gastric acid secretion in typhoid fever. *Postgraduate Medical Journal*: **69**, 121–3.

Majno, G. (1975) *The Healing Hand: Man and Wound in the Ancient World*. Harvard University Press, Cambridge, Massachusetts.

Managan, L. P., Sehulster, L. M., Chiarello, L., Simonds, D. N., Jarvis, W. R. (1998) Risk of infectious disease transmission from a common communion cup. *American Journal of Infection Control*: **26**, 538–9.

Marimon, J. M. and Bujanda, L. (1998) Antibacterial activity of wine against *Salmonella enteritidis*. pH or alcohol? *Journal of Clinical Gastroenterology*: **27**, 179–80.

Masquelier, M. J. and Jensen, H. (1953) Recherches sur l'action bactericide des vins rouges. *Bulletin de la Société de Pharmacie de Bordeaux*: **91**, 24–9.

Quennell, M. and Quennell, C. H. B. (1954) *Everyday Things in Ancient Greece*. B.T. Batsford, London.

Pickleman, J. (1990) A glass a day keeps the doctor . . . *American Surgeon*: **56**, 395–7.

Riberau-Gayon, J. and Peynaud, E. (1966) *Traité d'Oenologie*, 2nd edn. Librairie Polytechnique Beranger, Paris.

Roberts, P. R., Zaloga, S. J., Burney, J. D., Zaloga, G. P. (2000) Wine components inhibit cholera toxin-induced intestinal secretion in rats. *Journal of Intensive Care Medicine*: **15**, 48–52.

Rosenstock, S. J., Jorgensen, T., Andersen, S. P., Bonnevie, O. (2000) Association of *Helicobacter pylori* infection with lifestyle, chronic disease, body-indices and age at menarche in Danish adults. *Scandinavian Journal of Public Health*: **28**, 32–40.

Scannell, A. G., Ross, R. P., Hill, C., Arendt, E. K. (2000) An effective lacticin biopreservative in fresh pork sausage. *Journal of Food Protection*: **63**, 370–5.

Sheth, N. K., Wisniewski, T. R., Franson, T. R. (1988) Survival of enteric pathogens in common beverages: an in vitro study. *American Journal of Gastroenterology*: **83**, 658–60.

Singer, M. V. and Leffmann, C. (1988) Alcohol and gastric acid secretion in humans: a short review. *Scandinavian Journal of Gastroenterology*: **146** (supplement), 11–21.

Soleas, G. J., Diamandis, E. P., Goldberg, D. M. (1997) Wine as a biological fluid: history, production and role in disease prevention. *Journal of Clinical and Laboratory Analysis*: **11**, 287–313.

Sugita-Kunishi, Y., Hara-Kudo, Y., Iwamoto, T., Kondo, K. (2001) Wine has activity against entero-pathogenic bacteria in vitro but not in vivo. *Bioscience, Biotechnology and Biochemistry*: **65**, 954–7.

Sullivan, J. J., Hollingworth, T. A., Wekell, M. M., Meo, V. A., Moghadam, A. (1990) Determination of free (pH 2.2) sulfite in wines by flow injection analysis: a collaborative study. *Journal of the Association of Official Analytical Chemists*: **73**, 223–6.

Uemura, N., Okamoto, S., Yamamoto, S., Matsumura, N., Yamaguchi, S., Yamakido, M., Taniyama, K., Sasaki, N., Schlemper, R. J. (2001) *Helicobacter pylori* infection and the development of gastric cancer. *New England Journal of Medicine*: **345**, 784–9.

Weisse, M. E., Eberly, B., Person, D. A. (1995) Wine as a digestive aid: comparative antimicrobial effects of bismuth salicylate and red and white wine. *British Medical Journal*: **311**, 1657–60.

Yanez Ortega, J. L., Carraminana Martinez, I., Bayona Ponte, M. (2001) *Salmonella enteritidis* in a home for the aged. *Revista Española de Salud Publica*: **75**, 81–8.

# Index